Lecture Notes in Computer Science 15269

Founding Editors

Gerhard Goos
Juris Hartmanis

Editorial Board Members

Elisa Bertino, *Purdue University, West Lafayette, IN, USA*
Wen Gao, *Peking University, Beijing, China*
Bernhard Steffen ⓘ, *TU Dortmund University, Dortmund, Germany*
Moti Yung ⓘ, *Columbia University, New York, NY, USA*

The series Lecture Notes in Computer Science (LNCS), including its subseries Lecture Notes in Artificial Intelligence (LNAI) and Lecture Notes in Bioinformatics (LNBI), has established itself as a medium for the publication of new developments in computer science and information technology research, teaching, and education.

LNCS enjoys close cooperation with the computer science R & D community, the series counts many renowned academics among its volume editors and paper authors, and collaborates with prestigious societies. Its mission is to serve this international community by providing an invaluable service, mainly focused on the publication of conference and workshop proceedings and postproceedings. LNCS commenced publication in 1973.

Marcin Bieńkowski · Matthias Englert
Editors

Approximation and Online Algorithms

22nd International Workshop, WAOA 2024
Egham, UK, September 5–6, 2024
Proceedings

Editors
Marcin Bieńkowski ⓘ
University of Wrocław
Wrocław, Poland

Matthias Englert ⓘ
University of Warwick
Coventry, UK

ISSN 0302-9743   ISSN 1611-3349 (electronic)
Lecture Notes in Computer Science
ISBN 978-3-031-81395-5   ISBN 978-3-031-81396-2 (eBook)
https://doi.org/10.1007/978-3-031-81396-2

© The Editor(s) (if applicable) and The Author(s), under exclusive license to Springer Nature Switzerland AG 2025

This work is subject to copyright. All rights are solely and exclusively licensed by the Publisher, whether the whole or part of the material is concerned, specifically the rights of translation, reprinting, reuse of illustrations, recitation, broadcasting, reproduction on microfilms or in any other physical way, and transmission or information storage and retrieval, electronic adaptation, computer software, or by similar or dissimilar methodology now known or hereafter developed.
The use of general descriptive names, registered names, trademarks, service marks, etc. in this publication does not imply, even in the absence of a specific statement, that such names are exempt from the relevant protective laws and regulations and therefore free for general use.
The publisher, the authors and the editors are safe to assume that the advice and information in this book are believed to be true and accurate at the date of publication. Neither the publisher nor the authors or the editors give a warranty, expressed or implied, with respect to the material contained herein or for any errors or omissions that may have been made. The publisher remains neutral with regard to jurisdictional claims in published maps and institutional affiliations.

This Springer imprint is published by the registered company Springer Nature Switzerland AG
The registered company address is: Gewerbestrasse 11, 6330 Cham, Switzerland

If disposing of this product, please recycle the paper.

Preface

The 22nd Workshop on Approximation and Online Algorithms (WAOA 2024) focused on the design and analysis of algorithms for online and computationally hard problems. Both types of problems have a large number of applications in a wide variety of fields. The workshop was held in Egham, UK, on September 5–6, 2024, and was a success, providing many interesting presentations and opportunities for stimulating interactions. WAOA 2024 was part of ALGO 2024, which also hosted ESA, ALGOCLOUD, ALGOWIN, ATMOS, IPEC, and WABI.

WAOA 2024 invited submissions on algorithmic game theory, algorithmic trading, coloring and partitioning, competitive analysis, computational advertising, computational finance, cuts and connectivity, FPT approximation algorithms, geometric problems, graph algorithms, inapproximability results, mechanism design, network design, packing and covering, paradigms for designing and analyzing approximation and online algorithms, resource augmentation, and scheduling problems.

In response to the call for papers, we received 47 submissions. Each submission was reviewed by at least three referees in a double-blind review process. Submissions were judged primarily on originality, technical quality, clarity, and relevance to the conference topics. Based on the reviews, the program committee selected 15 papers. This volume contains the final revised versions of these papers, as well as an invited contribution from our plenary speaker, Rico Zenklusen. The EasyChair conference system was used to manage the electronic submissions, the review process, and the electronic program committee discussions.

We would like to thank all authors who submitted papers to WAOA 2024 and all participants of WAOA 2024, including the presenters of the accepted papers. Special thanks go to the plenary speaker Rico Zenklusen for accepting our invitation and giving a very nice talk. We would also like to thank the PC members and the external reviewers for their diligent work in evaluating the submissions and their contributions to the electronic discussions. Last but not least, we are grateful to all the local organizers, especially Argyrios Deligkas and Eduard Eiben, who served as local chairs of the organizing committee.

September 2024   Marcin Bieńkowski
                                                                                                 Matthias Englert

Organization

Program Committee Chairs

Marcin Bieńkowski — University of Wrocław, Poland
Matthias Englert — University of Warwick, UK

Steering Committee

Evripidis Bampis — Sorbonne Université, France
Thomas Erlebach — Durham University, UK
Christos Kaklamanis — University of Patras, Greece
Nicole Megow — Universität Bremen, Germany
Laura Sanità — Bocconi University, Italy
Martin Skutella — Technische Universität Berlin, Germany
Roberto Solis-Oba — University of Western Ontario, Canada

Program Committee

Spyros Angelopoulos — CNRS and Sorbonne University, France
Antonios Antoniadis — University of Twente, The Netherlands
Sujoy Bhore — IIT Bombay, India
Hans-Joachim Böckenhauer — ETH Zurich, Switzerland
Maike Buchin — Ruhr University Bochum, Germany
Panagiotis Charalampopoulos — Birkbeck, University of London, UK
Marek Chrobak — University of California, Riverside, USA
Christian Coester — University of Oxford, UK
Naveen Garg — IIT Delhi, India
Meng He — Dalhousie University, Canada
Martin Hoefer — Goethe Universität, Germany
Jochen Koenemann — University of Waterloo, Canada
Shi Li — Nanjing University, China
Alantha Newman — Université Grenoble Alpes, France
Zeev Nutov — Open University of Israel, Israel
Kevin Schewior — University of Southern Denmark, Denmark
Chris Schwiegelshohn — Aarhus University, Denmark
Hadas Shachnai — Technion, Israel

Joachim Spoerhase   University of Liverpool, UK
David Wajc   Technion, Israel
Prudence Wong   University of Liverpool, UK
Meirav Zehavi   Ben-Gurion University of the Negev, Israel

Additional Reviewers

Andreas Abels
Ishan Bansal
Ziyad Benomar
Bartłomiej Bosek
Elisabet Burjons
Martin Böhm
Sergio Cabello
Vincent Chau
Ilan Cohen
Syamantak Das
Minati De
Ilan Doron
Christoph Dürr
Franziska Eberle
Leah Epstein
Moran Feldman
Harmender Gahlawat
Tatsuya Gima
Daniel Gonçalves
Swati Gupta
Niklas Hahn
Martin Herold
Felix Hommelsheim
Lars Huth
Tobias Jacobs
Shahin Kamali
Tomohiro Koana
Yusuke Kobayashi
Guy Kortsarz
Ariel Kulik
Roie Levin

Fu-Hong Liu
Hsiang-Hsuan Liu
Nikolaos Melissinos
Gopinath Mishra
Pranabendu Misra
Eiji Miyano
Tobias Mömke
Daniel Neuen
Neel Patel
Kirk Pruhs
Sharath Raghvendra
Saladi Rahul
Runtian Ren
Matteo Russo
Harald Räcke
David Saulpic
Baruch Schieber
Jens Schlöter
Uwe Schwiegelshohn
Yongho Shin
Bertrand Simon
Satyam Singh
George Skretas
Moritz Stocker
Noam Touitou
Walter Unger
Rob van Stee
Philip Whittington
Xiaowei Wu
Karol Węgrzycki
Qiankun Zhang

Advances in Approximation Algorithms for Tree and Connectivity Augmentation (Invited Talk)

Rico Zenklusen

ETH Zürich, Switzerland

Augmentation problems are a fundamental class of Network Design problems. The goal is to find a cheapest way to increase a graph's (edge-)connectivity by adding edges from a given set of options. The Minimum Spanning Tree Problem is one of its most elementary examples, which can be interpreted as determining a cheapest way to increase the edge-connectivity of a graph from 0 to 1. The next step, to increase from 1 to 2, leads to the heavily studied Tree Augmentation Problem, and the more general problem of increasing the edge-connectivity of an arbitrary graph by 1 is known as Connectivity Augmentation. While these problems admit simple 2-approximations, obtaining approximation factors below 2 has proven challenging.

This talk has several goals, namely:

1. Providing a brief introduction to Tree and Connectivity Augmentation.
2. Discussing relevant algorithmic techniques that have recently led to better-than-2 approximations for these problems, including the Relative Greedy method and an interesting connection of it to local search procedures.
3. Demonstrating how these techniques can be leveraged to obtain better-than-2 approximations for (weighted) augmentation problems.

Contents

Bounding the Price-of-Fair-Sharing Using Knapsack-Cover Constraints
to Guide Near-Optimal Cost-Recovery Algorithms 1
    Sander Aarts, Jacob Dentes, Manxi Wu, and David B. Shmoys

Improved Online Scheduling with Restarts on a Single Machine 16
    Aflatoun Amouzandeh and Rob van Stee

Searching in Euclidean Spaces with Predictions 31
    Sergio Cabello and Panos Giannopoulos

Lower Bounds for Approximate (& Exact) k-Disjoint-Shortest-Paths 46
    Rajesh Chitnis, Samuel Thomas, and Anthony Wirth

Approximating δ-Covering 61
    Tim A. Hartmann and Tom Janßen

Fast Approximation Algorithms for Euclidean Minimum Weight Perfect
Matching 76
    Stefan Hougardy and Karolina Tammemaa

Approximation Algorithms for k-Scenario Matching 89
    *Danny Blom, Dylan Hyatt-Denesik, Afrouz Jabal Amelia,
    and Bart Smeulders*

Tight Approximation Bounds on a Simple Algorithm for Minimum
Average Search Time in Trees 104
    Svein Høgemo

Online Deterministic Minimum Cost Bipartite Matching with Delays
on a Line 119
    Tung-Wei Kuo

Maximizing Throughput for Parallel Jobs with Speed-Up Curves 135
    Kefu Lu and Mason Marchetti

Improved Approximation Algorithms for Covering Pliable Set Families
and Flexible Graph Connectivity 151
    Zeev Nutov

Small Additive Error for Unsplittable Multicommodity Flow
in Outerplanar Graphs 167
    Richard Shapley and David B. Shmoys

Complexity of Fixed Order Routing 183
    Steven Miltenburg, Tim Oosterwijk, and René Sitters

Approximate Min-Sum Subset Convolution 198
    Mihail Stoian

Online String Attractors: And Their Relation to the Lempel-Ziv
Factorization 213
    Philip Whittington

Author Index 229

Bounding the Price-of-Fair-Sharing Using Knapsack-Cover Constraints to Guide Near-Optimal Cost-Recovery Algorithms

Sander Aarts[(✉)], Jacob Dentes, Manxi Wu, and David B. Shmoys

School of Operations Research and Information Engineering, Cornell University, Ithaca, NY, USA
sea78@cornell.edu

Abstract. We consider the problem of fairly allocating the cost of providing a service among a set of users, where the service cost is formulated by an NP-hard *covering integer program (CIP)*. The central issue is to determine a cost allocation to each user that, in total, recovers as much as possible of the actual cost while satisfying a stabilizing condition known as the *core property*. The ratio between the total service cost and the cost recovered from users has been studied previously, with seminal papers of Deng, Ibaraki, & Nagomochi and Goemans & Skutella linking this *price-of-fair-sharing* to the integrality gap of an associated LP relaxation. Motivated by an application of cost allocation for network design for LPWANs, an emerging IoT technology, we investigate a general class of CIPs and give the first non-trivial price-of-fair-sharing bounds by using the natural LP relaxation strengthened with knapsack-cover inequalities. Furthermore, we demonstrate that these LP-based methods outperform previously known methods on an LPWAN-derived CIP data set. We also obtain analogous results for a more general setting in which the service provider also gets to select the subset of users, and the mechanism to elicit users' private utilities should be group-strategyproof. The key to obtaining this result is a simplified and improved analysis for a cross-monotone cost-allocation mechanism.

Keywords: Cost sharing · Covering integer programs · LoRaWAN

1 Introduction

Allocating costs fairly when providing some collective service to a set of users has been a mainstream topic within cooperative game theory in which the notion of the core, and the associated *core property*, plays a central role. Specifically, the core property requires that the payment from each user is such that each subset of users pays, in total, no more than the cost of providing service just to them. LP duality has long been known to play a fundamental role in such allocations, and the seminal work of Deng, Ibaraki & Nagomochi [7] and Goemans & Skutella

[12] proved the link between the integrality gap (or lack thereof) for several NP-hard discrete optimization settings and the fraction of the service cost that can be recovered by an allocation satisfying the core property. We term the ratio between optimal cost and the cost recoverable satisfying the core property as the *price-of-fair-sharing*.

For a natural subclass of *covering integer programs* (CIPs), known as multiset-multi-covering problems, the strongest known price-of-fair-sharing was not obtained using a dual LP-based framework [22]. We show that an LP-based approach can yield improved bounds and simplified analyses, not just for this special case considered by [22], but for general CIPs. More specifically, our main result is that by strengthening the natural LP relaxation for CIPs with knapsack-cover inequalities, the resulting dual LP does have the property that any feasible solution yields a cost-allocation that satisfies the core property. This reduces the problem of finding a cost-allocation to that of finding a strengthened-LP-relative approximation algorithm, allowing us to leverage approximation algorithms already known for these settings (e.g., [3–5, 18]). Not only do these existing algorithms yield bounds on the price-of-fair sharing for general CIPs, they also exploit existing bounds for families of sparse CIP instances. These existing algorithms provide improved cost-recovery ratios for families of sparse CIP instances.

We also obtain analogous results in a more general setting where the service provider selects a subset of users to receive the service based on their elicited private valuations of the service. The mechanism used to elicit users' private valuations needs to be group-strategyproof so that no subset of users have an incentive to misreport their valuations. The key to obtaining this result is a simplified and improved analysis for a cross-monotone cost-allocation mechanism. These results also rely on our knapsack-cover-strengthened LP relaxation.

Our investigation into these questions was directly motivated by challenges in the development of an emerging technology, LPWANs, a popular Internet of Things solution for wirelessly connecting devices to the internet [28]. CIPs emerge from the appropriate model for determining the location of gateways to provide LPWAN coverage to a collection of users, and our result is useful to determine how the cost of LPWANs should be shared among users in an incentive compatible manner. We complement our algorithmic and structural results with an empirical investigation into the effectiveness of our methods on relatively large-scale LPWAN data. Our empirical study demonstrates that, in comparison to several natural heuristic alternatives, our theoretically-motivated methods provide far superior results.

2 Optimizing Coverage for LPWAN Networks and IoT

The growth of the Internet of Things (IoT) is creating both new opportunities and challenges in wireless coverage provision and pricing. Low-power wide-area networks (LPWANs) are a ubiquitous technology for connecting devices, or things, to the internet. These networks use radio communication to transmit signals over long distances. Demodulating, and transferring these signals to the internet requires networks of physical wireless receivers. Many such networks are already deployed. In 2023 alone, LPWANs reportedly served over 1.3 billion IoT connections [1]. LoRaWAN is among the most popular LPWAN solutions, representing roughly 40% of all connections made outside of China. Globally, over 150 different network operators provide LoRaWAN coverage [29].

Sharing LoRaWAN coverage among multiple users is an effective yet challenging approach to better utilize wireless infrastructure and lower costs. Two key features make sharing LoRaWAN coverage especially beneficial. Firstly, the receivers, or *gateways*, can process orders of magnitude more wireless traffic than that produced by a typical single IoT application (see, e.g., the application in [2]). Secondly, the quality of coverage improves as the number of gateways increases. Transmitted LoRaWAN packets tend to fail at random, and coverage quality is measured by the *packet reception rate*. However, transmissions are *association free*; a device broadcasts to all nearby gateways, many of which may process the same packet [28]. A packet is lost if and only if all nearby gateways simultaneously fail to receive it. Thus, having multiple gateways to cover the transmitted signal can significantly improve the reception rate. This multi-coverage feature is a key aspect of our model and raises interesting computational challenges.

The problem of optimally providing wireless coverage to a group of users using multiple gateways can be modeled as a covering integer program. Let $\mathcal{U}$ denote a set of m users, and $\mathcal{F}$ a set of n potential gateway locations, or facilities. Each user $j \in \mathcal{U}$ has a service requirement $r_j > 0$, and each gateway has an opening cost $c_i > 0$. Requirements and costs are represented by vectors $\mathbf{r} = (r_1, r_2, \ldots, r_m)$ and $\mathbf{c} = (c_1, \ldots, c_n)$, respectively. Each facility-user pair is associated with a contribution parameter $a_{ij} \geq 0$ that specifies the quality of coverage that facility i provides user j if opened. Contributions are represented as an $n \times m$ matrix $\mathbf{A}$. The objective is to minimize cost, subject to providing sufficient coverage to each user.

$$c^* = \min\{\mathbf{c}^T \mathbf{x} : \mathbf{A}\mathbf{x} \geq \mathbf{r},\ \mathbf{x} \in \{0,1\}^n\} \tag{1}$$

Here, the binary decision variables $\mathbf{x} = (x_1, \ldots, x_n)$ indicate whether each facility i is opened or not. Details on how CIPs capture LoRaWAN coverage provision are given in an extended online version. (A more general formulation of CIPs permits the purchase of up to d_i integer copies of facility i. Our results extend to this setting in a straightforward way, but we omit this extension for simplicity and clarity of exposition.)

3 Cost-Sharing Fundamentals

The main objective in this work is to develop a principled algorithmic framework for finding fair and stable ways to share the cost of CIP solutions among the set of users. We follow a standard approach in mechanism design and define "fair and stable" using the *core property* [16,26]. An allocation satisfies the core property if no subset of users is allocated more cost than the minimum cost of providing coverage for this subset alone. This property can also be viewed as an absence of cross-subsidies; no sub-group of users pays more than what it would cost them to serve themselves, thereby not subsidizing other users [10].

More formally, a cost allocation is a vector $\xi = (\xi_1, \ldots, \xi_m)$, where $\xi_j \geq 0$ represents the dollar amount charged to user $j \in \mathcal{U}$. Allocations ξ_j are also called cost-shares. Following the standard definition, cost-shares ξ satisfy the core property if for each sub-group $J \subseteq \mathcal{U}$,

$$\sum_{j \in J} \xi_j \leq c_J^*, \quad \forall J \subseteq \mathcal{U}, \tag{2}$$

where c_J^* is the minimum cost of serving group J [16]. Cost allocations can vary in magnitude. We say cost allocation ξ is *budget-balanced* whenever the total amount paid by the served users, $\sum_{j \in \mathcal{U}} \xi_j$, equals the cost of serving them. The *core* is the set of budget-balanced cost allocations satisfying the core property [16]. In some problems, finding cost-shares in the core can be challenging.

Linear programming (LP) duality is the workhorse tool for finding cost-shares satisfying the core property, but it is not always possible to simultaneously satisfy both the core property and be budget-balanced. There are simple CIP instances that have no cost-allocations in the core (see, e.g., Li *et al.* [22]). Deng, Ibaraki, and Nagamochi [7] show that a generic covering problem has a non-empty core if and only if its natural LP-relaxation has no integrality gap. As such, it is common to maximize the cost recovered, subject to satisfying the core-property [12,16]; a cost-allocation ξ is β-budget-balanced if it recovers a fraction β of the cost.

For many optimization problems, every dual feasible solution is a cost allocation that satisfies the core property. This connection has important algorithmic implications. First, one can leverage duality to find cost-shares satisfying the core property by using (approximation) algorithms that produce feasible dual solutions as well as feasible integer solutions to the primal. If the integer solution costs at most α times the dual, we see that the corresponding cost allocation is $1/\alpha$-budget balanced. We shall say then that α is the *price-of-fair-sharing* (or equivalently, $1/\alpha$ has also been termed the cost-recovery ratio). This relationship has been frequently used to obtain strong results in many settings [7,8,12,13,19,21,23,25].

On the other hand, seminal work by Deng, Ibaraki, & Nagamochi [7] and Goemans & Skutella [12] shows that all cost allocations satisfying the core property are dual feasible solutions in the set cover problem and the facility location problem. Since the cost-allocation mechanism purchases an integer solution, but only allocates costs as a fractional solution, duality implies that price-of-fair-sharing is lower-bounded by the integrality gap of the problem. It is folklore that the price-of-fair-sharing is α whenever the integrality gap of the natural LP relaxation is α [16,23].

The relationship between the price-of-fair-sharing and the integrality gap of the natural IP/LP formulation does not, however, appear to hold for CIPs. The multi-cover constraints in CIPs render the natural linear programming relaxation ill-suited for cost-sharing. Even with just one user $|U| = 1$ and integral inputs, the integrality gap of the CIP is unbounded [4]. Naively, this would imply that the price-of-fair-sharing in CIPs is unbounded. However, Li, Sun, Wang, and Lou [22] show that a bound can be attained for the special case of CIPs with integer-valued $\mathbf{A}$ and $\mathbf{r}$. Their analysis, however, does not make explicit use of duality. Moreover, our IoT application yields CIPs that do not satisfy the assumption of integer-valued $\mathbf{A}$ and $\mathbf{r}$. This begs two questions: How do cost-sharing and duality relate in CIPs, and can there be an effective cost-recovery at all when data are non-integer-valued?

4 Knapsack-Cover Constraints to the Rescue

This paper presents a principled framework for finding cost-shares in CIPs using linear programming duality. Our approach makes use of a well-known strengthened LP formulation based on knapsack-cover inequalities [4]. Our main contribution is to show that every feasible dual solution in this strengthened LP, which we shall denote KC-LP, naturally induces cost-shares that satisfy the core property. This has significant algorithmic consequences. First, our results imply that any approximation algorithm that produces CIP solutions with cost at most α times the KC-LP optimum can be used to produce cost-shares that recover $1/\alpha$ of the cost; that is, a price-of-fair-sharing of α. There are many such algorithms [4–6,18]. More generally, our framework can be used to find cost-shares for any integer CIP solution by solving the strengthened dual linear program.

These methods yield the first cost-sharing algorithm with bounded price-of-fair-sharing for general CIPs. Furthermore, we prove that the implied price-of-fair-sharing bounds are tight. In addition, we showcase the efficacy of our framework by recovering up to 93% of the cost in semi-stylized LoRaWAN covering problems at scale; this is over twice the recovery of the next-best method. We also use our KC-LP approach to obtain analogous results for a more general setting in which the service provider also selects a subset of users to receive the service based on the users' private valuations elicited from a mechanism that is group-strategyproof. This reinforces the central message of this paper, affirming the powerful connection between an effective cost sharing mechanism and KC-LP dual solutions.

The knapsack-cover inequalities are introduced in the seminal work by Carr, Fleischer, Leung, and Phillips [4] to strengthen the LP so as to bound its integrality gap. It is helpful to first understand how the integrality gap of the natural LP relaxation is unbounded. Carr *et al.* provide the following instance: let $R > 0$ be a large integer requirement for a single user, and let there be two facilities; facility a provides $R - 1$ units of coverage at near-zero cost, whereas facility b provides R units of coverage at a cost of 1. Clearly, a feasible integer solution must include item b, with a total cost of 1. A fractional solution, however, can select a full unit of item a, and only a $1/R$ fraction of item b, with a cost of $1/R$. The resulting integrality gap is R, which can be arbitrarily large.

Carr *et al.* [4] derive a strong bound on the integrality gap by adding exponentially-many valid *knapsack-cover* (KC) inequalities to produce a strengthened LP relaxation. These inequalities represent residual coverage requirements at partial solutions. Suppose facilities $S \subseteq \mathcal{F}$ have been built. Now user j has a *residual requirement* $r_j^S = \max\left\{r_j - \sum_{i \in S} a_{ij}, 0\right\}$. To satisfy j's residual requirement, an additional contribution of at least r_j^S is needed from the remaining unbuilt facilities $\mathcal{F} \setminus S$. In addition, there is no benefit to exceeding r_j^S, so the original contributions a_{ij} can be clipped to r_j^S if they exceed this value. We call $a_{ij}^S = \min\left\{a_{ij}, r_j^S\right\}$ the *residual contribution*, and set it to zero if i is in S. This defines the *knapsack-cover inequality*:

$$\sum_{i \in \mathcal{F} \setminus S} a_{ij}^S \geq r_j^S. \tag{3}$$

There are $|\mathcal{U}| \times |2^{\mathcal{F}}| = m2^n$ KC inequalities. The KC inequalities are valid; their presence does not change the set of feasible *integer* solutions, and reduces the feasible set of the LP relaxation. The strengthened linear program is as follows:

$$\min_x \ \sum_{i \in \mathcal{F}} c_i x_i,$$
$$\text{s.t.} \ \sum_{i \in \mathcal{F} \setminus S} a_{ij}^S x_i \geq r_j^S, \qquad \forall (j,S) \in \mathcal{U} \times 2^{\mathcal{F}}, \qquad \text{(KC-LP)}$$
$$x_i \geq 0, \qquad \forall i \in \mathcal{F}.$$

Note that the constraints $x_i \leq 1$ have been dropped. (Multiplicity constraints are implicit, in that $a_{ij}^S = 0$ whenever $i \in S$.) Associated with this primal is the knapsack-cover dual program (KC-DP), referred to as the KC-LP dual. The dual has one constraint for each facility, and one variable for every KC inequality:

$$\max_y \ \sum_{j \in \mathcal{U}} \sum_{S \subseteq \mathcal{F}} r_j^S y_j^S,$$
$$\text{s.t.} \ \sum_{j \in \mathcal{U}} \sum_{S \subseteq \mathcal{F} \setminus \{i\}} a_{ij}^S y_j^S \leq c_i, \qquad \forall i \in \mathcal{F}, \qquad \text{(KC-DP)}$$
$$y_j^S \geq 0, \qquad \forall (j,S) \in \mathcal{U} \times 2^{\mathcal{F}}.$$

There are known approximation algorithms for CIPs that yield integer solutions with costs within a multiplicative factor of the KC-LP optimum. The worst-case performance of these algorithms depend on the column and row sparsity of the contributions $\mathbf{A}$. Let $\Delta = \max_i \{\sum_{j \in \mathcal{U}} \mathbf{1}[a_{ij} > 0]\}$ denote the column sparsity, and let $\Gamma = \max_j \{\sum_{i \in \mathcal{F}} \mathbf{1}[a_{ij} > 0]\}$ denote the row sparsity. Carr *et al.* [4] propose a Γ-approximation algorithm based on rounding a KC-LP solution; Kolliopoulos and Young [18] design an alternative KC-LP rounding procedure to yield a $\mathcal{O}\left(\log(1+\Delta)\right)$-approximation algorithm. Together, these provide upper bounds on the strengthened integrality gap of Γ and $\mathcal{O}\left(1+\log\Delta\right)$). Some algorithms attain improved guarantees at the cost of small violations of the multiplicity constraints [5,6]. Common to these LP-rounding rounding methods is that they first require obtaining an optimal solution to the KC-LP.

The strengthened linear program and its dual can be solved to near-optimality in polynomial time. Chekuri and Quanrud [5] develop a multiplicative weights method that returns approximately-optimal primal and dual solutions in near linear time, but with an $\mathcal{O}\left(1/\epsilon^5\right)$ dependence on the relative error ϵ. Their approach builds upon earlier work of Plotkin, Shmoys, and Tardos [27], and the solution method outlined by Carr et al. [4]. The returned dual solutions are feasible, lie within a $\frac{1}{1-\epsilon}$-factor of the optimum, and have a polynomially-bounded number of non-zero variables [5]. In practice, the primal and dual problems can be solved exactly using column generation for quite large instances; the problem of finding a *most violated inequality* can be reduced to solving a sequence of pseudopolynomially-many minimum-cost knapsack problems, each of which admits a pseudopolynomial exact algorithm (e.g., [20]). In our case study, we solve the KC-LP dual to optimality using this approach on large-scale problem instances with thousands of users and thousands of facilities.

5 Cost-Shares and the Strengthened LP

Our main technical contribution is to show that any feasible solution to the strengthened dual (KC-DP) produces a cost-allocation that satisfies the core property. This reconciles the apparent inconsistency for the CIP with respect to the folk understanding cost-shares and duality. The cost-shares themselves are remarkably intuitive: each user pays the part of the dual objective associated with her share of service utilization.

Theorem 1. *Let* $\mathbf{y} = (y_j^S)_{S \subseteq \mathcal{F}, j \in \mathcal{U}}$ *be a KC-LP dual-feasible solution. Then, cost-shares*

$$\xi_j := \sum_{S \subseteq \mathcal{F}} r_j^S y_j^S, \qquad \forall j \in \mathcal{U}, \tag{4}$$

satisfy the core property. We say the cost-shares in (4) are induced by $\mathbf{y}$.

Clearly, by summing the cost-shares over all users, we recover the dual objective in (KC-DP). The theorem is simple, but its consequences are profound. In particular, it can be used to exploit approximation algorithms that find integer solutions with costs bounded in terms of a corresponding dual-feasible solution.

Corollary 1. *Let* $X \subseteq \mathcal{F}$ *be a feasible solution to the CIP, and* $\mathbf{y}$ *a KC-LP dual feasible solution. If* X *is a KC-LP relative* α-*approxmationmate solution, i.e.,*

$$c(X) \equiv \sum_{i \in X} c_i \leq \alpha \sum_{j \in U} \sum_{S \subseteq \mathcal{F}} r_j^S y_j^S,$$

then the cost-shares induced by $\mathbf{y}$ *have cost-of-fair-sharing at most* α.

This implies that the Carr et al. [4] algorithm yields cost-allocations with cost-of-fair-sharing at most $\Gamma \leq m$, and that the rounding approach of Kolliopoulos and Young [18] produces a cost-of-fair-sharing of $\mathcal{O}(\ln \Delta)$, where $\Delta \leq n$.

Moreover, if small violations of the multiplicity bounds are tolerated, this is improved further on sparse instances [5,6]. These are the first price-of-fair-sharing bounds for general CIPs. Furthermore, even on the more restricted multi-set multi-cover problem, i.e., CIPs with integer inputs, our work yields improvements. The Γ-bound is new, and the latter $\mathcal{O}(\ln \Delta)$ bound dominates the existing $\mathcal{O}(\max_i \sum_j a_{ij})$ bound of Li *et al.* [22]. Finally, our theorem implies that one can also find cost-shares directly by solving the KC-LP dual. This decouples the problem of finding an integer CIP solution from that of finding cost-shares. Our case study shows this can have great value in practice.

Proof of Theorem 1. The proof is relatively simple; it naturally uses dual feasibility, and a rearrangement of summations that is standard in the analysis of primal-dual schema (see, e.g., [31]). Fix a KC-LP dual feasible solution $\mathbf{y}$, and let ξ be the induced cost-shares. The main burden of proof is to show that no subset of users has incentive to act separately from the others. In other words, we need to show that for any $J \subseteq \mathcal{U}$, we have $c_J^* \geq \sum_{j \in J} \xi$, where c_J^* is the minimum cost of an integer solution serving group J only; this is the minimum cost to the CIP in which only constraints associated with users J are included.

We prove this by considering an optimal solution to the problem of serving the smaller set of users. Let $X_J^* \subseteq \mathcal{F}$ be a minimum-cost solution for serving group J. Using (KC-DP) feasibility of $\mathbf{y}$, we see that

$$c_J^* = \sum_{i \in X_J^*} c_i \geq \sum_{i \in X_J^*} \left(\sum_{j \in \mathcal{U}} \sum_{S \subset \mathcal{F} \setminus \{i\}} a_{ij}^S y_j^S \right) \geq \sum_{i \in X_J^*} \sum_{j \in J} \sum_{S \subset \mathcal{F} \setminus \{i\}} a_{ij}^S y_j^S,$$

where the last inequality is due to the fact that all variables are non-negative.

The last part can be shown using a standard change in the order of summation. The right-hand-side expression above counts, for every item $i \in X_J^*$, the sum over subsets $S \subseteq \mathcal{F}$ that do not contain i. Equivalently, one can sum, for each subset $S \subseteq \mathcal{F}$, every item in X_J^* not in S, i.e.,

$$\sum_{i \in X_J^*} \sum_{j \in J} \sum_{S \subset \mathcal{F} \setminus \{i\}} a_{ij}^S y_j^S = \sum_{j \in J} \sum_{S \subseteq \mathcal{F}} y_j^S \left(\sum_{i \in X_J^* \setminus S} a_{ij}^S \right). \tag{5}$$

Recall that X_J^* is a feasible solution to the sub-problem of serving users in set J only. The residual demand of each subset S is satisfied by X_J^* for all users in J:

$$\sum_{i \in X_J^* \setminus S} a_{ij}^S \geq r_j^S, \quad \forall j \in J. \tag{6}$$

By applying the inequality (6) to (5), we arrive at the cost allocation to group J under the cost-shares ξ. In summary, we have shown that

$$c(X_J^*) \geq \sum_{j \in J} \sum_{S \subseteq \mathcal{F}} r_j^S y_j^S = \sum_j \xi_j. \tag{7}$$

This proves that each group $J \subseteq \mathcal{U}$ prefers to accept the cost-shares ξ over forming a group on their own. In other words, any (KC-DP) dual solution **y** produces cost-shares satisfying the core property. □

A natural question is whether there are cost-allocations, perhaps not derived from KC-LP dual variables, that have lower worst-case prices-of-fair-sharing. For CIP the answer is *no*; our bounds are provably tight, at least with respect to parameters $\varGamma$ and $\varDelta$. In particular, in the special case in which all contributions and requirements are binary, the KC-LP and its dual reduce to the standard set cover linear programs. Here, the folk theorem applies, and the cost-recovery ratio is provably upper-bounded by the integrality gap [7]. For Set Cover, and hence CIP as well, the integrality gap can be as large as $\varGamma$ and $\ln \varDelta$ [11,30]. Hence, the lower bounds of our cost-recovery ratios for CIP are tight.

6   A Case Study on LoRaWAN Coverage

We conduct a case study to evaluate the effectiveness of the KC-LP cost-sharing framework on a practical coverage-sharing problem for LoRaWAN. The study is based on a scenario in which the goal is to provide coverage over Brooklyn, NY, a relatively large and densely built urban area. The challenge is to find a gateway placement that provides sufficient coverage throughout the area at minimal cost, and to allocate the cost of the gateways across the users, subject to the core constraint, while recovering as much of the cost as possible. This is the main problem motivating our work. Our results are based on the average performance over a sequence of random problem instances. Our results suggest the KC-LP framework performs well in practice; by solving both the IP and KC-LP dual to optimality, we recover on average 93% of the cost of the gateways.

The set of problem instances is derived from a combination of real-world data and assumptions made in the absence of available data. Each instance is a CIP with random contribution matrix **A**, requirements **r**, and facilities **f**. The demand points, or users $\mathcal{U}$, are defined as a regular gird of 7,808 points over the study-area. Each instance uses a sub-sample of size 2,000. Facility locations are derived from the corners of real building footprints, and sub-sampled down to 4,380 per instance. This mimics a practical constraint faced by operators who do not have access to every site due to the high access cost. Next, each facility i is associated with a cost c_i uniformly distributed between 0 and 1. For the contribution matrix **A**, we use a distance-based Okumura-Hata model with normal noise to generate random connection qualities a_{ij} between each gateway-demand point pair (i,j) [14]. Finally, to ensure feasibility, each demand point has a requirement equal to the value of building all gateways, divided by a geometrically distributed random variable, sampled independently for each user. A total of 10 instances are used.

For each instance, we solve both the CIP and the KC-LP dual to optimality. The CIPs are solved using an off-the-shelf IP solver; the KC-LP dual is solved using the column generation routine, similar to [4]. This produces the optimal cost-shares, in the sense that the price-of-fair-sharing matches the instance-specific integrality gap exactly. To reiterate, the ability to find cost-shares by solving the KC-LP dual is a valuable consequence of our work. We are not aware of another equally tractable method for finding cost-shares of this quality in practice. We also consider additional benchmark algorithms.

The KC-LP solver is compared against two natural approximation algorithms that produce cost-shares: the PRIMALDUAL algorithm and the GREEDY algorithm. The PRIMALDUAL algorithm, or dual-ascent algorithm, incrementally grows both a feasible KC-LP dual solution, and an infeasible integer CIP solution. The algorithm terminates as soon as the integer solution is feasible. The main idea behind this algorithm is standard (see e.g. [31]), and analogous to that of [3] in a multi-user setting. Next, we also employ an existing GREEDY algorithm that also produces both an integer solution to the CIP, as well as cost-shares [22]. This algorithm produces cost-shares via dual fitting; it greedily selects gateways to add, and amortizes the per-coverage cost of selected facilities into an infeasible KC-LP dual solution. Finally, the dual variables are scaled down by $\log(n)$ to ensure feasibility. We also introduce an improved variant of this algorithm, called GREEDY+. By viewing the greedy-generated cost-shares as KC-LP dual variables, these need to be scaled down by worst-case bound, but only the minimal amount to make them KC-LP dual feasible.

One caveat of both greedy algorithms is that they only seem to work well on CIPs with integer requirements $\mathbf{r}$ and contributions $\mathbf{A}$ [9]. As such, for these algorithms only, we modify the instances by multiplying the inputs $\mathbf{A}$ and $\mathbf{r}$ by a large constant, and then rounding the connections up, and the requirements down, to the nearest integer. This modification never compromises feasibility, and does not increase the minimum cost. However, solutions to the rounded CIPs need not be feasible for the original real-valued instance. This is a potential weakness of using greedy algorithms on real-valued CIPs.

The results are summarized in Fig. 1. Each set of bars along the x-axis represent one instance; the bar heights represent costs. The GREEDY algorithm performs remarkably well, with its solution cost exceeding the minimum by only 10% on average, whereas the primal-dual algorithm performs worse, averaging nearly 30% above optimal. More remarkably, the OPTIMAL method recovers 93% of the cost on average; over twice as much as the next best algorithm. GREEDY recovers relatively little cost, even after the improved dual-fitting GREEDY+; PRIMALDUAL recovers considerably more – 45% of the minimum cost.

Overall, the KC-LP framework finds near perfect cost-shares when sharing the cost of LoRaWAN coverage. Using column generation to solve the KC-LP dual, one can recover nearly all of the cost of reasonably realistic coverage provision instances. If the instances are larger such that solving the IP and KC-LP to optimality is prohibitive, one can use both GREEDY and PRIMAL-DUAL; the former to solve the CIP, the latter to find cost-shares. Alternatively, one can solve the KC-LP to near-optimality using the algorithms methods of [4,6]. The consistency of performance across this data set (along with other results that achieved analogous performance on variants in which the costs were Gaussian rather than uniform) demonstrate the effectiveness of this theoretically-inspired approach to deliver significant results in our application.

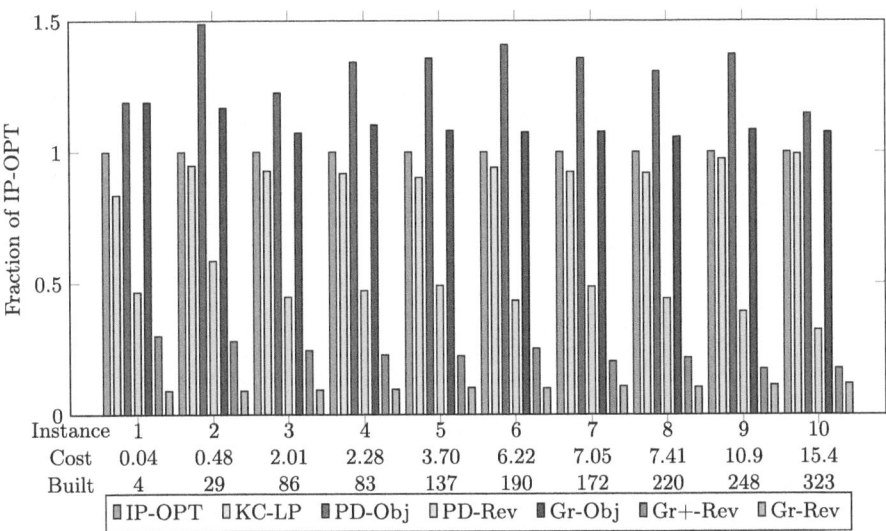

Fig. 1. Costs incurred and recovery for 10 instances. IP-OPT is the cost of the IP optimum, KC-LP the cost of the KC-LP optimum. Prefixes PD, Gr, and GR+, represent the PRIMALDUAL, GREEDY, and GREEDY+, respectively. Suffixes Obj and Rev represent the integer objective cost, and cost-share revenue, respectively.

7 Group-Strategyproof KC-LP Cost-Shares

So far it has been assumed that the group of users to be served is given; sometimes this choice also falls on the service provider. In an extended setting, the goal of the service provider is to elicit private user preferences and design a mechanism for choosing who to serve, in addition to allocating costs to those served. See Jain and Mahdian [16] for more details and definitions. Assume that each user $j \in \mathcal{U}$ has a private utility u_j for receiving service, and has the option of opting out for a utility of 0. When utilities are unknown, they must be elicited by a mechanism. This creates a challenge, because users and groups of users may be able to strategically misreport their utilities. A mechanism is said to be *strategyproof* if individual users cannot benefit from misreporting their utility, and *group-strategyproof* if no group of users can benefit by colluding to misreport their utilities. The goal of a mechanism in this setting is to elicit preferences in a group-strategyproof manner, decide who to serve, and allocate-costs in a way that maximizes the cost-recovery ratio.

To find a group-strategyproof mechanism, Moulin and Shenker [24] prove that it suffices to have cross-monotonic cost-allocation mechanism. Cost-shares $\xi : \mathcal{U} \times 2^{\mathcal{U}} \to \mathbf{R}^{|\mathcal{U}|}$ are *cross-monotonic* if the cost allocated to user j does not increase when more users are served, i.e., when for all $j \in J \subseteq U$, $\xi_j(U) \leq \xi_j(J)$. Not all cost-shares that satisfy the core property are cross-monotonic. Pál and Tardos [25] provided a general approach for using primal-dual algorithms to find cross-monotonic cost-shares in the facility location problem and the rent-or-buy problem. This approach has been widely used [17,19,21]. Interestingly, imposing cross-monotonicity usually lowers the best attainable cost-recovery ratio (or equivalently, increases the price-of-fair-sharing) [25]. Indeed, Immorlica, Mahdian, and Mirrokni [15] upper bounded the cost-recovery ratio for cross-monotonic cost allocations. In particular, a cross-monotonic cost allocation for set cover can recover at most $1/\Delta$ of the cost, where Δ is the size of the largest set. Moreover, even if all elements are covered by at most two sets, the cost recovery ratio is $\mathcal{O}\left((2+\epsilon)/n^{1/3}\right)$. Li et al. [22] describe a cross-monotonic cost allocation for the CIP that recovers $1/(2n)$ of the cost without explicit use of duality; we simplify their proof using the strengthened dual, generalize the algorithm to general CIP, and improve this to $1/(2\Delta)$.

Our strengthened linear programming approach can be used to find cross-monotonic cost-shares. The approach uses a primal-dual algorithm, in line with the general framework of Pál and Tardos [25]. The algorithm is the same as in Li et al. [22], but our analysis is different, in that it uses the KC-LP dual.

Theorem 2. *Fix users $U \subseteq \mathcal{U}$. The primal-dual mechanism [22] produces a feasible solution X for users U, and cross-monotone cost-shares $\xi_j(J)$ that satisfy $\sum_{j \in J} \xi_j(J) \geq \left(\frac{1}{2\Delta}\right) \cdot c(X)$.*

The main algorithmic idea is to let each user independently select facilities in complete isolation from the other users. Cross-monotonicity is enforced by preventing any interactions between users' dual variables. Moreover, the problem faced by an individual user is a minimum cost-knapsack problem, each of which can be solved by a primal-dual algorithm [3]. This produces, for each user $j \in U$, a selection of facilities X_j, and a KC-LP dual solution $\mathbf{y}_j$ that is feasible for the individual problem with constraints for user j only, and no variables corresponding to other users. Finally, our mechanism selects the union of all selected facilities $(X_j)_{j \in U}$, and scales down the individual dual variables $\mathbf{y}_j$ by the column sparsity Δ. The procedure is summarized in Algorithm 1, in which MinCostKnapsackPrimalDual is given in [3,22] (and our online version).

Algorithm 1: A cross-monotonic primal-dual

Input: $(\mathcal{F}, \mathcal{U}, \mathbf{r}, \mathbf{c}, \mathbf{A})$
for *all users $j \in J$ independently* **do**
$\quad X_j, \mathbf{y}'_j \leftarrow$ MinCostKnapsackPrimalDual$(\mathcal{F}, r_j, c, a_j)$
$\quad \mathbf{y}_j \leftarrow \mathbf{y}'_j/\Delta$
$X \leftarrow \cup_{j \in \mathcal{U}} X_j$
$\xi_j(U) \leftarrow \frac{1}{2\Delta} \sum_{S \subseteq F} y_j^S$ for all $j \in U$
return X, ξ

Carnes and Shmoys [3] develop and analyze a primal-dual algorithm for the minimum-cost knapsack problem based on the KC-LP formulation. Fix a single user $j \in \mathcal{U}$ and let $\mathbf{a}_j = (a_{1j}, \ldots, a_{nj})$ denote their contributions. The user starts with an all-zero dual solution $\mathbf{y}_j$, and an empty selection $X = \emptyset$. While the residual demand r_j^X is positive, they increase the dual variable y_j^X. Eventually, some constraint $\sum_{S \subseteq F - \{i\}} a_{ij}^X y_j^X \leq c_i$ becomes tight for a facility i. This facility is added to the selection X, and the process repeats. The algorithm returns a selection of facilities X_j and a feasible KC-LP dual solution. Critically, the cost of X_j is at most twice the KC-LP dual objective under $\mathbf{y}_j$.

Theorem 3 (Carnes and Shmoys [3]). *Let $X_j \subseteq \mathcal{F}$ and $\mathbf{y}_j$ be a selection of facilities, and the corresponding dual solution returned by the primal-dual algorithm for min-cost knapsack. These dual variables are feasible, and satisfy*

$$\sum_{i \in X_j} c_i \leq 2 \sum_{S \subseteq \mathcal{F}} r_j^S y_j^S.$$

This result is used in two parts of our proof. First, we use the dual feasibility of the individual dual variables $\mathbf{y}_j$ to construct a feasible dual solution to the master CIP, which gives us the core-property via Theorem 1. Secondly, the approximation ratio is used to derive our cost-recovery ratio.

Proof of Theorem 2. To prove this result, we need to argue for cross-monotonicity, the core-property, and cost recovery. We argue for cross-monotonicity first. Clearly, the dual variables $\mathbf{y}'_j$ of each user are independent of other users. Meanwhile, the maximum number of users in U served by any facility, Δ, is monotonically increasing in the size of U, so the dual variables $\mathbf{y}_j = \mathbf{y}'_j/\Delta$ are monotonically decreasing in U, as are the induced cost-shares.

The core property is easy to prove using dual feasibility and our Theorem 1. Consider some selected facility $i \in X$. Then,

$$\sum_{j \in \mathcal{U}} \sum_{S \subseteq F \setminus \{i\}} a_{ij}^S y_j^S = \frac{1}{\Delta} \sum_{\{j \in \mathcal{U}: a_{ij} > 0\}} \left(\sum_{S \subseteq \mathcal{F} \setminus \{i\}} a_{ij}^S y_j'^S \right) \leq \frac{1}{\Delta} \sum_{\{j \in J: a_{ij} > 0\}} c_i \leq c_i.$$

The first equality follows from dropping users not served by facility i. The second equality uses the definition of $\mathbf{y}_j = \mathbf{y}'_j/\Delta$. The following inequality uses the fact that the dual variables are individually feasible to the min-cost knapsack LP of each user j. The final inequality follows from the definition of Δ. The above shows that dual variables $(\mathbf{y}_j)_{j \in \mathcal{U}}$ are feasible for the CIP induced by users U, and thus Theorem 1 implies that the core-property is satisfied by the accompanying cost-shares $(\xi_j(U))_{j \in \mathcal{U}}$.

Finally, the cost-recovery recovery ratio follows from the approximation ratio of the minimum-cost knapsack algorithm. In particular, observe that

$$\sum_{i \in X} c_i \leq \sum_{j \in \mathcal{U}} \sum_{i \in X_j} c_i \leq 2 \sum_{j \in \mathcal{U}} \sum_{S \subseteq \mathcal{F}} r_j^S y_j'^S = 2\Delta \sum_{j \in \mathcal{U}} \sum_{S \subseteq \mathcal{F}} r_j^S y_j^S.$$

The first inequality is obvious; the second follows from applying Theorem 3 to each user j in U individually. The equality follows from the definition of $\mathbf{y}_j$. This proves that the cost-of-fair-sharing is at most 2Δ, as claimed. □

Finally, the proof suggests there is potential for improvement. In fact, whenever the contributions **A** are binary, the min-cost knapsack algorithm is exact, in which case the cost-recovery ratio is $1/\Delta$ [22]. Moreover, if we know that each selected facility X always selected by at least two users, it also follows that the cost-of-fair-sharing is a factor 2 smaller, i.e. Δ. On the other hand, no group-strategyproof mechanism can recover more than $1/\Delta$ of the cost in general [15]. Whether the 2Δ cost-of-fair-sharing is tight when contributions are non-binary, however, remains an open problem [22].

Acknowledgements. This material is based on work supported by the NSF under Grant CNS-1952063.

Disclosure of Interests. The authors have no competing interests to declare that are relevant to the content of this article.

References

1. LPWAN Market. https://iot-analytics.com/lpwan-market/. Accessed 29 Jun 2024
2. Aarts, S., Shmoys, D.B., Coy, A.: An interpretable determinantal choice model for subset selection (2023)
3. Carnes, T., Shmoys, D.: Primal-dual schema for capacitated covering problems. In: 13th International Conference on Integer Programming and Combinatorial Optimization, IPCO 2008, Bertinoro, Italy, 26–28 May 2008 Proceedings 13, pp. 288–302. Springer (2008)
4. Carr, R.D., Fleischer, L.K., Leung, V.J., Phillips, C.A.: Strengthening integrality gaps for capacitated network design and covering problems. In: Proceedings of the Eleventh Annual ACM-SIAM Symposium on Discrete Algorithms, SODA '00, pp. 106–115. Society for Industrial and Applied Mathematics, USA (2000)
5. Chekuri, C., Quanrud, K.: On Approximating (Sparse) Covering Integer Programs. In: Proceedings of the Thirtieth Annual ACM-SIAM Symposium on Discrete Algorithms, pp. 1596–1615. SIAM (2019)
6. Chen, A., Harris, D.G., Srinivasan, A.: Partial resampling to approximate covering integer programs. Random Struct. Algorithms **58**(1), 68–93 (2021)
7. Deng, X., Ibaraki, T., Nagamochi, H.: Algorithmic aspects of the core of combinatorial optimization games. Math. Oper. Res. **24**(3), 751–766 (1999)
8. Devanur, N.R., Mihail, M., Vazirani, V.V.: Strategyproof cost-sharing mechanisms for set cover and facility location games. In: Proceedings of the 4th ACM Conference on Electronic Commerce, pp. 108–114 (2003)
9. Dobson, G.: Worst-case analysis of greedy heuristics for integer programming with nonnegative data. Math. Oper. Res. **7**(4), 515–531 (1982)
10. Faulhaber, G.R.: Cross-subsidization: pricing in public enterprises. Am. Econ. Rev. **65**(5), 966–977 (1975)
11. Feige, U.: A threshold of ln n for approximating set cover. J. ACM (JACM) **45**(4), 634–652 (1998)
12. Goemans, M.X., Skutella, M.: Cooperative facility location games. J. Algorithms **50**(2), 194–214 (2004)
13. Gupta, A., Kumar, A., Pál, M., Roughgarden, T.: Approximation via cost-sharing: a simple approximation algorithm for the multicommodity rent-or-buy problem. In: Proceedings of the 44th Annual IEEE Symposium on Foundations of Computer Science, 2003, pp. 606–615. IEEE (2003)

14. Hata, M.: Empirical formula for propagation loss in land mobile radio services. IEEE Trans. Veh. Technol. **29**, 317–325 (1980)
15. Immorlica, N., Mahdian, M., Mirrokni, V.S.: Limitations of cross-monotonic cost-sharing schemes. ACM Trans. Algorithms (TALG) **4**(2), 1–25 (2008)
16. Jain, K., Mahdian, M.: Cost Sharing, pp. 385-410. Cambridge University Press (2007)
17. Jain, K., Vazirani, V.V.: Equitable cost allocations via primal-dual-type algorithms. SIAM J. Comput. **38**(1), 241–256 (2008)
18. Kolliopoulos, S.G., Young, N.E.: Approximation algorithms for covering/packing integer programs. J. Comput. Syst. Sci. **71**(4), 495–505 (2005)
19. Könemann, J., Leonardi, S., Schäfer, G., van Zwam, S.H.: A group-strategyproof cost sharing mechanism for the Steiner forest game. SIAM J. Comput. **37**(5), 1319–1341 (2008)
20. Lawler, E.L.: Fast approximation algorithms for knapsack problems. In: 18th Annual Symposium on Foundations of Computer Science, SFCS 1977, pp. 206–213. IEEE (1977)
21. Leonardi, S., Schäfer, G.: Cross-monotonic cost sharing methods for connected facility location games. Theor. Comput. Sci. **326**(1), 431–442 (2004). https://www.sciencedirect.com/science/article/pii/S0304397504004876
22. Li, X.Y., Sun, Z., Wang, W., Lou, W.: Cost sharing and strategyproof mechanisms for set cover games. J. Comb. Optim. **20**(3), 259–284 (2010)
23. Meir, R., Bachrach, Y., Rosenschein, J.S.: Minimal subsidies in expense sharing games. In: Algorithmic Game Theory: Third International Symposium, SAGT 2010, Athens, Greece, 18–20 October 2010, Proceedings 3, pp. 347–358. Springer (2010)
24. Moulin, H., Shenker, S.: Strategyproof sharing of submodular costs: budget balance versus efficiency. Econ. Theor. **18**, 511–533 (2001)
25. Pál, M., Tardos, É.: Group strategyproof mechanisms via primal-dual algorithms. In: Proceedings of the 44th Annual IEEE Symposium on Foundations of Computer Science, 2003, pp. 584–593. IEEE (2003)
26. Parsons, S.G.: Cross-subsidization in telecommunications. J. Regul. Econ. **13**, 157–182 (1998)
27. Plotkin, S.A., Shmoys, D.B., Tardos, É.: Fast approximation algorithms for fractional packing and covering problems. Math. Oper. Res. **20**(2), 257–301 (1995)
28. Semtech: LoRa® and LoRaWAN®: a technical overview. https://lora-developers.semtech.com/documentation/tech-papers-and-guides/lora-and-lorawan/. Accessed 07 Sept 2021
29. Semtech: Network Providers. https://www.semtech.com/lora/ecosystem/networks. Accessed 29 Jun 2024
30. Singh, M.: Integrality gap of the vertex cover linear programming relaxation. Oper. Res. Lett. **47**(4), 288–290 (2019)
31. Williamson, D.P., Shmoys, D.B.: The Design of Approximation Algorithms. Cambridge University Press (2011)

Improved Online Scheduling with Restarts on a Single Machine

Aflatoun Amouzandeh[✉][iD] and Rob van Stee[iD]

University of Siegen, Siegen, Germany
{aflatoun.amouzandeh,rob.vanstee}@uni-siegen.de

Abstract. We consider the problem of minimizing the total completion time on a single online machine using restarts. Although restarts can potentially be very beneficial in the context of online scheduling, there has been relatively little research on this topic up until now.

We present a very simple online algorithm which is better than 1.4568-competitive. The basic rule of the algorithm is to run jobs in order of increasing size. For possible restarts, the algorithm only considers whether the completion time of an incoming job is more than a factor of 1.4568 higher if we first let the running job finish compared to the case where we start the new job immediately. If this is the case, we interrupt the running job and start running the new job. All other existing or past jobs are ignored for this decision.

The analysis of the algorithm has become significantly easier and shorter than for the previous best result which was 3/2. We hope that this result can lead to further research in this interesting topic.

Keywords: Online Algorithms · Single Machine Scheduling · Completion Time

1 Introduction

Online scheduling of jobs on a single machine is a fundamental problem in online algorithms and its study dates back to 1966, when Graham [7] considered load balancing (that paper mentions the concept of online algorithms by name, although they were not widely studied until the early 1990s). An online scheduling algorithm determines whether to process a job or not without knowing the future events. Despite many years of study, many online scheduling problems are not (completely) resolved, and even for the standard load balancing problem a small gap remains.

We study online algorithms for the single machine scheduling problem, using restarts with the objective of minimizing the total completion time. The problem is denoted by $1|r_j, online, restart| \sum C_j$. Each job has a size which is the amount

---

This research was financially supported by the Deutsche Forschungsgemeinschaft (DFG, German Research Foundation) - Project Number 451155613.

© The Author(s), under exclusive license to Springer Nature Switzerland AG 2025
M. Bieńkowski and M. Englert (Eds.): WAOA 2024, LNCS 15269, pp. 16–30, 2025.
https://doi.org/10.1007/978-3-031-81396-2_2

of time it needs to be run in order to complete. Allowing restarts means that the processing of a job may be interrupted and in this case, the interrupted job loses all its previous progress and must be started again later, until it is completed without interruptions. The scheduling decisions must be made online, without prior knowledge of future job arrivals. The effectiveness of an online algorithm is measured by its competitive ratio [2,5,10]. Given an instance I of the problem, we denote the objective values of the online and optimal offline algorithms by $C_{ALG}(I)$ and $C_{OPT}(I)$, respectively. The competitive ratio $\mathcal{R}$ of an online algorithm ALG is defined as

$$\mathcal{R} = \sup_I \frac{C_{ALG}(I)}{C_{OPT}(I)}.$$

This means that if an algorithm has a competitive ratio of $\mathcal{R}$, we have the guarantee that no matter how bad the input (from the point of view of the algorithm), the algorithm will never have cost more than $\mathcal{R}$ times the optimal cost. Note that the focus here is entirely on the lack of knowledge about the input; the online algorithm may use as much computation as it needs to make its decisions.

The online single machine scheduling with the objective to minimize the total completion time is the same as minimizing the average completion time of the jobs: we consider how much time jobs need to complete on average, and compare this to the optimal solution. Several 2-competitive algorithms exist, due to Phillips et al. [12], Stougie and Hoogeveen and Vestjens [9]. Hoogeveen and Vestjens [9] introduced the delayed SPT (DSPT) algorithm, showed that the ratio of 2 is optimal and that all three algorithms can perform twice as bad as any of the others. Another important and well-studied objective is to minimize the total weighted completion time, which is the weighted versions of this function [1,6,8,11,15].

A very important model which has not received so far the attention it deserves and will be a central focus of this paper, is allowing restarts. In this model, it is possible to interrupt a running job in case a more urgent job arrives (a smaller one, or a heavier one in the case of weighted jobs), but in this case the job that was running must be restarted from the beginning: the work done on it is lost. This model is also called preemption with restarts and is different from the more commonly studied preemption with resume model, where an interrupted job can simply be resumed from the point of interruption. The concept of restarts was initially introduced by Shmoys et al. [13] for makespan minimization. Van Stee and La Poutré in [14] introduce the RSPT algorithm that achieves a deterministic competitive ratio of $3/2$ for minimizing total completion time on a single machine with using restarts, surpassing the lower bound of $e/(e-1) \approx 1.582$ for randomized non-preemptive online algorithms [3]. The algorithm RSPT is based on a very simple rule that the job currently running is only interrupted if there exists another job that can complete before $2/3$ times the completion time of the running job (if left to run). The current lower bound for this problem is 1.21 and is due to Epstein and van Stee [4]. It uses an exponentially growing (but relatively short) sequence of jobs interspersed with jobs of (essentially) zero size which arrive at inopportune moments, forcing restarts in each step.

In our study, we consider the power of restarts to achieve a competitive ratio below 3/2. The primary contribution of this paper lies in presenting an online algorithm for the problem which achieves a competitive ratio better than 1.4568 (the real root of the equation $6 + 3\rho^2 = 4\rho^3$). The analysis of our algorithm is tight. This is the first improvement for this problem in twenty years and the analysis has become significantly easier and shorter than in [14] for the previous best result.

2 Algorithm

We begin with some terminology and notation. We denote by r_i and p_i, the release date and the processing time of job J_i, respectively. Given a specific set of jobs $S = \{J_1, \ldots, J_n\}$, we use r_i^S and p_i^S to denote the release date and the processing time of job J_i ($i = 1, \ldots, n$), respectively. Throughout the paper, we use σ to denote the schedule of the algorithm and π^* to denote an optimal offline schedule. The completion time of J_i in a schedule ψ will be denoted by ψ_i. The relative completion time of a job J_i in σ compared to a feasible schedule ψ is $rct_i^\psi = \sigma_i / \psi_i$. We propose an improved version of the RSPT algorithm. In the original RSPT algorithm, the currently running job J_i which started most recently at time s, is only interrupted if there exists an incoming job J_q that can complete before 2/3 times the completion time of J_i. This interruption condition can be represented by the equation

$$r_q + p_q \leq \frac{2}{3}(s + p_i).$$

Improved RSPT Algorithm (IRSPT). Our new algorithm introduces a more nuanced interruption condition that taken into account the size of the incoming jobs more. At any time when the machine is idle or the algorithm completes a job, the algorithm starts processing the smallest available job. If there is a choice, it takes the job with the smallest release date. Consider a job J_i that started (most recently) at time s and an incoming job J_q. There are two conditions for interrupting a job. The incoming job must be smaller than the running job and the algorithm interrupts J_i based on the following equation

$$r_q + p_q < \frac{1}{R}(s + p_i + p_q), \tag{1}$$

where R is the real root of the equation $6 + 3\rho^2 = 4\rho^3$. For any incoming job that satisfies (1), the algorithm interrupts J_i and starts processing of J_q at time r_q. The interrupted job loses all its previous progress and must be started again later.

In (1), the value $r_q + p_q$ is the completion time of J_q if it starts at time r_q, whereas the value $s + p_i + p_q$ is the completion time of J_q if it starts immediately after J_i completes. This inequality implies that an incoming job J_q will only cause an interruption if it can complete before time $\frac{1}{R}(s+p_i+p_q)$ after starting at its release date. This means we start J_q immediately if its relative completion time compared to the optimal schedule could otherwise be larger than R. An interruption based solely on this inequality is not always beneficial, as there may be incoming jobs larger than p_i that also satisfy the inequality. Therefore, as mentioned we refine our rule with a second condition, so that the algorithm only restarts for incoming jobs smaller than p_i. Figure 1 illustrates an example of our restarting rule.

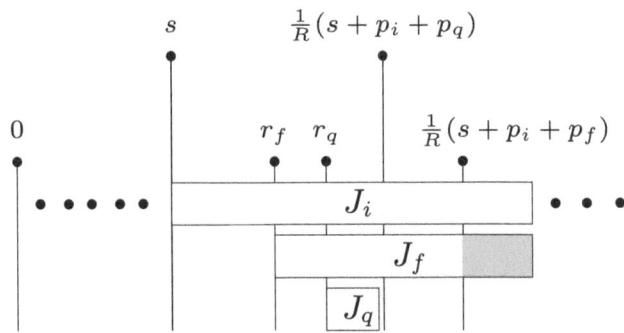

Fig. 1. Illustration of incoming jobs J_f and J_q. The job J_f does not cause an interruption of J_i, but the job J_q does.

2.1 IRSPT-Like Algorithms

It can be seen that the competitive ratio of IRSPT does not improve beyond R: consider a job of size 1 that arrives at time 0, and N jobs of size 0 that arrive at time $1/R+\varepsilon$. IRSPT will run these jobs in order of release date and have a total completion time of $N+1$. However, it is possible to achieve a total completion time of $(1/R+\varepsilon)(N+1)+1$ by running the jobs of size 0 first. Therefore, as N grows and ε tends to 0, the competitive ratio tends to R. Generally, we define a family of algorithms for the problem $1|r_j, online, restart| \sum C_j$. Any IRSPT-like algorithm behaves exactly like IRSPT algorithm, but with using a value ρ instead of R.

We show in the next theorem that no algorithm from the IRSPT-like family outperforms IRSPT for any value of ρ.

Theorem 1. *For the problem* $1|r_j, online, restart| \sum C_j$, *any IRSPT-like algorithm with* $1 < \rho < 2$ *has a competitive ratio of at least* R *(that is, the real root of the equation* $6 + 3\rho^2 = 4\rho^3$*).*

Proof. Consider an IRSPT-like algorithm ALG and the following job sequence: a job J_1 of size $1+\varepsilon$ arrives at time 0, a job J_2 of size 1 at time $\frac{2-\rho}{\rho}$, and a job

J_3 of size 0 at time $\frac{2-\varepsilon}{\rho^2}$. Both J_2 and J_3 cause interruptions upon their release dates: J_2 interrupts J_1 and J_3 in turn interrupts J_2. As ε approaches 0, if $\rho < \sqrt{2}$ then the optimal cost for this sequence tends to $2 + \frac{4}{\rho^2}$ (using the order J_1, J_3, J_2 and noting that $r_3 > 1 + \varepsilon$ for ε small enough and $\rho^2 < 2$), while the cost of ALG tends to $\frac{6}{\rho^2} + 3$ (using the order J_3, J_2, J_1). Therefore, $\mathcal{R}(ALG) \geq 1.5$. Otherwise, if $\rho \geq \sqrt{2}$ then the optimal cost for this sequence tends to 4 (using the order J_1, J_3, J_2), while the cost of ALG still tends to $\frac{6}{\rho^2} + 3$ (using the order J_3, J_2, J_1). Hence, $\mathcal{R}(ALG) \geq \max\left(\rho, \frac{6+3\rho^2}{4\rho^2}\right) \geq R$, where R is the real root of the equation $6 + 3\rho^2 = 4\rho^3$. □

In this paper we have used jobs of size 0. If it is required that job sizes are strictly positive, our upper bound of course still holds, and our lower bounds also continue to hold by using jobs of arbitrarily small size instead of size 0.

3 Analysis of IRSPT Algorithm

To simplify the analysis, we will focus exclusively on the IRSPT algorithm. In order to prove that IRSPT has a competitive ratio R, we assume there exists a smallest counterexample consisting of a minimum number of jobs, and show by contradiction that such a counterexample does not exist. We partition the processing of this input into two events, which we will analyze separately.

- Job completions. While the completed job was running, other jobs may have arrived. We will need to show that these jobs complete early enough (on average) and that they do not delay other waiting jobs too much.
- Job restarts. While the interrupted job was running, other jobs may have arrived earlier without causing a restart. The input may also contain larger jobs that get delayed as well due to this interruption.

To better understand the behavior of the IRSPT algorithm, we explore several properties. In our analysis, we will use these properties to upper bound the competitive ratio of the IRSPT algorithm.

Observation 2. *Any interruption of a job J_i starting at time s occurs before time $\frac{1}{R}(s + p_i)$ and job J_i cannot be interrupted at $\frac{p_i}{R-1}$ or later.*

Proof. If an interruption of J_i happens by J_q, then

$$r_q + p_q < \frac{1}{R}(s + p_i + p_q) = \frac{1}{R}(s+p_i) + \frac{1}{R}p_q \Leftrightarrow r_q + \left(1 - \frac{1}{R}\right)p_q < \frac{1}{R}(s+p_i). \quad (2)$$

Hence, $r_q < \frac{s+p_i}{R}$, meaning that any interruption of J_i is before time $\frac{1}{R}(s+p_i)$. Additionally, $r_q < \frac{s+p_i}{R} < \frac{r_q+p_i}{R}$ implies $r_q < \frac{p_i}{R-1}$, so J_i cannot be interrupted at $\frac{p_i}{R-1}$ or later. □

Observation 3. *The algorithm interrupts the processing of J_i for another job J_q if and only if less than $\frac{1}{R}$ of J_q will remain unfinished at time $\frac{1}{R}(s + p_i)$.*

Proof. By 2, an interruption happens if more than $1 - \frac{1}{R}$ of J_q will be completed strictly before time $\frac{1}{R}(s+p_i)$, when J_q is started at time r_q. Therefore, less than $\frac{1}{R}$ of J_q will remain unfinished at time $\frac{1}{R}(s+p_i)$. □

Observation 4. *A job J_i that starts at time s and is still running at time $\frac{1}{R}(s+p_i)$ will not be restarted.*

Proof. This follows immediately from Observation 2. □

Lemma 1. *Suppose the algorithm ALG interrupts a job J_i at time t. Then, at time t, OPT has not completed any job of size less than p_i that ALG has not completed.*

Proof. At time s, ALG has completed all jobs of size less than p_i that have arrived because the algorithm always prefers smaller available jobs. All jobs that arrived in the time interval (s,t) are unfinished at time t in any feasible schedule by Observation 3, since they do not cause a restart. (In fact, more than $1/R$ of each such job is unfinished) □

We explain several structural properties of a smallest counterexample (with the minimal number of jobs), denoted by I. Let σ be the schedule produced by IRSPT algorithm ALG for I. The schedule σ starts at a nonnegative time. Since I is a smallest counterexample and the algorithm does not base its decision on anything that happened before the last gap, there are no gaps in σ after the time of the last interruption, and no gaps at all if there was no interruption.

Observation 4 implies that for every job of nonzero size, there is an interval of length $\min(p_i, (1-1/R)(s+p_i))$ in which it will not be restarted. We use this to simplify the input as follows. If a job J_q arrives exactly at a time t when a job of nonzero size completes, we do the following. Let J_i be the unique job of nonzero size that completes at time t. We decrease r_q to some value in the open interval $(\max(s, (s+p_i)/R), s+p_i)$ without affecting the schedule σ created by IRSPT. The optimal cost can only decrease as a result. This adjustment ensures that each job after the last interruption arrives while another job J_i of nonzero size is running (strictly before J_i completes, but after the incoming job could cause an interruption).

We next demonstrate how the instance I can be modified to achieve a simpler and more specific structure.

Lemma 2. *Let I be a smallest counterexample for IRSPT algorithm ALG. Suppose that the algorithm restarted at least once. Another smallest counterexample I' can be constructed such that no job arrives after the last interruption.*

Proof. Consider the last release date in I, we denote the job being processed at that time in σ by J_i. We will show that without increasing the number of jobs, we can construct a new instance I' by modifying I so that no jobs arrive while J_i is processing, and the competitive ratio of modified instance does not decrease.

Let s denote the (final) starting time of J_i and the set of jobs that arrive while J_i is processing is S. Let $p_{\max}^S$ represent the largest size among the jobs in S. If there are multiple jobs with size $p_{\max}^S$ in S, we can adjust their release dates so that all these jobs arrive at the earliest release date among jobs of size $p_{\max}^S$ in S without changing the processing order in σ (because ALG did not interrupt J_i with the earliest job of size $p_{\max}^S$) and this does not increase the optimal cost. Let $r_{\max}^S$ represent the earliest release date among the jobs of size $p_{\max}^S$ in S. In the next step, when we change the value of $p_{\max}^S$, we change the size of all the jobs of size $p_{\max}^S$ in S (so that all these jobs continue to have size $p_{\max}^S$).

If $p_{\max}^S \geq p_i$, we can set the release date of the job(s) with size $p_{\max}^S$ to time r_i without changing the behavior of the algorithm (apart from possibly the order in which some jobs of size p_i are being run) or increasing the optimal cost. As a result, such jobs are no longer in S. Therefore, without loss of generality we have $p_{\max}^S < p_i$.

We define $\delta_{\max}^i$ as the smallest possible processing time such that any job with a smaller processing time that arrives at $r_{\max}^S$ would cause ALG to interrupt J_i. Consider the cost of the algorithm and the optimal cost as functions of parameter $p_{\max}^S$ as it varies in within the size range $[x_0, p_i]$, where x_0 is the maximum of $\delta_{\max}^i$ and the second largest size in S. We observe that in the entire size range $[x_0, p_i]$, the cost of the algorithm is linearly increasing because there are no gaps in its schedule. The optimal cost is a piecewise linear function (in each piece, a fixed permutation of the jobs is used) with a slope that does not increase with $p_{\max}^S$, thus the optimal cost function is concave. The worst-case instance (the highest competitive ratio) is therefore achieved on one of the endpoints of this size range $[x_0, p_i]$. Therefore, we can obtain an instance with a worse competitive ratio by choosing the value of $p_{\max}^S$ to be either minimal or maximal. In the following, we show that by repeating this instance reduction procedure as often as needed, the lemma follows. Recall that s denote the (final) starting time of J_i.

Case 1. If $s \geq \frac{p_i}{R-1}$, meaning that J_i would never be restarted, no matter what other jobs arrived (Observation 2). Then we can move the release date of all jobs in S to time $s + \varepsilon$ for some very small ε. This can only decrease the optimal cost and does not change the schedule of ALG. Now all jobs in S have the same release date and will be processed in the same order by ALG and OPT, since no further jobs arrive after time $s + \varepsilon$.

Now, if $p_{\max}^S$ is maximal in the worst-case instance, we change the value of $p_{\max}^S$ to p_i and then the release date of the job(s) with size $p_{\max}^S$ can be moved back (decreasing it) to time r_i. Alternatively, if $p_{\max}^S$ is minimal in the worst-case instance, then we change the value of $p_{\max}^S$ to x_0 and repeat the procedure for the new instance with the updated value of $p_{\max}^S$. These adjustments do not change the processing order in σ and do not increase the optimal cost. By repeating the above instance reduction procedure based on $p_{\max}^S$ as often as needed, eventually all jobs in S have the same size. The relative completion time of all jobs in S is at most R. Therefore, it cannot be the case that any of these jobs have size 0, because a job of size 0 could be removed without affecting the schedule and this would imply that I is not a smallest counterexample. Consequently, all jobs in

S should have size p_i and can arrive earlier in another instance with the same processing order as in σ, without decreasing the competitive ratio.

Case 2. If $s < \frac{p_i}{R-1}$, then consider a job J_j^S and suppose that ALG and OPT start to run this job at time t_j and t_j^*, respectively. Let d_j be the number of unfinished jobs in ALG schedule at time $c_j = t_j + p_j$ and d_j^* be the number of unfinished jobs in OPT schedule at time $c_j^* = t_j^* + p_j$. Consider the jobs in S, where $|S| = n$, numbered according to their positions in σ. We define the above variables analogously for the job J_n^S with size $p_{\max}^S$ that ALG starts to run last (among all the jobs in S). It is clear that the relative completion time of J_n^S is less that or equal to R when both ALG and OPT complete J_n^S last, because job J_i is not interrupted for any job in S. Additionally, we can assume J_n^S is the last arriving job, and then the contribution of J_n^S to the online cost is $c_n + d_n p_{\max}^S$ and its contribution to the offline cost is $c_n^* + d_n^* p_{\max}^S$. This immediately means that $d_n/d_n^* > R$ because $\frac{c_n + d_n p_{\max}^S}{c_n^* + d_n^* p_{\max}^S} > R$, since the instance is a smallest counterexample. Therefore, the size of J_n^S in the worst-case instance should be maximized and set equal to the size of J_i and then it can be moved back to the release date of J_i. Now we are in the same situation as before, without changing the processing order in σ and without increasing the optimal cost, but with one fewer job in S.

If OPT does not run J_n^S last, let J_f be the job in S that OPT starts first. Then we can modify the instance I by letting all jobs that arrive before OPT completes J_f to the optimal finishing time of J_f (or to $s + p_i - \varepsilon$ for some very small ε if that is earlier), without increasing the optimal cost. Furthermore, all jobs that arrive after OPT completes J_f but before ALG completes J_i can be shifted to arrive at the time at which OPT finishes J_f (or to $s + p_i - \varepsilon$). Now, all jobs in S except for J_f arrive at the same time and are processed in nondecreasing order after this time in OPT schedule. Therefore, in the schedule of OPT there is only one job in S, that is out of sequence, and that one job (if it exists) is run first. Hence, if OPT does not run J_n^S last, thus we know that it starts this job first.

Let J_ℓ denote the job in S that OPT starts last and assume OPT starts the first job (among all the jobs in S) at time t^*. The variables for J_ℓ^S are defined analogously as before. Because job J_i is not interrupted by any job in S, the following inequality holds:

$$\frac{c_\ell}{c_\ell^*} = \frac{s + p_i + \sum_{j=1}^{n-1} p_j^S}{t^* + p_{\max}^S + \sum_{j=1}^{n-1} p_j^S} \leq \frac{s + p_i + \sum_{j=1}^{n-1} p_j^S}{r_n^S + \frac{1}{R} p_{\max}^S + \sum_{j=1}^{n-1} p_j^S} \leq R.$$

Therefore, the relative completion time of J_ℓ is less than or equal to R. Without loss of generality, we can assume that J_ℓ is the last arriving job, then the contribution of J_ℓ^S to the online cost is $c_\ell + d_\ell p_\ell$ and its contribution to the offline cost is $c_\ell^* + d_\ell^* p_\ell$. This immediately means that $d_\ell/d_\ell^* > R$ because $\frac{c_\ell + d_\ell p_\ell}{c_\ell^* + d_\ell^* p_\ell} > R$, since the instance is a smallest counterexample. Hence, if the size of J_ℓ^S is not equal to J_n^S, it should be maximized and set equal to the size of J_n^S and then it can be moved back to the release date of J_n^S. Therefore, J_n^S and J_ℓ^S with same size have same release date. This allows us to exchange them in the OPT schedule without any change in the optimal cost. As a result, both ALG and OPT complete J_n^S last, and then the release date of this job can be decreased to time r_i as described above.

Consequently, we have shown that the largest job(s) in S can be moved back while keeping a counterexample. Repeating this as often as needed (finitely many times) means that S contains no jobs at all, and more generally that no jobs arrive after the last interruption. □

In the following proofs, we will use the concept of credit. The credit of a job indicates how much more it could (additionally) be delayed without violating the competitive ratio. For instance, a job of size 1 that arrives at time 0 has credit $R-1$ (if it is the only job in the input), because it cannot complete before time 1, so the competitive ratio would still be maintained if it completed at time R, but ALG will complete it at time 1 as long as no other jobs arrive.

To show that our algorithm is R-competitive, it is then sufficient to show that the set of jobs as a whole has nonnegative credit at the end of processing, combined of course with the fact that the credit of each job is calculated properly when it arrives, based on the optimal schedule. If a job J_i is currently scheduled to complete at time σ_i and we know that OPT completes it at time ψ_i, then this job has credit of $R\psi_i - \sigma_i$. Of course, job J_i may be delayed further because of restarts and/or other (smaller) jobs arriving. Therefore, when a job arrives we generally do not know when the online algorithm will complete it nor when the offline algorithm will complete it. To deal with these problems, we will show that for each event separately, it is possible to reassign credits so that every job has nonnegative credit (not just the set as a whole), no matter what the optimal schedule is, and given that all jobs that already existed at the previous event had nonnegative credit after that event. We distinguish cases based on the possible optimal schedules and the starting time of the completed or interrupted job. The R-competitiveness then follows by combining this with using the technique of a smallest counterexample.

We showed that any smallest counterexample can be modified so that no job arrives after the last interruption without decreasing the implied competitive ratio. It is then sufficient to show that each job can have nonnegative credit immediately after the last interruption; it follows from the minimality of the counterexample that credit could previously be assigned such that all jobs have nonnegative credit.

Let D_j denote the set containing J_j and the jobs at least as large as J_j that ALG will complete after it (and have already arrived). Let S_j denote the set containing all jobs smaller than J_j that arrived in the time interval $(s, t]$. The jobs in D_j and S_j are numbered according to their position in the schedule of ALG, respectively. We say a job J_j in an instance I delays the processing of another job (or jobs) if that other job (or jobs) starts processing earlier in the ALG schedule of the instance $I \setminus \{J_j\}$.

Consider the job J_i started running (most recently) at time s before being interrupted by J_q at time t and we denote the starting time of J_i in OPT schedule by s^*. We define the above sets analogously for the job J_i and J_q. We consider all possible cases for the schedule of OPT based on the number of jobs in D_i that OPT completes before J_q. Therefore, it is sufficient to consider the possible optimal schedules for the jobs that have arrived by time t. The optimal schedule might vary if some jobs arrive later, but obviously the total cost of the existing jobs by time t cannot decrease in such a scenario.

Observation 2 indicates that interruptions do not occur after time $\frac{p_i}{R-1}$. Therefore, since $\frac{p_i}{R-1} \approx 2.1891 p_i$, OPT can complete at most three jobs from D_i before it starts processing J_q. This is because J_q is smaller than every job in D_i, and OPT processes jobs after time t in non-decreasing order of size, under the condition that no other jobs arrive. Hence, we only need to consider a few cases.

Lemma 3. *If no job from the set D_i is completed before J_q by OPT, then immediately after an interruption of J_i at time t, we can reassign credit such that all jobs maintain nonnegative credit.*

Proof. We distinguish two cases depending on the job that OPT runs first.
Case 1. OPT processes the job that causes the interruption first. In this scenario, the same set of unfinished jobs is processed in non-decreasing order of size after time t in both of OPT's and ALG's schedules, since we analyze the schedule immediately after the interruption. Therefore, the credit of J_i is $(R-1)(t + \sum_j p_j^{S_i} + p_i)$, ensuring that all jobs maintain nonnegative credit.
Case 2. OPT runs another job first, say a job J_f that arrives at time t_1 and does not satisfy (1). In this case, the jobs in S_f delay the completion time of the other jobs except J_f by the same amount in the schedule of ALG and OPT. We have that OPT completes the ℓth job in S_f no earlier than at time $t_1 + p_f + \sum_{j=1}^{\ell} p_j^{S_f}$, whereas ALG completes the ℓth job in S_f at time $t + \sum_{j=1}^{\ell} p_j^{S_f}$. This means the credit of the ℓth job in S_f is at least $R(t_1 + p_f) - t + (R-1) \sum_{j=1}^{\ell} p_j^{S_f}$. Additionally, we have $R(t_1 + p_f) - p_f \geq s + p_i > t$, because J_f does not cause an interruption. Therefore, $p_f + (R-1) \sum_{j=1}^{\ell} p_j^{S_f} = (R-1)(\frac{1}{R-1} p_f + \sum_{j=1}^{\ell} p_j^{S_f}) \geq (R-1)(\frac{1}{R-1} p_f + p_\ell^{S_f}) \geq p_\ell^{S_f}$, since $p_\ell^{S_f} \leq p_f$. The delay of J_f caused by the ℓth job is also at most $p_\ell^{S_f}$, and we have shown that we have enough credit for the delay of J_f. Hence, in the worst-case scenario, S_f only contains the job that caused the interruption (J_q). Now, we consider the jobs in $D_f \setminus \{J_f\}$. We have

$$\frac{t + p_q + 2p_f}{t_1 + p_q + 2p_f} < \frac{s + p_i + p_q + 2p_f}{t_1 + p_q + 2p_f} \leq R,$$

holds because the job J_f does not cause an interruption. Therefore, the relative completion time of $J_2^{D_f}$ is less than R. Hence, for the jobs in $D_f \setminus \{J_f\}$, the relative completion time is at most R because both ALG and OPT process these jobs in a same order. Consequently, in the worst-case instance (the highest competitive ratio), D_f comprises J_f, the interrupted job (J_i) and the large jobs (at least as large as J_i) that caused delays before time s to the processing of at least one other job. The ratio between the completion time of J_f in ALG schedule and completion time of J_q in OPT schedule is less than R, since J_f does not satisfy (1). This implies the existence of a credits reassignment where all jobs have nonnegative credit. □

Lemma 4. *If OPT completes one job from D_i (the interrupted job J_i or a job that is at least as large as it), then immediately after an interruption of J_i at time t, we can reassign credit such that all jobs maintain nonnegative credit.*

Proof. In this scenario, the interrupted job J_i is delayed by $t + \sum_j p_j^{S_i} - s$ because of the restart. We first consider the jobs smaller than J_i that are available at the time of interruption (the set S_i). ALG will complete these jobs sequentially from smallest to largest, while OPT might use a different order. Additionally, OPT may run a job different from J_i (but at least as large as it) first. However, if OPT completes a job at least as large as J_i first, the completion time of all jobs will increase, thus the credits of all jobs increase or remain the same (if the size matches with the size of J_i, we can simply swap the labels and the cost of OPT remains unchanged). Therefore, it is sufficient to only consider the case where OPT runs the interrupted job J_i first. Here, we begin by analyzing the jobs in S_i that did not cause an interruption to J_i. We consider two scenarios based on the interruption time t.

For $t \leq p_i$, under the condition that OPT completes J_i before J_q, OPT completes the ℓth job in S_i no earlier than at time $p_i + \sum_{j=1}^{\ell} p_j^{S_i}$, whereas ALG completes the ℓth job in S_i at most at time $t + \sum_{j=1}^{\ell} p_j^{S_i}$. This implies the credit of the ℓth job in S_i is at least $Rp_i - t + (R-1)\sum_{j=1}^{\ell} p_j^{S_i}$. The additional delay of J_i caused by the ℓth job in S_i is at most $p_\ell^{S_i}$. Since $Rp_i - t \geq Rp_i - p_i = (R-1)p_i$, the job J_i can receive $(R-1)(p_i + \sum_{j=1}^{\ell} p_j)$ credit from the ℓth job. This credit is greater than $p_\ell^{S_i}$ under the assumption that $p_\ell^{S_i} \leq \frac{R-1}{2-R} p_i$.

For $t > p_i$, OPT may choose to run another job J_f in S_i before J_q. Suppose J_f has size f and OPT starts it at time t^*. Other jobs in S_i completed by ALG before J_f gain more credit in this scenario. Consider the case that S_i only contains J_q and J_f. It is important to note that J_f did not cause a restart, so its completion time is after $t + p_q$. Therefore, the difference in credit for J_q, compared to when OPT completes J_q first, is $R(t^* + f + p_q) - R(t + p_q) = R(t^* - t + f) > t^* - t + p_q$. Hence, we only need to consider the case where OPT also completes J_q first. Thus, the credit of the ℓth job in S_i is at least $(R-1)(t + \sum_{j=1}^{\ell} p_j^{S_i})$. It follows that the interrupted job can receive $(R-1)(t + \sum_{j=1}^{\ell} p_j^{S_i})$ credit from the ℓth job. This credit is greater than $p_\ell^{S_i}$ under the assumption that $p_\ell^{S_i} \leq \frac{R-1}{2-R} p_i$ as above. Consequently, there can only be one job in $S_i \setminus \{p_q\}$ and at most $(1 - \frac{R-1}{2-R})p_i$ credit is needed for the delay caused by S_i to J_i.

Next, we consider the jobs in $(D_i \setminus \{J_i\})$, which are at least as large as J_i. It is sufficient to analyze the case where OPT processes the jobs in D_i sequentially from smallest to largest. Additionally, note that the jobs in S_i delay the jobs in

D_i by the same amount in both ALG's and OPT's schedules: their total size. Therefore, the only additional delay for all of these jobs is $t-s$. Ignoring the sizes of the jobs in S_i, which only improve the relative completion times of jobs in $D_i \setminus \{J_i\}$, we analyze two cases: if $t \leq p_i$, OPT completes the ℓth job in $D_i \setminus \{J_i\}$ no earlier than at time $p_i + \sum_{j=2}^{\ell} p_j^{D_i}$. Alternatively, if $t > p_i$, OPT completes the ℓth job in D_i no earlier than at time $t + \sum_{j=2}^{\ell} p_j^{D_i}$. ALG completes this job at time $t + p_i + \sum_{j=2}^{\ell} p_j^{D_i}$ in both cases. Therefore, the relative completion time of the ℓth job in $D_i \setminus \{J_i\}$ is at most R, where $\ell = 3, \ldots, k$ and each job in D_i has a size of at least p_i.

Consequently, we define the worst-case instance where the set S_i consist of job J_q, which caused the interruption, and potentially another job $J_2^{S_i}$ that does not receive enough credit proportional to its size. The set D_i comprises the current interrupted job J_i and other interrupted jobs (If any) that ALG will complete after J_i, where these jobs are at least as large as J_i and delayed the processing of other jobs before time s. Moreover, D_i may include an additional job with a relative completion time greater than R under the constraint that S_i exclusively contains J_q.

In the following, we analyze the interruption at time t in the worst-case instance. We consider the following scenarios based on the interruption time t. If $t \leq p_i$, ALG completes J_q at time $t + p_q$, and OPT completes J_q no earlier than at time at time $s^* + p_i + p_q$, meaning that $(R-1)p_q - t + R(s^* + p_i)$ credit remains available thereafter, which can be allocated to the other jobs. Alternatively, if $t > p_i$ OPT completes J_q no earlier than at time $t + p_q$, ALG completes this job exactly at time $t + p_q$, meaning that $(R-1)(p_q + t)$ credit remains available which can be allocated to the other jobs.

In the case that the number of jobs in $D_i \setminus \{J_i\}$ that delayed processing of at least one other job (before the time s) is $k \neq 0$, we can say that any job in D_i arrives during the last running of its next job in D_i; otherwise, the larger job that delayed running of another job, can not run in the ALG schedule before time s. Additionally, the algorithm interrupts with every incoming job before time $(2-R)(s+p_i)$, that can potentially be completed before the currently running job. Therefore, we have $s^* \geq k(2-R)p_i$ in every schedule of OPT. Hence, there is no job $J_2^{D_i}$ with relative completion time greater than R, and $J_2^{S_i}$ does not exist in the worst-case instance. Now if $t \leq s^* + p_i$, we need to check that $k((1-R)p_q + t - s) \leq (R-2)p_q - 2t + R(s^* + p_i) + s$, and if $t > s^* + p_i$, then we show $k((1-R)p_q + t - s) \leq (R-2)(p_q + t) + s$. Based on our rule of restarting when $s^* \geq k(2-R)p_i$, both cases can be proven by induction. Hence, the credit of every job in D_i is always nonnegative. Now, consider the case that the number of jobs in $D_i \setminus \{J_i\}$ that delayed processing of at least one other job (before the time s) is 0. Then, it is straightforward to verify that the credit of every large job does not decrease after the interruption when $s \geq \frac{2-R}{R-1}p_i \approx 1.18914 p_i$. If $s < \frac{2-R}{R-1}p_i$, the additional credit which can be given to the large jobs (denote this additional credit by K_i) does not decrease to below $0.18 p_i$. Therefore, to show that the credits of the large jobs are always nonnegative, we only have the following two cases that can be easily checked. If $t \leq p_i$ we have that $t + p_q - s - \min((R(p_i + p_q +$

$p_2^{D_i}) - (t + p_q + p_i + p_2^{D_i})), 0) \leq (R-1)p_q - t + Rp_i + K_i$. Otherwise, when $t > p_i$ we have $t + p_q - s - \min((R(t + p_q + p_2^{D_i}) - (t + p_q + p_i + p_2^{D_i})), 0) \leq (R-1)(p_q + t) + K_i$. Hence, there exists an assignment of credits such that all jobs have nonnegative credit. □

Lemma 5. *If OPT completes at least two jobs from D_i before J_q, then immediately after an interruption of J_i at time t, we can reassign credit such that all jobs maintain nonnegative credit.*

Proof. Let T denote the set containing two jobs that OPT completes from D_i before J_q. Without loss of generality we can assume that two jobs in T are J_i and another job $J_2^{D_i}$ that is completed immediately after it. Again, if OPT completes a job at least as large as J_i first, then all the bounds below increase or stay the same. The jobs in S_i delay the jobs in $D_i \setminus T$ by the same amount in the schedule of ALG and OPT. Therefore, these additional delays decrease the relative completion times of these jobs. No matter what order OPT uses for S_i and the larger jobs, under the condition that it completed at least two jobs from D_i before J_q, OPT completes the ℓth job in S_i no earlier than at time $2p_i + \sum_{j=1}^{\ell} p_j^{S_i}$, whereas ALG completes the ℓth job in S_i at time $t + \sum_{j=1}^{\ell} p_j^{S_i}$. This means that its credit is at least $2Rp_i - t + (R-1)\sum_{j=1}^{\ell} p_j^{S_i}$. Therefore, the credit of the ℓth job in S_i is enough for the additional delay caused by the ℓth job which is at most $p_\ell^{S_i}$. As a result, we ignore the sizes of the jobs in S_i which only improve the relative completion times of jobs in $D_i \setminus T$. Hence, every job in $D_i \setminus T$ is only additionally delayed by $t - s$ (relative to the difference between OPT and the schedule of ALG for the instance without interruption).

Suppose that OPT starts J_q at time t^*. Given that $t^* \geq 2p_i$, $t < 2p_i$ and $p_j^{D_i} \geq p_i$ for every j, we have

$$\frac{ALG(J_5^{D_i})}{OPT(J_5^{D_i})} = \frac{t + \sum_{j=1}^{5} p_j^{D_i}}{t^* + \sum_{j=3}^{5} p_j^{D_i}} < \frac{2p_i + \sum_{j=1}^{5} p_j^{D_i}}{2p_i + \sum_{j=3}^{5} p_j^{D_i}} \leq \frac{2+5}{2+3} < R.$$

Therefore, after the second job ($J_4^{D_i}$) in $D_i \setminus T$, subsequent jobs in $D_i \setminus T$ will have a relative completion time smaller than R. This is because ALG and OPT will process these jobs sequentially from smallest to largest, starting after $J_4^{D_i}$.

Consequently, we define the worst-case instance as follows: S_i contains only the job J_q that caused the interruption. D_i consists of the current interrupted job J_i and other jobs (at least as large as J_i) that are completed by ALG after J_i, under the condition that these jobs have relative completion times greater than R, delayed the processing of other jobs before time s, or satisfy both properties.

Now, consider the case that the number of jobs in $D_i \setminus \{T\}$ that delayed processing of at least one other job (before the time s) is 0. We show that enough credits for $J_1^{D_i}, J_2^{D_i}, J_3^{D_i}$ and $J_4^{D_i}$ are available. OPT competes J_q at the earliest at time $2p_i + p_q$. ALG completes this job at the latest at time $t + p_q$, meaning that $(R-1)p_q - t + 2Rp_i$ credit remains, which can be allocated to $J_1^{D_i}$, $J_2^{D_i}, J_3^{D_i}$ and $J_4^{D_i}$, then we have the following inequality which is true because of our interruption rule, $(R-1)p_q - t + 2Rp_i \geq 2(t + p_q - s) + (t - s) + \min(t + (1 - R)(p_q + \sum_{j=1}^{4} p_j^{D_i}), 0)$.

In the case that the number of jobs in $D_i \setminus \{T\}$ that delayed processing of at least one other job (before the time s) is $k \neq 0$, the condition $s^* \geq k(2 - R)p_i$ holds because the algorithm interrupts with every incoming job before time $(2 - R)(s + p_i)$ that can potentially be completed before the currently running job. We therefore need to verify that $k((1-R)p_q + t - s) \leq (2R-4)p_q - 4t + R(s^* + 2p_i) + 3s$, given $t > s^* + p_i$. This can be easily proven by induction and based on our rule of interruption. Consequently, an appropriate credit assignment exists where all jobs maintain nonnegative credit. □

Theorem 5. *The $IRSPT$ algorithm is R-competitive for the scheduling problem $1|r_j, online, restart| \sum C_j$.*

Proof. Suppose there is no restart in the schedule. By Lemma 2, this implies that all jobs arrive at a time when the machine is idle in the smallest counterexample (all at the same time). However, in this case the competitive ratio of our algorithm is actually 1, as it processes the jobs in their optimal sequence. It follows that an instance which is a smallest counterexample contains at least one restart. However, as shown in Lemmas 3, 4 and 5, all jobs maintain nonnegative credit after the last interruption. This contradicts the assumption that the given instance is a counterexample. Hence, the $IRSPT$ algorithm is R-competitive for $1|r_j, online, restart| \sum C_j$. □

Disclosure of Interests. The authors have no competing interests to declare that are relevant to the content of this article.

References

1. Anderson, E.J., Potts, C.N.: Online scheduling of a single machine to minimize total weighted completion time. Math. Oper. Res. **29**(3), 686–697 (2004). https://doi.org/10.1287/MOOR.1040.0092
2. Borodin, A., El-Yaniv, R.: Online Computation and Competitive Analysis. Cambridge University Press (1998)
3. Chekuri, C., Motwani, R., Natarajan, B., Stein, C.: Approximation techniques for average completion time scheduling. In: Saks, M.E. (ed.) Proceedings of the Eighth Annual ACM-SIAM Symposium on Discrete Algorithms, 5–7 January 1997, New Orleans, Louisiana, USA, pp. 609–618. ACM/SIAM (1997). http://dl.acm.org/citation.cfm?id=314161.314396
4. Epstein, L., van Stee, R.: Lower bounds for on-line single-machine scheduling. Theor. Comput. Sci. **299**(1–3), 439–450 (2003). https://doi.org/10.1016/S0304-3975(02)00488-7
5. Fiat, A., Woeginger, G.J.: Online Algorithms: The State of the Art, vol. 1442. Springer (1998)
6. Goemans, M.X., Queyranne, M., Schulz, A.S., Skutella, M., Wang, Y.: Single machine scheduling with release dates. SIAM J. Discret. Math. **15**(2), 165–192 (2002). https://doi.org/10.1137/S089548019936223X

7. Graham, R.L.: Bounds for certain multiprocessing anomalies. Bell Syst. Tech. J. **45**(9), 1563–1581 (1966)
8. Hall, L.A., Schulz, A.S., Shmoys, D.B., Wein, J.: Scheduling to minimize average completion time: off-line and on-line approximation algorithms. Math. Oper. Res. **22**(3), 513–544 (1997). https://doi.org/10.1287/MOOR.22.3.513
9. Hoogeveen, J.A., Vestjens, A.P.A.: Optimal on-line algorithms for single-machine scheduling. In: Cunningham, W.H., McCormick, S.T., Queyranne, M. (eds.) IPCO 1996. LNCS, vol. 1084, pp. 404–414. Springer, Heidelberg (1996). https://doi.org/10.1007/3-540-61310-2_30
10. Komm, D.: An Introduction to Online Computation - Determinism, Randomization, Advice. Texts in Theoretical Computer Science. An EATCS Series. Springer, Cham (2016). https://doi.org/10.1007/978-3-319-42749-2
11. Ma, R., Tao, J., Yuan, J.: Online scheduling with linear deteriorating jobs to minimize the total weighted completion time. Appl. Math. Comput. **273**, 570–583 (2016). https://doi.org/10.1016/J.AMC.2015.10.058
12. Phillips, C., Stein, C., Wein, J.: Scheduling jobs that arrive over time. In: Akl, S.G., Dehne, F., Sack, J.-R., Santoro, N. (eds.) WADS 1995. LNCS, vol. 955, pp. 86–97. Springer, Heidelberg (1995). https://doi.org/10.1007/3-540-60220-8_53
13. Shmoys, D.B., Wein, J., Williamson, D.P.: Scheduling parallel machines on-line. In: McGeoch, L.A., Sleator, D.D. (eds.) On-Line Algorithms, Proceedings of a DIMACS Workshop. DIMACS Series in Discrete Mathematics and Theoretical Computer Science, New Brunswick, New Jersey, USA, 11–13 February 1991, vol. 7, pp. 163–166. DIMACS/AMS (1991). https://doi.org/10.1090/DIMACS/007/13
14. van Stee, R., La Poutré, H.: Minimizing the total completion time on-line on a single machine, using restarts. J. Algorithms **57**(2), 95–129 (2005)
15. Tao, J., Chao, Z., Xi, Y., Tao, Y.: An optimal semi-online algorithm for a single machine scheduling problem with bounded processing time. Inf. Process. Lett. **110**(8–9), 325–330 (2010). https://doi.org/10.1016/J.IPL.2010.02.013

Searching in Euclidean Spaces with Predictions

Sergio Cabello[1,2]($\boxtimes$) and Panos Giannopoulos[3]

[1] Faculty of Mathematics and Physics, University of Ljubljana, Ljubljana, Slovenia
[2] Institute of Mathematics, Physics and Mechanics, Ljubljana, Slovenia
sergio.cabello@fmf.uni-lj.si
[3] Department of Computer Science, City St George's, University of London, London, UK
panos.giannopoulos@city.ac.uk

Abstract. We study the problem of searching for a target at some unknown location in $\mathbb{R}^d$ when additional information regarding the position of the target is available in the form of predictions. In our setting, predictions come as approximate distances to the target: for each point $p \in \mathbb{R}^d$ that the searcher visits, we obtain a value $\lambda(p)$ such that $|pt| \leq \lambda(p) \leq c \cdot |pt|$, where $c \geq 1$ is a fixed constant, $\boldsymbol{t}$ is the position of the target, and $|pt|$ is the Euclidean distance of p to $\boldsymbol{t}$. The cost of the search is the length of the path followed by the searcher. Our main positive result is a strategy that achieves $(12c)^{d+1}$-competitive ratio, even when the constant c is unknown. We also give a lower bound of roughly $(c/16)^{d-1}$ on the competitive ratio of any search strategy in $\mathbb{R}^d$.

Keywords: search games · predictions · Euclidean space

1 Introduction

The problem of searching for a target positioned at some unknown location in some region is a classic search game problem that has been well-studied in both fields of Computational Geometry and Operations Research. The problem comes in many different versions including, linear search (i.e., searching for a target on a line) [6,7,9,15], searching in the plane [4,8,16,20], searching in concurrent rays [4,14,15], and searching inside polygonal regions [21,26]. The book by Alpern and Gal [1] provides an extensive overview of general search games, while Ghosh and Klein [17] survey search problems in planar domains.

The searcher starts from some given position, follows some path according to some strategy until the target, usually a point, is reached or detected, for some appropriate definition of "detection". In 1-dimensional settings, e.g., an infinite line, one usually requires that the searcher passes through the target point. In the plane one usually requires that the searcher is within some distance of the target, or sees the target, if there are obstacles, or that the target lies on the segment connecting the searcher's current and starting positions [16,17,23].

The cost of the search is the length of the path followed by the searcher and the objective is to find an efficient search strategy. Efficiency is usually measured by the **competitive ratio**, which, in this setting, is the ratio of the length of the path of the searcher to the actual Euclidean distance of the starting position to the target.

In this work, we consider the problem of finding a target point t in Euclidean space $\mathbb{R}^d$ when additional information regarding the position of the target is available in the form of **predictions**. Here, the predictions are the approximate distance to t for all points visited during the search, e.g., a value between, say, $|pt|$ and $2|pt|$ for each point p visited. Such an estimate could be obtained for example in a scenario where one takes into account the strength of a signal broadcasted by the target.

Algorithms with predictions is a concept that has been introduced relatively recently; see the survey by Mitzenmacher and Vassilvitskii [25]. The general idea is that on top of the usual input data we are also given additional and possibly inaccurate (noisy) information, the prediction, that should assist the algorithm to be more effective. The improvement in performance depends on the accuracy of the prediction.

We continue this section with the problem setup, a summary of our contribution, a short discussion on our predictions model, and related work.

1.1 Problem Setup

We consider the following search problem in $\mathbb{R}^d$. Assume that there is a fixed but unknown **target** point $t \in \mathbb{R}^d$. Without loss of generality, we start the search at the origin, which we denote by o. We want to find a curve γ that starts at o and ends at t. The **cost** of the search is the Euclidean length of the curve γ.

As we search for the target, we have approximate information about the distance to it from each point that we have visited in so far. More precisely, we assume that there is a constant $c \geq 1$ and an unknown function

$$\lambda \colon \mathbb{R}^d \to \mathbb{R}_{\geq 0} \text{ such that } \forall p \in \mathbb{R}^d \ |pt| \leq \lambda(p) \leq c \cdot |pt|. \tag{1}$$

We refer to such a function λ as a c-**prediction** for the target t. The constant c is the **prediction factor** of λ. See Fig. 1 for an example in $d = 1$. Note that for $c = 1$, the function λ gives the exact distance to the target.

For each point p along the search path we have traversed so far, we obtain the value $\lambda(p)$, and the search strategy decides how to continue the search depending on that information. We know when we have reached the target because $\lambda(p) = 0$ holds only when $p = t$.

As it is common in search games, we are interested in the competitive ratio of the search strategy: how does the length of the search path compares to the straight-line distance from the origin to the target? To formalize this, for each target $t \in \mathbb{R}^d$ and each constant $c \geq 1$, we consider the family $\Lambda(t,c)$ of c-predictions for t, that is, functions satisfying condition (1). The family of c-prediction functions is $\cup_{t \in \mathbb{R}^d} \Lambda(t,c)$.

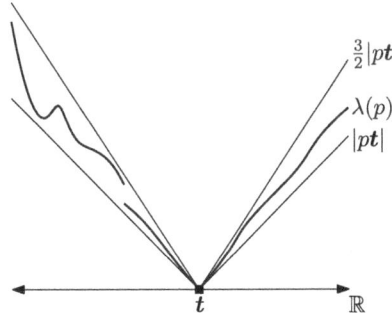

Fig. 1. Example of c-prediction function $\lambda(p)$ for $c = 3/2$. In this example the function is not monotone in $|pt|$ and it is not continuous.

A search strategy S is $\alpha = \alpha(S, c, d)$ **competitive** if for all $t \in \mathbb{R}^d$ and all $\lambda \in \Lambda(t, c)$, the length of the path defined by S to reach t from o is at most $\alpha |ot|$. Note that we have two possible regimes, depending on whether c is known or unknown to the search strategy.

1.2 Our Contribution

Our main contributions are the following:

- We introduce a natural, new search problem in $\mathbb{R}^d$, under a predictions model where we have approximate information about the distance to the target.
- We show that for each dimension d and each constant prediction factor $c \geq 1$ there is a search strategy with competitive ratio $(6c)^{d+1}/3$. To achieve this, we use ε-nets from metric spaces, also known as r-nets, and provide a path of finite length but an infinite number of pieces. This result holds assuming that we know the prediction factor c. For *unknown* prediction factor, a similar statement holds with a slightly different search strategy and larger competitive ratio. This result is given in Sect. 3.
- We show that for $c > 4$, any deterministic search strategy in $\mathbb{R}^d$ with c-predictions will have a competitive ratio of at least $(1/4) \cdot (c/16)^{d-1} \cdot \min\{\sqrt{\pi/d}, 1\}$. For this result we again use ε-nets as a basic tool and we employ techniques developed for the Euclidean Traveling Salesperson with Neighbourhoods. With a slightly worse constant, the lower bound holds also for randomized search strategies. This lower bound is shown in Sect. 4.

Additionally, we give several basic properties of the model. For example, we show that without a prediction function throughout the whole search, we cannot find the target using a path of bounded length, and having an infinite number of pieces (segments) in the search path is unavoidable. We also show that we may assume that the prediction function is continuous. We also note that when $c = 1$, i.e., we have exact information about the distance to the target, the target can be reached with competitive ratio arbitrarily close to 1. A detailed discussion and proofs of these results can be found in the full version of the paper.

To our knowledge, this is the first search problem in Euclidean spaces with $d \geq 2$ where a search point reaches a target point with constant competitive ratio; see the discussion below about related work. For our upper and lower bounds we use bounds on the cardinality of ε-nets. For the sake of simplicity in the calculations, we used suboptimal but simple-to-parse estimates in the simplifications and the cardinality of ε-nets. In any case, our upper and lower bounds for the best competitive ratio are still far apart.

Our model, as presented in the continuous setting, is scale-free and general. In the discrete setting, where the target has integral coordinates or where the target can be detected when the searcher is within a given distance from it, our strategies can be easily modified to have a finite number of steps. The lower bound holds in the discrete setting too, albeit with a slightly worse constant. Moreover, by having an approximate distance to the target as the prediction, the model can also accommodate underestimates of the actual distance (in addition to the default overestimates), which can be of interest in practical scenarios. These as well as other extensions are discussed in Sect. 3.1.

1.3 Related Work

For *linear* search, when the exact distance to the target is known, one can easily find the target by walking at most three times this distance. When the distance to the target is unknown, Beck and Newman [7] and later Baeza-Yates at al. [4] showed that a simple doubling strategy has competitive ratio 9; here, a lower bound on the distance to the target is assumed, otherwise there is no search strategy with constant competitive ratio. There are also other similar strategies with the same competitive ratio [10,17]. Moreover, various approaches [4,7,19,22] show that 9 is the best possible competitive ratio for the problem.

Gal [14,15] introduced the problem of searching for a target on multiple *rays* that are concurrent at the starting position and gave an optimal strategy for the case where the distance to the target is unknown. This result was rediscovered by Baeza-Yates et al. [4].

Baeza-Yates et al. [4] also considered the problem of finding a target with *integer coordinates* in the plane and presented various search strategies. When the distance to the target is known, it is also easy to get an optimal strategy. However, when the distance is unknown, and with no additional information available, no search strategy can have constant competitive ratio as there are $\Theta(n^2)$ integral points within distance at most n from the origin and any search strategy has to visit all of them in some order. For the natural extension of the problem in $\mathbb{R}^d$, the latter generalizes to any $d \geq 3$, as there are $\Theta(n^d)$ integral points within distance at most n from the origin. When we are in $\mathbb{R}^d$ and the distance to the target is known, we hit a classical problem in Number Theory: on how many ways can we express a positive integer as sum of d squares of integers. For $d \geq 5$, there is no search strategy with constant competitive ratio as there are there are $\Omega(n^{d/2-1})$ integral points at distance exactly n from the origin; see, for example, Vaughan and Wooley [27].

In another variant of the problem by Gal [16], the searcher travels along a path until the target lies on the *segment* connecting the searcher's current and starting positions, essentially sweeping around its starting position with an infinitely elastic cord until the target is swept. For this, Gal [16] gave a spiral search strategy with 17.289 competitive ratio, which Langetepe [23] showed to be optimal.

Hipke et al. [19] considered linear search when the target is at *distance* at least 1 and at most $D \geq 1$ from the starting point, where D is known at the start of the search. Bose et al. [10] provided a more careful analysis using the roots of a recursive sequence of polynomials and gave better lower and upper bounds on the competitive ratio with dependence on D. López-Ortiz and Schuierer [24] considered also this setting, for the case of concurrent rays. Compared to these works, there are two main differences in our work. Firstly and most importantly, we have a prediction all the way through the search, while they have a prediction only at the start. Secondly, we consider the problem in more general settings, namely in $\mathbb{R}^d$ for arbitrary d. An upper bound at the start does not suffice to find a point when $d \geq 2$. (If $d = 2$ and the exact distance is known, the problem can be easily solved since the target has to lie on a known circle, and that is the only additional instance that is solvable.)

Banerjee et al. [5] considered the problem of finding a target in a *graph* with information about the distance to the target. In their model, the target is at one vertex of the graph and at each vertex we have a value stored that is made available only when we are adjacent to the vertex. For most vertices, the value stored at a vertex is the true distance from the vertex to the target, but for some vertices the value is wrong. Contrary to our setting, they do not assume a bounded error for the information at each vertex, but that the information is wrong at a bounded number of nodes. The bound on the number of nodes with wrong information then appears in the bound for the length of the search path.

Finally, Angelopoulos [2,3] gave strategies for linear and multiple-ray search under a different model where a one-off, possibly erroneous hint or prediction on the target's position is given at the start of the search. The prediction can be positional, directional, or, in general, a k-bit string encoding answers to k binary queries and the measure of the performance of a strategy is a trade-off between the competitive ratio under error-free prediction and that under erroneous prediction.

2 Notation and Preliminaries

Since the dimension d is always fixed and clear from the context, we drop in the notation the dependency on d.

The **ball** centered at $p \in \mathbb{R}^d$ with radius r is $B(p,r) = \{q \in \mathbb{R}^d \mid |pq| \leq r\}$. We will also consider the **spherical shells** $S(p, r_1, r_2) = \{q \in \mathbb{R}^d \mid r_1 \leq |pq| \leq r_2\}$. A spherical shell in the plane is an annulus. For a c-prediction function λ, whenever we are at a point $p \in \mathbb{R}^d$, we get a prediction $\lambda(p)$ and we deduce that the target point t lies in the spherical shell $S(p, \lambda(p)/c, \lambda(p))$. See Fig. 2.

Note that we have made a modeling decision, namely we have assumed that there is an unknown function λ such that at each point p we get the prediction $\lambda(p)$. More generally, it could happen that we visit the same point p multiple times and at each time we get a different estimate of the distance from p to the target. However, getting a different estimate can only help, as it provides more information: if we get two different c-predictions λ and λ' at different times at the same point p, then we know that the target lies in the spherical shell $S(p, \max\{\lambda, \lambda'\}/c, \min\{\lambda, \lambda'\})$. This is more information than what we get if $\lambda' = \lambda$ because then we can only conclude that the target lies in $S(p, \lambda/c, \lambda)$, which is strictly larger. Thus, when searching for an optimal search strategy, we can assume that each time we visit the same point we get the same prediction. In particular, a search strategy could simply ignore the information obtained in the second and subsequent visits to the same point.

A **path** in $\mathbb{R}^d$ is a continuous function $\pi\colon [0,1] \to \mathbb{R}^d$. The paths in our search strategies will consist of an infinite number of straight-line segments and will exhibit a Zeno-like phenomena: they make an infinite number of turns in finite time and length. To show that the paths we define reach the target, we will use the following property, whose proof is an standard argument in continuity.

Lemma 1. *Let $\pi\colon [0,1] \to \mathbb{R}^d$ be a path. Assume that there is a point $p \in \mathbb{R}^d$ with the following property: for each $\varepsilon > 0$ there exists some $\delta \in (0,1]$ such that the subpath $\pi([1-\delta, 1])$ is contained in $B(p, \varepsilon)$. Then $\pi(1) = p$, that is, p is the endpoint of the path π.*

We will use ε-nets from metric space theory: An ε-**net** for the ball $B(p,r)$ is a subset N of points from $B(p,r)$ such that (i) each point of $B(p,r)$ is at distance at most ε from some point of N, and (ii) each two distinct points of N are at distance at least ε. Condition (i) is equivalent to $B(p,r) \subseteq \bigcup_{q \in N} B(q, \varepsilon)$. Condition (ii) is equivalent to the balls $B(q, \varepsilon/2)$, where $q \in N$, being pairwise interior disjoint.

The following bound follows from a well-known technique using volumes.

Lemma 2. *In $\mathbb{R}^d$, for each $\varepsilon \leq r$, a ball of radius r has a ε-net with at least $(r/\varepsilon)^d$ and at most $(3r/\varepsilon)^d$ elements.*

3 Upper Bound

In this section we provide search strategies to reach the target in $\mathbb{R}^d$ when we have a c-prediction. We first provide the key lemma that tells us how to get a sequence of points whose λ-values decreases geometrically. We then provide a strategy when the prediction factor is known, and then discuss how to handle the case for unknown prediction factor. In this setting, we adapt the notation so that c^* is the true prediction factor, while c is a guess for the true prediction factor. At the end of the section we discuss some extensions.

Lemma 3. *Assume that we are at point p_i and the prediction factor c^* is perhaps unknown. Let c be a guess for c^*. Using a search through a path γ_{i+1} of length at most $2(6c)^d \cdot \lambda(p_i)$ we get to one of the following outcomes:*

- *we move from p_i to a point p_{i+1} such that $\lambda(p_{i+1}) \leq \lambda(p_i)/2$, or*
- *we come back to p_i and correctly deduce that $c < c^*$.*

Moreover, all points of the path γ_{i+1} are at distance at most $2\lambda(p_i)$ from t.

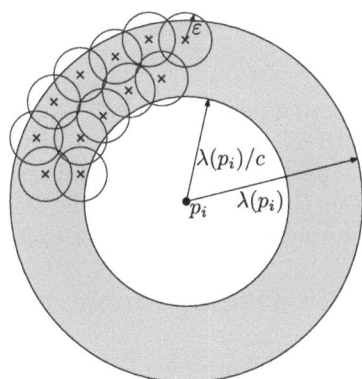

Fig. 2. Spherical shell $S(p_i, \lambda(p_i)/c, \lambda(p_i))$ where t must lie and (part of) a net.

Proof. Recall that $|p_i t| \leq \lambda(p_i)$. Set $\varepsilon = \frac{\lambda(p_i)}{2c}$ and let N be a ε-net for the ball $B(p_i, \lambda(p_i))$ by Lemma 2. Since the target is contained in $B(p_i, \lambda(p_i))$ and N is a ε-net, there is some point $p_* \in N$ such that $|p_* t| \leq \varepsilon = \lambda(p_i)/2c$. If $c^* \leq c$, then we have $\lambda(p_*) \leq c^* \cdot |p_* t| \leq c^* \cdot \lambda(p_i)/2c \leq \lambda(p_i)/2$. If $c^* > c$, then we have no guaranteed useful bound for $\lambda(p_*)$.

We take a path that goes through the points of N in arbitrary order and, at each point of N, we query for the prediction $\lambda(\cdot)$. We finish the path as soon as we reach some point $q_* \in N$ such that $\lambda(q_*) \leq \lambda(p_i)/2$. If for all the points p of N we have $\lambda(p) > \lambda(p_i)/2$, then we go back to the point p_i. This finishes the description of the path γ_{i+1}.

In the first case, we set $p_{i+1} = q_*$ and we have moved to a point p_{i+1} with $\lambda(p_{i+1}) \leq \lambda(p_i)/2$. When $c^* \leq c$, we have to be in this case since the point p_* satisfies the stopping condition $\lambda(p_*) \leq \lambda(p_i)/2$. Thus, if we do not have such a point, we can conclude that $c^* > c$.

To bound the length of γ_{i+1}, we note that any two points in $B(p_i, \lambda(p_i))$ are at distance at most $2\lambda(p_i)$. We further note that the first edge has length at most $\lambda(p_i)$ and, if γ_{i+1} comes back to p_i, also the last edge has length at most $\lambda(p_i)$. Using the bound of Lemma 2 we get that the path γ_{i+1} has length at most

$$\lambda(p_i) + (|N| - 1) \cdot (2\lambda(p_i)) + \lambda(p_i) \leq \left(\frac{3\lambda(p_i)}{\varepsilon}\right)^d \cdot 2\lambda(p_i) = 2(6c)^d \cdot \lambda(p_i).$$

Finally, we note that the whole path γ_{i+1} is contained in $B(p_i, \lambda(p_i))$, which is contained in $B(\boldsymbol{t}, 2\lambda(p_i))$ because $\boldsymbol{t} \in B(p_i, \lambda(p_i))$.

Note: It is known [13] that for any set of n points in the d-dimensional ball $B(p,r)$ there is a tour of length $r \cdot O(n^{\frac{d-1}{d}})$ visiting them. This implies that we can also use a path γ_{i+1} of length $\lambda(p_i) \cdot O(|N|^{\frac{d-1}{d}}) = O((6c)^{d-1}) \cdot \lambda(p_i)$. With this, the dependency on c is slightly better at the expense of having more ugly-looking constants.

Theorem 1 (Known c). *Consider the search with predictions problem in $\mathbb{R}^d$ where the prediction factor $c^* > 1$ is known. There is a search strategy to reach the target with competitive ratio $2 \cdot 6^d \cdot (c^*)^{d+1}$.*

Proof Let $p_0 = \boldsymbol{o}$ be the starting point and recall that $\lambda(p_0) \le c^* \cdot |\boldsymbol{ot}|$.

For $i = 0, 1, 2 \ldots$ iteratively, we use Lemma 3 with the guessed prediction factor $c = c^*$ to obtain a path γ_{i+1}. As the prediction factor is correct, we always have the outcome in the first item: γ_{i+1} finishes at a point p_{i+1} with $\lambda(p_{i+1}) \le \lambda(p_i)/2$. It follows by induction that for each $i \in \mathbb{N}$ we have $\lambda(p_i) \le \lambda(p_0)/2^i$ and therefore $\operatorname{len}(\gamma_i) \le 2(6c^*)^d \cdot \lambda(p_i) \le 2(6c^*)^d \cdot \lambda(p_0)/2^i$.

Let γ be the concatenation of the paths $\gamma_1, \gamma_2, \ldots$. Then

$$\operatorname{len}(\gamma) = \sum_{i=1}^{\infty} \operatorname{len}(\gamma_i) \le \sum_{i=1}^{\infty} 2(6c^*)^d \cdot \frac{\lambda(p_0)}{2^i} = 2(6c^*)^d \cdot \lambda(p_0)$$

$$\le 2 \cdot 6^d (c^*)^{d+1} \cdot |\boldsymbol{ot}|.$$

The path makes an infinite number of straight-line moves. Since for each $i \in \mathbb{N}$, the suffix of the path γ after p_{i+1} is at distance at most $2 \cdot \lambda(p_{i+1}) \le 2 \cdot \lambda(p_0)/2^{i+1} = \lambda(p_0)/2^i$ from $\boldsymbol{t}$, Lemma 1 implies that $\boldsymbol{t}$ is the endpoint of the path γ.

Theorem 2 (Unknown c). *Consider the search with predictions problem in $\mathbb{R}^d$ where the prediction factor c^* is unknown. There is a search strategy to reach the target with competitive ratio $(12c^*)^{d+1}$.*

Proof The basic idea is using an exponential search for the constant c. At each step we use Lemma 3 to either move to a point with smaller predicted distance to the target or to detect that our guess for c is too small and double it. Index j parameterizes the current guess $c = 2^j$, and $p_i^{(j)}$ denotes a point during that guess. At each step we will increase either i or j.

We start setting $j = 1$, $i = 0$ and $p_0^{(j)} = p_0^{(1)} = \boldsymbol{o}$.

From the current point $p_i^{(j)}$, we use Lemma 3 with the guessed prediction factor $c = 2^j$ to obtain a path $\gamma_{i+1}^{(j)}$. If the outcome is given by the first item of Lemma 3, then we get to a point $p_{i+1}^{(j)}$ with $\lambda(p_{i+1}^{(j)}) \le \lambda(p_i^{(j)})/2$; in this case we increase i. If the outcome is given by the second item of Lemma 3, then we get back to $p_i^{(j)}$; in this case we set $p_i^{(j+1)} = p_i^{(j)}$, increase j and, from this point on, we will use the new guessed prediction factor 2^j, which is twice larger than before. See Fig. 3 for an schematic view of the sequence of points.

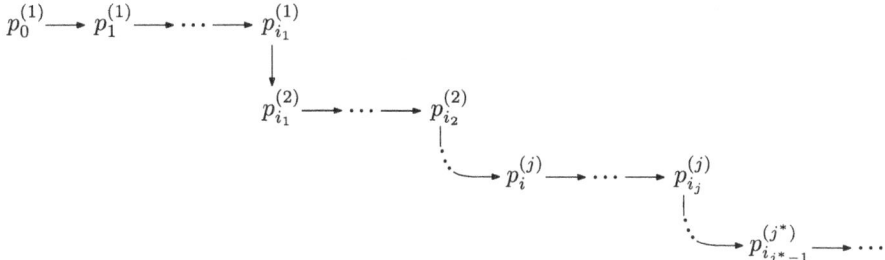

Fig. 3. Visualizing the sequence of points $\left((p_i^{(j)})_{i \in I_j}\right)_{j \in J}$. When increasing i we move right, when increasing j we move down. Here, we are using $j^* = \max(J)$ and $i_j = \max(I_j)$ for all $j \in J$.

Let j^* be the largest value that index j takes through the procedure. Note that $j^* \leq \lceil \log_2 c^* \rceil$ since $2^{\lceil \log_2 c^* \rceil}$ is an overestimate to c^*. If we arrive to $j^* = \lceil \log_2 c^* \rceil$, from that point on we will always extend the search path using the outcome in the first item of Lemma 3. Set $J = \{1, \ldots, j^*\}$, which is the set of values that index j takes through the procedure.

For each $j \in J$, let I_j be the set of indices i such that the point $p_i^{(j)}$ is defined and let $i_j = \max(I_j)$. Note that i_{j^*} is undefined because I_{j^*} is infinite, but i_j is defined for all $j < j^*$ because I_j is finite for all $j < j^*$. By construction, the index i_j is the first element of I_{j+1}, for all $j < j^*$. It may happen that, for some $j < j^*$, the set I_j contains a single element. This happens when j is increased in successive steps and thus $p_i^{(j)} = p_i^{(j+1)} = p_i^{(j+2)}$.

It follows by induction that for all $j \in J$ and all $i \in I_j$ we have $\lambda(p_i^{(j)}) \leq \lambda(p_0^{(1)})/2^i = \lambda(o)/2^i$. Note that this bound is independent of the index j. Therefore

$$\forall j \in J, \; i \in I_j: \quad \text{len}(\gamma_{i+1}^{(j)}) \leq 2(6 \cdot 2^j)^d \cdot \lambda(p_i^{(j)}) \leq 2(6 \cdot 2^j)^d \cdot \frac{\lambda(o)}{2^i}.$$

Let γ be the concatenation of the paths $\gamma_{i+1}^{(j)}$, in the same order as we constructed them: first $\gamma_{i+1}^{(1)}$ for increasing $i \in I_1$, then $\gamma_{i+1}^{(2)}$ for increasing $i \in I_2$, and so on until we reach the infinite sequence $\gamma_{i+1}^{(j^*)}$ for increasing $i \in I_{j^*}$. We then have

$$\text{len}(\gamma) = \sum_{j=1}^{j^*} \sum_{i \in I_j} \text{len}(\gamma_{i+1}^{(j)}) \leq \sum_{j=1}^{j^*} \sum_{i \in I_j} 2(6 \cdot 2^j)^d \cdot \frac{\lambda(o)}{2^i}.$$

Using that for all $j < j^*$ the sets I_j and I_{j+1} have only $i_j = \max(I_j)$ in common, that $2^j \le c^*$ for all $j < j^*$, and that $2^{j^*} < 2c^*$, we get

$$\begin{aligned}
\text{len}(\gamma) &\le \sum_{j=1}^{j^*-1} 2(6 \cdot 2^j)^d \cdot \frac{\lambda(\mathbf{o})}{2^{i_j}} + \sum_{i=0}^{\infty} 2(6 \cdot 2c^*)^d \cdot \frac{\lambda(\mathbf{o})}{2^i} \\
&\le 2 \cdot 6^d \cdot \lambda(\mathbf{o}) \cdot \sum_{j=1}^{j^*-1} (2^d)^j + 4(12 \cdot c^*)^d \cdot \lambda(\mathbf{o}) \\
&\le 2 \cdot 6^d \cdot \lambda(\mathbf{o}) \cdot \frac{(2^d)^{j^*}}{2^d - 1} + 4(12 \cdot c^*)^d \cdot \lambda(\mathbf{o}) \\
&\le 4 \cdot 3^d \cdot \lambda(\mathbf{o}) \cdot (2c^*)^d + 4(12 \cdot c^*)^d \cdot \lambda(\mathbf{o}) \\
&\le 8 \cdot 12^d \cdot (c^*)^d \cdot \lambda(\mathbf{o}) \le 12^{d+1} \cdot (c^*)^{d+1} \cdot |\mathbf{ot}|.
\end{aligned}$$

The path makes an infinite number of straight-line moves. Since for each $i \in \mathbb{N}$, the suffix of the path γ after $p_{i+1}^{(j)}$, for j such that $i+1 \in I_j$, is at distance at most $2 \cdot \lambda(p_{i+1}^{(j)}) \le 2 \cdot \lambda(\mathbf{o})/2^{i+1} = \lambda(\mathbf{o})/2^i$ from $\mathbf{t}$, Lemma 1 implies that $\mathbf{t}$ is the endpoint of the path γ.

3.1 Extensions

First, the same approach works for spaces of bounded doubling dimension as long as there is a concept of path to connect points whose length is the same as the distance or we change the setting so that the cost of moving from one point to another is the distance between the points. For this, one has to use ε-nets in spaces of bounded doubling dimension [18].

When the target is known to have integral coordinates, we can modify the search so that it has a finite number of segments. Indeed, as soon as we reach a point p_i such that $\lambda(p_i) < 1/2$, we know that the target is at distance smaller than $1/2$ from p_i and there is a unique point with integral coordinates in $B(p_i, \lambda(p_i))$. We can then just move that point. A similar approach works when the target is known to have coordinates with bounded resolution by scaling the setting. In the case where we can detect the target when we are within a given distance δ, we can finish the search when we reach a point p_i with $\lambda(p_i) \le \delta$, thus, also bounding the number of steps (which will depend on the initial estimate $\lambda(p_0)$).

Our strategies work also in the case where predictions may additionally underestimate the actual distance to the target by, say, some factor $c' \le 1$, i.e., $c' \cdot |pt| \le \lambda(p)$. Then, when both factors c, c' are known, this is equivalent to scaling up the prediction by $1/c'$ to get a new one with factor c/c'. When the factors are unknown, the strategy in Theorem 2 still visits points with geometrically decreasing predictions and since for each point, say p_i, we now have that $|p_i t| \le \lambda(p_0)/(2^i \cdot c')$ the path converges to the target (as c' is constant).

One can consider the version when the target is an unknown k-flat F in $\mathbb{R}^d$. The same strategies work, also if we do not know the prediction factor nor the dimension, k, of the flat. Indeed, we are constructing a path as a concatenation of segments such that the distance to F gets arbitrarily small for each suffix of the path. Formally, for each $\varepsilon > 0$, there is a suffix of the path such that all the points on the path are at distance ε from F. Since the path is continuous and has bounded length, the endpoint of the path has to be at distance 0 from F.

Finally, note that for small d there are tighter bounds on the size of ε-nets translating to better constants in our approach. For $c \approx 1$ one can also exploit that we need an ε-net of the spherical shell $S(\mathbf{o}, r/c, r)$, which is much smaller than the whole ball $S(\mathbf{o}, r)$. However, this improvement is expected to be small since even the ε-nets for spheres are not much better.

4 Lower Bound

In this section we provide a lower bound for the search problem with c-predictions. Our lower bounds are meaningful when we assume that c is large enough. The idea is to construct many c-predictions for several different targets that are indistinguishable, unless we are quite close to the target. For each such prediction λ, there is a small ball around the target where the value of λ is different from the other predictions, but all the other predictions have the same value on that small ball. Then, the searcher has to visit *all* the small balls around the targets, because in the worst case the target is going to be in the last small ball visited.

First, using the techniques by Elbassioni, Fishkin and Sitters [12] developed for the Euclidean Traveling Salesperson with Neighbourhoods, one can show the following bound. See Dumitrescu and Tóth [11] for an improvement over [12] where the same idea is reused.

Lemma 4. *Let $\mathbb{B}$ be a set of n congruent balls of radius δ in $\mathbb{R}^d$ that are pairwise interior disjoint. Each path in $\mathbb{R}^d$ that contains at least one point from each ball of $\mathbb{B}$ has length at least $(n/2^d - 1) \cdot \delta \cdot \sqrt{\frac{\pi}{d}}$.*

Next, fix a value $c > 2$. For each $\mathbf{t} \in B(\mathbf{o}, 1/2 - 1/c)$, let λ_t be the function defined as follows; see Figs. 4 and 5 for intuition.

$$\lambda_t(p) = \begin{cases} c \cdot |p\mathbf{t}|, & \text{if } x \in B(\mathbf{t}, 1/c), \\ 1, & \text{if } x \in B(\mathbf{o}, 1/2) \setminus B(\mathbf{t}, 1/c), \\ 2 \cdot |p\mathbf{o}|, & \text{if } x \notin B(\mathbf{o}, 1/2). \end{cases} \quad (2)$$

Lemma 5. *Assume that $c > 2$ and $|\mathbf{ot}| \leq \frac{1}{2} - \frac{1}{c}$. Then the function λ_t is a c-prediction for the target $\mathbf{t}$. Moreover, for any two distinct $\mathbf{t}, \mathbf{t}'$ satisfying the hypothesis, the functions λ_t and $\lambda_{t'}$ agree on all points outside $B(\mathbf{t}, 1/c) \cup B(\mathbf{t}', 1/c)$.*

Theorem 3. *Assume that $c \geq 4$. There is family of c-predictions for targets in $B(\mathbf{o}, 1/4)$ such that any search path in $\mathbb{R}^d$ that uses c-predictions to find the target has length at least $(c^{d-1}/16^d) \cdot \min\{\sqrt{\pi/d}, 1\}$.*

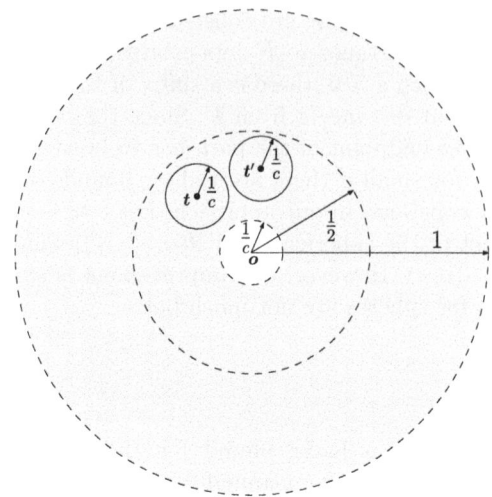

Fig. 4. Parts in the domains for λ_t and $\lambda_{t'}$ for two targets when $d = 2$.

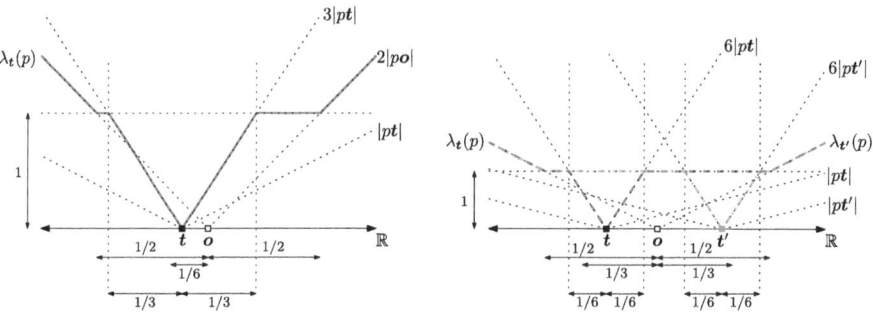

Fig. 5. Examples of functions λ_t for $d = 1$. Note that the axes have different scales. Left: example for $c = 3$; the target has to be at distance at most $1/6$ from the origin. Right: two functions for $c = 6$; the target has to be at distance at most $1/3$ from the origin.

Proof. Set $\varepsilon = 2/c$. Because of Lemma 2, the ball $B(o, 1/4)$ has a ε-separated set N with

$$|N| \geq \left(\frac{1/4}{2/c}\right)^d = (c/8)^d$$

points. Note that the balls $B(q, 1/c)$, where $q \in N$, are pairwise interior disjoint.

For each point $t \in N$, consider the function λ_t defined in Eq. (2). Note that $\frac{1}{4} \leq \frac{1}{2} - \frac{1}{c}$ because $c \geq 4$, and therefore $B(o, 1/4) \subseteq B(o, 1/2 - 1/c)$, as required for the definition. Lemma 5 shows that the function λ_t is a c-prediction for each $t \in N$. Consider the set of prediction functions $\Lambda = \{\lambda_t \mid t \in N\}$. Let $\mathbb{B} = \{B(t, 1/c) \mid t \in N\}$ and note that these balls are pairwise interior disjoint. All the functions $\lambda \in \Lambda$ take the same value on each point p outside

⋃ 𝔹. Moreover, to distinguish between two functions $\lambda_t, \lambda_{t'} \in \Lambda$ we have to evaluate them at some point of $B(t, 1/c) \cup B(t', 1/c)$.

Consider any search strategy when the c-prediction function is selected from Λ by an *adversary*. It may use that the c-prediction is from Λ. In particular, we know beforehand that the target t is a point of N, selected by the adversary. We claim that the search path has to visit the interior of all the balls of $\mathbb{B}$. Indeed, while there are two distinct points $t, t' \in N$ such that the balls $B(t, 1/c)$ and $B(t', 1/c)$ are not visited by the search path, we have $\lambda_t(p) = \lambda_{t'}(p)$ for all points p along the path, and therefore the information collected cannot discern whether the target is t or t'. Thus, in the worst case, the search path has to visit all the balls of $\mathbb{B}$ but one, to identify which point of N is the target, and then still may have to move to that last ball. That is, we may assume that in the worst case the adversary chooses $\lambda_t \in \Lambda$ such that $B(t, 1/c)$ is the *last* ball of $\mathbb{B}$ visited by the search strategy. After the searcher deduces at which point of N the target is, they still have to move there.

We conclude that the search path has to visit all $|N| \geq (c/8)^d$ balls of $\mathbb{B}$, and for the last ball it still has to travel $1/c$ to the center. Since those balls are pairwise interior disjoint, Lemma 4 implies that the search path has length at least

$$\frac{1}{c} + \left(\frac{|N|}{2^d} - 1\right) \cdot \left(\frac{1}{c}\right) \cdot \sqrt{\frac{\pi}{d}} \geq \frac{1}{c} + \left(\frac{|N|}{2^d} - 1\right) \cdot \left(\frac{1}{c}\right) \min\{\sqrt{\pi/d}, 1\}$$
$$\geq \frac{|N|}{2^d} \cdot \left(\frac{1}{c}\right) \cdot \min\{\sqrt{\pi/d}, 1\}$$
$$\geq \frac{(c/8)^d}{2^d} \cdot \left(\frac{1}{c}\right) \cdot \min\{\sqrt{\pi/d}, 1\}$$
$$= \frac{c^{d-1}}{16^d} \cdot \min\{\sqrt{\pi/d}, 1\}.$$

For the *competitive ratio* we have to compare the length of the search path to the distance to the target, which in the construction is at most $1/4$. We then obtain the following bound, which is interesting when $d \geq 2$ and c is large enough.

Corollary 1. *Consider the search with predictions problem in $\mathbb{R}^d$ with prediction factor $c \geq 4$. Any search strategy to reach the target has competitive ratio at least $(1/4) \cdot (c/16)^{d-1} \cdot \min\{\sqrt{\pi/d}, 1\}$.*

The lower bound still holds in the case where the target has integral coordinates or is detected when we are within a given distance, with a slightly worse constant in the basis of the exponential function. For this, we can just scale the input such that the balls in the construction are large enough or the epsilon-net can be assumed to have points with integer coordinates. Then, we still have to visit all balls in order to detect which one has the target.

A similar lower bound holds for randomized search strategies. In our construction, the searcher has to visit a certain number of disjoint regions (balls) and they can deduce something only when visiting the ball that contains the target. In the worst-case (or adversary) model, the target is going to be in the last ball that is visited. In a randomized setting, all balls would be equally likely to contain the target, and thus the searcher has to visit in expectation half of the balls because visiting the wrong ball does not provide any information. Thus, we have a TSP where you have to visit (in expectation) half of the pairwise disjoint balls, and a similar lower bound applies because the bound is linear in the number of balls.

Acknowledgements. The authors would like to thank Kostas Tsakalidis for suggesting the area of algorithmic problems with predictions and Konrad Swanepoel for providing a simple argument on why a 2-dimensional disk cannot be covered by a curve of bounded length.

This research was funded in part by the Slovenian Research and Innovation Agency (P1-0297, J1-2452, N1-0218, N1-0285), in part by the Erdős Center, and in part by the European union (ERC grant KARST, 101071836; ERC grant 882971, GeoScape). Views and opinions expressed are however those of the authors only and do not necessarily reflect those of the European Union or the European Research Council. Neither the European Union nor the granting authority can be held responsible for them.

References

1. Alpern, S., Gal, S.: The Theory of Search Games and Rendezvous. International Series in Operations Research and Management Science, vol. 55. Kluwer (2003). https://doi.org/10.1007/b100809
2. Angelopoulos, S.: Online search with a hint. In: Lee, J.R. (ed.) 12th Innovations in Theoretical Computer Science Conference, ITCS 2021, 6–8 January 2021, Virtual Conference. LIPIcs, vol. 185, pp. 51:1–51:16. Schloss Dagstuhl - Leibniz-Zentrum für Informatik (2021). https://doi.org/10.4230/LIPICS.ITCS.2021.51
3. Angelopoulos, S.: Competitive search in the line and the star with predictions. In: Leroux, J., Lombardy, S., Peleg, D. (eds.) 48th International Symposium on Mathematical Foundations of Computer Science, MFCS 2023. LIPIcs, 28 August–1 September 2023, Bordeaux, France, vol. 272, pp. 12:1–12:15. Schloss Dagstuhl - Leibniz-Zentrum für Informatik (2023). https://doi.org/10.4230/LIPICS.MFCS.2023.12
4. Baezayates, R.A., Culberson, J.C., Rawlins, G.J.E.: Searching in the plane. Inf. Comput. **106**(2), 234–252 (1993). https://doi.org/10.1006/inco.1993.1054
5. Banerjee, S., Cohen-Addad, V., Gupta, A., Li, Z.: Graph searching with predictions. In: Kalai, Y.T. (ed.) 14th Innovations in Theoretical Computer Science Conference, ITCS 2023, 10–13 January 2023, MIT, Cambridge, Massachusetts, USA. LIPIcs, vol. 251, pp. 12:1–12:24. Schloss Dagstuhl - Leibniz-Zentrum für Informatik (2023). https://doi.org/10.4230/LIPICS.ITCS.2023.12
6. Beck, A.: On the linear search problem. Israel J. Math. **2**, 221–228 (1964). https://doi.org/10.1007/BF02759737
7. Beck, A., Newman, D.J.: Yet more on the linear search problem. Israel J. Math. **8**, 419–429 (1970). https://doi.org/10.1007/BF02798690
8. Bellman, R.: Minimization problem. Bull. Amer. Math. Soc. **62**, 270 (1956)
9. Bellman, R.: Problem 63-9, an optimal search. SIAM Rev. **5**(3), 274–274 (1963)

10. Bose, P., Carufel, J.D., Durocher, S.: Searching on a line: a complete characterization of the optimal solution. Theor. Comput. Sci. **569**, 24–42 (2015). https://doi.org/10.1016/J.TCS.2014.12.007
11. Dumitrescu, A., Tóth, C.D.: The traveling salesman problem for lines, balls, and planes. ACM Trans. Algorithms **12**(3), 43:1–43:29 (2016). https://doi.org/10.1145/2850418
12. Elbassioni, K.M., Fishkin, A.V., Sitters, R.: Approximation algorithms for the Euclidean traveling salesman problem with discrete and continuous neighborhoods. Int. J. Comput. Geom. Appl. **19**(2), 173–193 (2009). https://doi.org/10.1142/S0218195909002897
13. Few, L.: The shortest path and the shortest road through n points. Mathematika **2**(2), 141–144 (1955). https://doi.org/10.1112/S0025579300000784
14. Gal, S.: A general search game. Israel J. Math. **12**, 32–45 (1972)
15. Gal, S.: Minimax solutions for linear search problems. SIAM J. Appl. Math. **27**(1), 17–30 (1974)
16. Gal, S.: Search Games. Academic Press, New York (1980)
17. Ghosh, S.K., Klein, R.: Online algorithms for searching and exploration in the plane. Comput. Sci. Rev. **4**(4), 189–201 (2010). https://doi.org/10.1016/J.COSREV.2010.05.001
18. Har-Peled, S., Mendel, M.: Fast construction of nets in low-dimensional metrics and their applications. SIAM J. Comput. **35**(5), 1148–1184 (2006). https://doi.org/10.1137/S0097539704446281
19. Hipke, C.A., Icking, C., Klein, R., Langetepe, E.: How to find a point on a line within a fixed distance. Discret. Appl. Math. **93**(1), 67–73 (1999). https://doi.org/10.1016/S0166-218X(99)00009-8
20. Isbell, J.R.: An optimal search pattern. Nav. Res. Logist. Q. **4**, 357–359 (1957)
21. Klein, R.: Algorithmische Geometrie, 2nd edn. Springer-Verlag, Heidelberg (2005). https://doi.org/10.1007/3-540-27619-X
22. Kupavskii, A., Welzl, E.: Lower bounds for searching robots, some faulty. Distrib. Comput. **34**(4), 229–237 (2021). https://doi.org/10.1007/S00446-019-00358-Y
23. Langetepe, E.: On the optimality of spiral search. In: Charikar, M. (ed.) Proceedings of the Twenty-First Annual ACM-SIAM Symposium on Discrete Algorithms, SODA 2010, Austin, Texas, USA, 17–19 January 2010, pp. 1–12. SIAM (2010). https://doi.org/10.1137/1.9781611973075.1
24. López-Ortiz, A., Schuierer, S.: The ultimate strategy to search on m rays? Theor. Comput. Sci. **261**(2), 267–295 (2001). https://doi.org/10.1016/S0304-3975(00)00144-4
25. Mitzenmacher, M., Vassilvitskii, S.: Algorithms with predictions. In: Roughgarden, T. (ed.) Beyond the Worst-Case Analysis of Algorithms, pp. 646–662. Cambridge University Press (2020). https://doi.org/10.1017/9781108637435.037
26. Schuierer, S.: On-line searching in simple polygons. In: Christensen, H.I., Bunke, H., Noltemeier, H. (eds.) Sensor Based Intelligent Robots, International Workshop, Dagstuhl Castle, Germany, 28 September–2 October 1998, Selected Papers. LNCS, vol. 1724, pp. 220–239. Springer (1998). https://doi.org/10.1007/10705474_12
27. Vaughan, R.C., Wooley, T.D.: The asymptotic formula in Waring's problem: higher order expansions. J. für die reine und angewandte Mathematik (Crelles J.) **2018**(742), 17–46 (2018). https://doi.org/10.1515/crelle-2015-0098

Lower Bounds for Approximate (& Exact) k-Disjoint-Shortest-Paths

Rajesh Chitnis[1], Samuel Thomas[1,2(✉)], and Anthony Wirth[2,3]

[1] School of Computer Science, The University of Birmingham, Birmingham, UK
samuelthomascs@gmail.com
[2] School of Computing and Information Systems, The University of Melbourne, Melbourne, Australia
[3] School of Computer Science, The University of Sydney, Sydney, Australia
anthony.wirth@sydney.edu.au

Abstract. Given a graph $G = (V, E)$ and a set $\mathcal{T} = \{(s_i, t_i) : 1 \leq i \leq k\} \subseteq V \times V$ of k pairs, the k-VERTEX-DISJOINT-PATHS (resp. k-EDGE-DISJOINT-PATHS) problem asks to determine whether there exist k pairwise vertex-disjoint (resp. edge-disjoint) paths $P_1, P_2, \ldots, P_k$ in G such that, for each $1 \leq i \leq k$, P_i connects s_i to t_i. Both the edge-disjoint and vertex-disjoint versions in undirected graphs are famously known to be FPT (parameterized by k) due to the Graph Minor Theory of Robertson and Seymour.

Eilam-Tzoreff [DAM '98] introduced a variant, known as the k-DISJOINT-SHORTEST-PATHS problem, where each path is further required to be a shortest path connecting its pair. They showed that the k-DISJOINT-SHORTEST-PATHS problem is NP-complete on both directed and undirected graphs; this holds even if the graphs are planar and have unit edge lengths. We focus on four versions of the problem, corresponding to considering edge/vertex disjointness, and to considering directed/undirected graphs. Building on the reduction of Chitnis [SIDMA '23] for k-EDGE-DISJOINT-PATHS on planar DAGs, we obtain the following *inapproximability lower bound* for each of the four versions of k-DISJOINT-SHORTEST-PATHS on n-vertex graphs:

– Under the gap version of the Exponential Time Hypothesis (Gap-ETH), there exists a constant $\delta > 0$ such that for any constant $0 < \varepsilon \leq \frac{1}{2}$ and any computable function f, there is no $(\frac{1}{2} + \varepsilon)$-approximation in $f(k) \cdot n^{\delta \cdot k}$ time.

We provide a single, **unified framework** to obtain lower bounds for *each of the four versions* of k-DISJOINT-SHORTEST-PATHS. We are able to further strengthen our results by restricting the structure of the input graphs in the lower bound constructions as follows:

– Directed: The inapproximability lower bound for edge-disjoint (resp. vertex-disjoint) paths holds even if the input graph is a planar (resp. 1-planar) DAG with max in-degree and max out-degree at most 2.
– Undirected: The inapproximability lower bound for edge-disjoint (resp. vertex-disjoint) paths hold even if the input graph is planar (resp. 1-planar) and has max degree 4.

This extended abstract does not contain any proofs; rather, it gives a high-level sketch of the techniques used to obtain our lower bounds. For a full version of the paper, containing all the proofs, see the Arxiv version [9].

© The Author(s), under exclusive license to Springer Nature Switzerland AG 2025
M. Bieńkowski and M. Englert (Eds.): WAOA 2024, LNCS 15269, pp. 46–60, 2025.
https://doi.org/10.1007/978-3-031-81396-2_4

The reductions outlined in this paper produce graphs in which half of the terminal pairs are trivially satisfiable, so any improvement of our $(\frac{1}{2} + \varepsilon)$ inapproximability factor requires a different approach.

As a byproduct of our reductions, we also show that the exact versions of each problem is $W[1]$-hard and give a $f(k) \cdot n^{o(k)}$-time lower bound for them under ETH. This exact lower bound shows that the $n^{O(k)}$-time algorithms of Bérczi and Kobayashi [ESA '17] for Directed-k-EDSP and Directed-k-VDSP are tight.

Keywords: disjoint shortest paths · directed acyclic graphs · planar graphs · 1-planar graphs · exponential time hypothesis · lower bounds · FPT inapproximability

1 Introduction

The k-DISJOINT-PATHS problem is one of the oldest and best-studied in graph theory: given a graph on n vertices and a set of k terminal pairs, the question is to determine whether there exists a collection of k pairwise-disjoint paths where each path connects one of the given terminal pairs. This paper focuses on a variant called the k-DISJOINT-SHORTEST-PATHS problem, where there is an additional requirement that each of the paths must be a shortest path for the terminal pair that it connects. There are four versions of the k-DISJOINT-SHORTEST-PATHS problem, depending on whether we require edge-disjointness or vertex-disjointness and whether the input graph is directed or undirected. This problem was introduced by Eilam-Tzoreff [13], who provided hardness results for both directed versions. The k-DISJOINT-SHORTEST-PATHS problem arises in several real-world scenarios, such as effective packet switching [22,24] and integrated circuit design [15,23].

1.1 Organization of the Paper

We first briefly survey some of the known results for k-DISJOINT-SHORTEST-PATHS on directed graphs (Sect. 1.2) and undirected graphs (Sect. 1.3), before outlining our results in Sect. 2. Due to space constraints, we focus on a sketch for the proofs of our lower bounds (Theorem 1(I) and Corollary 2(I)) for the EDGE-DISJOINT-SHORTEST-PATHS problem here, and defer our remaining results to the full version. Note that the full version contains all relevant results and proofs, whereas here we outline the key ideas of the proofs.

All of our results are obtained by reductions from known hardness results for the k-*Clique* problem (Sect. 3). For each of the four versions of the k-DISJOINT-SHORTEST-PATHS problem, a similar template is followed: first by obtaining an intermediate graph from an instance of k-*Clique*, before then applying an operation to vertices of that graph. Section 4 demonstrates a high-level overview of our reduction technique to obtain Theorem 1(I) and Corollary 2(I) and briefly describes how to modify the reductions to obtain our other results.

1.2 Prior Work on k-DISJOINT-SHORTEST-PATHS on directed graphs

The EDGE-DISJOINT-SHORTEST-PATHS problem on directed graphs is defined as follows, and the VERTEX-DISJOINT-SHORTEST-PATHS problem can be defined analogously:

DIRECTED-k-EDGE-DISJOINT-SHORTEST-PATHS (**Directed-k-EDSP**)
Input: An integer k, a directed graph $G = (V, E)$ with non-negative edge-lengths, and a set $\mathcal{T} = \{(s_i, t_i) : 1 \leq i \leq k\} \subseteq V \times V$ of k terminal pairs.
Question: Does there exist a collection of k paths $P_1, P_2, \ldots, P_k$ in G such that

- P_i is a shortest $s_i \rightsquigarrow t_i$ path in G for each $1 \leq i \leq k$, and
- for each $1 \leq i \neq j \leq k$, the paths P_i and P_j are edge-disjoint?

By setting all edge-lengths to be 0, the hardness for k-DISJOINT-SHORTEST-PATHS on digraphs follows from that of the k-DISJOINT-PATHS problem on digraphs. Eilam-Tzoreff [13] showed that both Directed-k-VDSP and Directed-k-EDSP are NP-hard when k is part of the input, even when the input digraph is planar and all edge-lengths are 1. Bérczi and Kobayashi [5] designed $n^{O(k)}$-time algorithms for Directed-k-VDSP on planar digraphs, and for Directed-k-VDSP and Directed-k-EDSP on DAGs by modifying an earlier algorithm of Fortune et al. for the k-Disjoint-Paths problem on DAGs [14]. When each edge-length is positive, Bérczi and Kobayashi [5] also showed that Directed-2-VDSP and Directed-2-EDSP can be solved in $n^{O(1)}$ time.

Amiri and Wargalla [2] showed a tight lower bound for Directed-k-EDSP on planar DAGs: under the Exponential Time Hypothesis (ETH)[1], there is no computable function, f, such that Directed-k-EDSP on planar DAGs admits an $f(k) \cdot n^{o(k)}$-time algorithm. Our results are reached by advancing Chitnis's technique for obtaining an exact lower bound for EDGE-DISJOINT-PATHS on planar DAGs [8]. Although not explicitly mentioned in their paper, the hardness reduction for Undirected-k-VDSP on general graphs by Bentert et al., [4, Proposition 3], also holds for 1-planar graphs and for DAGs if one were to orient all edges from either left-to-right or bottom-to-top, and can also be adapted to hold for a bounded max degree of four.

1.3 Prior Work on k-DISJOINT-SHORTEST-PATHS on Undirected Graphs

The two versions of the k-DISJOINT-SHORTEST-PATHS problem on undirected graphs, being Undirected-k-EDSP and Undirected-k-VDSP, are defined analogously to their directed counterparts. Eilam-Tzoreff [13] designed an $O(n^8)$-time algorithm for Undirected-2-VDSP and Undirected-2-EDSP in the case when all edge costs are guaranteed to be positive. Akhmedov [1] improved this to $O(n^7)$ when the costs are positive and further to $O(n^6)$ when all costs are 1. Gottschau et al. [16] and Kobayashi and Sako [19] independently gave $n^{O(1)}$-time algorithms for Undirected-2-VDSP and Undirected-2-EDSP when edge costs are non-negative.

The complexity of Undirected-k-VDSP and Undirected-k-EDSP for $k \geq 3$ was a long-standing open problem until Lochet [20] designed an XP algorithm running in

---

[1] The Exponential Time Hypothesis (ETH) states that n-variable m-clause 3-SAT cannot be solved in $2^{o(n)} \cdot (n+m)^{O(1)}$ time [17, 18].

$n^{O(k^{5^k})}$ time for general k on VDSP. Bentert et al. improved the running time of this algorithm to $n^{O(k!k)}$ using some geometric ideas [4, Theorem 2], and also showed that there is no $f(k) \cdot n^{o(k)}$-time algorithm (for any computable function f) under ETH [4, Proposition 3].

2 Our Results

In this paper, we obtain lower bounds on the running time of exact and approximate algorithms for the edge-disjoint and vertex-disjoint versions of k-DISJOINT-SHORTEST-PATHS on undirected and directed graphs. Note that because we assign each edge a uniform cost of 1, we could equivalently measure path length by counting the number of unit-cost vertices along it and we choose the latter to simplify some of our arguments. Additionally, by considering each vertex to have non-zero cost, we cannot exploit the known hardness results for k-DISJOINT-PATHS (a special case of k-DISJOINT-SHORTEST-PATHS with all vertex costs 0).

We now define our results for the Directed-k-EDSP problem. Theorem 1 and Corollary 2 assume the Exponential Time Hypothesis (ETH) and Gap Exponential Time Hypothesis[2] (GAP-ETH), respectively. Although we sketch the proofs of Theorem 1(I) and Corollary 2(I) in this paper, the remainder of our proofs are deferred the full version of this paper [9].

Theorem 1 *(inapproximability). Assuming Gap-ETH, for each $0 < \varepsilon \leq \frac{1}{2}$ there exists a constant $\zeta > 0$ such that no $f(k) \cdot n^{\zeta k}$ time algorithm can distinguish between the following two cases in a k-DISJOINT-SHORTEST-PATHS problem instance:*

- *All k pairs can be satisfied*
- *At most $(\frac{1}{2} + \varepsilon) \cdot k$ pairs can be satisfied*

Here f is any computable function, n is the number of vertices and k is the number of terminal pairs. This bound holds for the following problems and graph classes:

(I) DIRECTED-k-EDGE-DISJOINT-SHORTEST-PATHS on planar DAGs with max in-degree and out-degree 2.

(II) DIRECTED-k-VERTEX-DISJOINT-SHORTEST-PATHS on 1-planar DAGs with max in-degree and out-degree 2.

(III) UNDIRECTED-k-EDGE-DISJOINT-SHORTEST-PATHS on planar graphs with max degree 4.

(IV) UNDIRECTED-k-VERTEX-DISJOINT-SHORTEST-PATHS on 1-planar graphs with max degree 4.

---

[2] The Gap-ETH [11,21] states that there exists a constant $\delta > 0$ such that there is no $2^{o(n)}$-time algorithm which, given an instance of 3-SAT on n variables, can distinguish between the case when all clauses are satisfiable versus the case when every assignment to the variables leaves at least δ-fraction of the clauses unsatisfied. We refer the interested reader to further discussions about the plausibility of Gap-ETH [6,11].

Corollary 2 *(exact lower bound). The following k-DISJOINT-SHORTEST-PATHS problems on the corresponding graph classes are W[1] hard parameterized by the number of terminal pairs k. Moreover, under the ETH, there is no computable function f which solves them in $f(k) \cdot n^{o(k)}$ time.*

(I) DIRECTED-k-EDGE-DISJOINT-SHORTEST-PATHS *on planar DAGs with max in-degree and out-degree 2.*
(II) DIRECTED-k-VERTEX-DISJOINT-SHORTEST-PATHS *on 1-planar DAGs with max in-degree and out-degree 2.*
(III) UNDIRECTED-k-EDGE-DISJOINT-SHORTEST-PATHS *on planar graphs with max degree 4.*
(Iv) UNDIRECTED-k-VERTEX-DISJOINT-SHORTEST-PATHS *on 1-planar graphs with max degree 4 (Table 1).*

Table 1. A compendium of results that we obtain. Full theorem statements are in Sect. 2. Throughout the table, f represents any computable function. Note that all our EDSP and VDSP results hold even for planar and 1-planar graphs respectively. Furthermore, the directed results hold if the input graph is a DAG and has both max in-degree and max-degree upper bounded by 2. The undirected results hold even if the max degree of the input graph is upper bounded by 4.

Problem	Inapproximability Factor	Hypothesis	Lower Bound	Reference
Directed-k-EDSP	$\left(\frac{1}{2}+\varepsilon\right)$ for each $0<\varepsilon\leq\frac{1}{2}$	Gap-ETH	$f(k)\cdot n^{\zeta k}$ for some $\zeta>0$	Theorem 1(I)
Directed-k-EDSP	Exact	ETH	$f(k)\cdot n^{o(k)}$	Corollary 2(I)
Directed-k-VDSP	$\left(\frac{1}{2}+\varepsilon\right)$ for each $0<\varepsilon\leq\frac{1}{2}$	Gap-ETH	$f(k)\cdot n^{\zeta k}$ for some $\zeta>0$	Theorem 1(II)
Directed-k-VDSP	Exact	ETH	$f(k)\cdot n^{o(k)}$	Corollary 2(II)
Undirected-k-EDSP	$\left(\frac{1}{2}+\varepsilon\right)$ for each $0<\varepsilon\leq\frac{1}{2}$	Gap-ETH	$f(k)\cdot n^{\zeta k}$ for some $\zeta>0$	Theorem 1(III)
Undirected-k-EDSP	Exact	ETH	$f(k)\cdot n^{o(k)}$	Corollary 2(III)
Undirected-k-VDSP	$\left(\frac{1}{2}+\varepsilon\right)$ for each $0<\varepsilon\leq\frac{1}{2}$	Gap-ETH	$f(k)\cdot n^{\zeta k}$ for some $\zeta>0$	Theorem 1(IV)
Undirected-k-VDSP	Exact	ETH	$f(k)\cdot n^{o(k)}$	Corollary 2(IV)

Placing Our Lower Bounds in the Context of Prior Work: Our inapproximability results are *tight* for our specific reductions because in all of our reductions of DISJOINT-SHORTEST-PATHS it is trivially possible to satisfy half the pairs[3]. To obtain stronger inapproximability results, one therefore needs ideas quite different from those introduced in this paper such as those given by Bentert et al. [3]. Their paper provides

---

[3] For example, in Fig. 1, selecting a shortest $c_i \leadsto d_i$ for every $1 \leq i \leq k$ or a shortest $a_j \leadsto b_j$ for every $1 \leq j \leq k$ provides k pairs that are necessarily pairwise edge-disjoint.

an $o(k)$-factor inapproximability lower bound in $f(k) \cdot \text{poly}(n)$ time under Gap-ETH for VERTEX-DISJOINT-SHORTEST-PATHS and EDGE-DISJOINT-SHORTEST-PATHS graphs for which the terminal vertices have a degree of at most 2 and every other vertex has degree at most 3. Our inapproximability results for EDGE-DISJOINT-SHORTEST-PATHS (Theorem 1(I) Theorem 1(III)) and VERTEX-DISJOINT-SHORTEST-PATHS (Theorem 1(II) Theorem 1(IV)), however, hold even if the input graph is planar or 1-planar respectively.

The framework presented in this paper provides one reduction technique for achieving the aforementioned inapproximability bounds, in addition to a number of exact hardness results (Corollary 2(I), Corollary 2(II), Corollary 2(III) and Corollary 2(IV)). Chitnis [8] focused on planar graphs in their hardness proof for k-EDGE-DISJOINT-PATHS, and we take this further in analysing how we can obtain analogous lower bounds on other specific graph classes.

Chitnis [8] used a reduction from GRID-TILING-$\leq$ and we note that Corollary 2(I) could also be obtained by reducing from GRID-TILING, although we present a reduction from k-Clique here. Corollary 2(I) was also obtained independently by Amiri & Wargalla [2], although without the added restriction of bounded in-degree and out-degree.

Although we obtain our results by adapting Chitnis's technique for obtaining lower bounds for EDGE-DISJOINT-PATHS on DAGs [8], we note that one can obtain Corollary 2(IV) by analysing Bentert et al.'s graph, [4, Proposition 3], for hardness on general undirected graphs to be 1-planar. Subsequently, we can obtain Corollary 2(II) by orienting all edges in their graph either left-to-right or bottom-to-top. To the best of our knowledge, Corollary 2(III) is the first result showing a lower bound for Undirected-k-EDSP on planar graphs.

3 Known Exact and Inapproximate Lower Bounds for k-CLIQUE

All our lower bounds are obtained using reductions from k-Clique, which is known to be W[1]-hard [12]. We now define k-Clique, before giving Chen et al.'s asymptotically-tight lower bound [7]:

---
k-Clique
Input: Integer k, and an undirected graph $G = (V, E)$, where $V = \{v_1, v_2, \ldots, v_N\}$.
Question: Does there exist a set $Z \subseteq V$ of size k such that for all $x \neq y \in Z$, $x - y \in E$?

Theorem 3 (Theorem 5.5, [7]). *Under the Exponential Time Hypothesis (ETH), the k-Clique problem on graphs with N vertices cannot be solved in $f(k) \cdot N^{o(k)}$ time for any computable function f.*

We use Theorem 3 to show all of our exact lower bounds. To obtain our approximability lower bounds, we use the stronger Gap-ETH assumption, under which hardness of approximating k-Clique is known. Theorem 4 paraphrases the hardness of k-Clique under Gap-ETH as given in [6, Theorem 18].

Theorem 4 (Theorem 18, [6]). *Assuming Gap-ETH, there exist constants $\delta, r_0 > 0$ such that, for any computable function g and for any positive integers $q \geq r \geq r_0$, there* is no algorithm that, given a graph G', can distinguish in $g(q,r) \cdot N^{\delta r}$ time, where $N = |V(G')|$ and CLIQUE(G') denotes the maximum size of a clique in G', between the cases where CLIQUE(G') $\geq q$, or CLIQUE(G') $< r$.

4 Proof Sketch of Theorem 1(I) and Corollary 2(I)

This section provides a sketch for the proof of our lower bounds for the k-EDSP problem. We begin by constructing an intermediate graph D_{int} and describing some of its key properties in Sect. 4.1 and Sect. 4.2 respectively. Section 4.3 then demonstrates how to modify D_{int} to obtain the final graph D_{edge} and Sect. 4.4 concludes this section by sketching how these components can be used to prove Theorem 1(I) and Corollary 2(I).

4.1 Construction of the Intermediate Graph D_{int}

Given an instance $G = (V,E)$ of k-*Clique* with $V = \{v_1, v_2, \ldots, v_N\}$, we now build an instance of an intermediate digraph D_{int} (Fig. 1). This section gives a sketch of this construction. D_{int} is later modified to obtain the final graphs, D_{edge} (Sect. 4.3) and D_{vertex} (Sect. 5), which are used to obtain exact and approximate lower bounds. We note that the D_{int} graph is essentially the same as the graph that was constructed to show the W[1]-hardness of DIRECTED-k-EDGE-DISJOINT-PATHS (via reduction from GRID-TILING-$\leq$) by Chitnis [8].

We first define the following sets for a given instance, G, of k-*Clique*:

$$\text{For each } i \in [k], \text{ let } S_{i,i} := \{(a,a) : 1 \leq a \leq N\}$$
$$\text{For each } 1 \leq i \neq j \leq k, \text{ let } S_{i,j} := \{(a,b) \mid v_a \text{ and } v_b \text{ form an edge in } G\} \quad (1)$$

1. **Origin**: The origin is marked at the bottom left corner of D_{int}. This is defined just so we can view the naming of the vertices as per the usual $X - Y$ coordinate system: increasing horizontally towards the right, and vertically towards the top.
2. **Grid (black) vertices and edges**: For each $1 \leq i, j \leq k$, introduce a (directed) $N \times N$ grid $D_{i,j}$ where the column numbers increase from 1 to N as we go from left to right, and the row numbers increase from 1 to N as we go from bottom to top. For each $1 \leq q, \ell \leq N$ the unique vertex which is the intersection of the qth column and ℓ^{th} row of $D_{i,j}$ is denoted by $\mathbf{w}_{i,j}^{q,\ell}$.
3. **Red edges for connections**: Red edges are used to form a direct perfect matching between neighbouring $k \times k$ grids.
4. **Green (terminal) vertices and magenta edges**: For each $i \in [k]$, add four sets of (terminal) vertices (shown in Fig. 1 using green colour). Then, these are connected using magenta edges to and from the relevant $k \times k$ grids. To reduce the degree of these vertices, without loss of generality, we can convert them into a directed binary tree of depth $\log_2 N$.

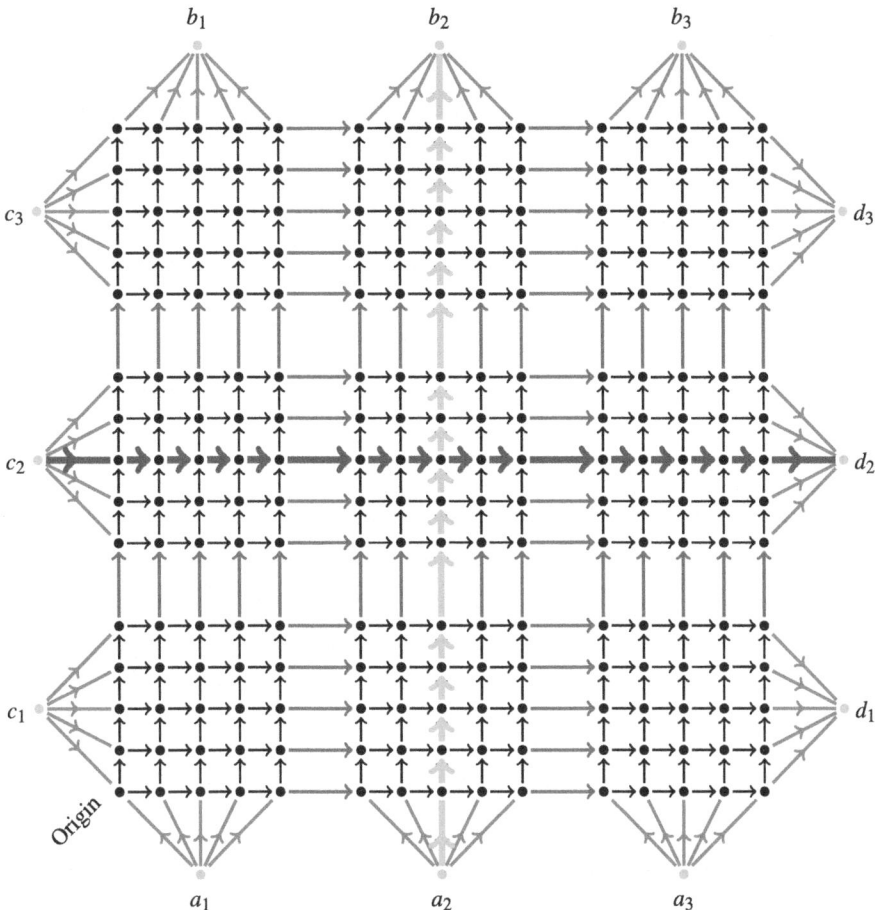

Fig. 1. The intermediate directed graph D_{int} constructed from an instance (G, k) of k-*Clique* (with $k = 3$ and $N = 5$) via the construction described in Sect. 4.1. The highlighted **blue** and yellow paths are examples of horizontal and vertical canonical paths respectively, which are described informally in Definition 6. (Color figure online)

4.2 Characterizing Shortest Paths Between Terminal Pairs in D_{int}

In this section, we introduce a high-level description of the structure of shortest paths between terminal pairs in D_{int} (Fig. 1).

Definition 5 (*costs of vertices in D_{int}*)**.** *Each vertex in D_{int} has a cost of 1.*

Definition 6 (*informal description of horizontal (resp., vertical) canonical paths in D_{int}*)**.** *Fix any $j \in [k]$. We define the horizontal canonical path $c_j \leadsto d_j$ (resp., $a_j \leadsto b_j$) in D_{int} to be that beginning with any* magenta *edge outgoing from c_j (resp., a_j), then taking only horizontal (resp., vertical) edges before ending with the available* magenta edge ingoing to d_j (resp., b_j) and can be visualised as a blue (resp., yellow) path in Fig. 1.

It is straight-forward to observe in Fig. 1 that each canonical path (vertical or horizontal) visits exactly $kN+2$ vertices and thus has a cost of $kN+2$.

The next lemma shows that if $j \in [k]$ then any shortest $c_j \rightsquigarrow d_j$ path in D_{int} must be a horizontal canonical path and vice versa.

Lemma 7 (equivalence of horizontal canonical paths and horizontal paths between terminal pairs in D_{int}). Let $j \in [k]$. The horizontal canonical paths in D_{int} satisfy the following two properties:

(i) Every horizontal canonical path from $c_j \rightsquigarrow d_j$ is a shortest $c_j \rightsquigarrow d_j$ path in D_{int}.
(ii) If P is a shortest $c_j \rightsquigarrow d_j$ path in D_{int}, then P must be a canonical $c_j \rightsquigarrow d_j$ path.

The analogous equivalent to Lemma 7 holds for vertical canonical paths. Thus, we have characterized shortest paths between terminal pairs in D_{int}: each is either a vertical canonical path or a horizontal canonical path.

4.3 Obtaining the Graph D_{edge} from D_{int} via the Splitting Operation

Section 4.1 makes clear that the structure of D_{int} depends only on the values of n and k, and not on the internal details of the k-Clique instance. Our splitting operation defined here ensures that the resultant graph is a sound and complete encoding of the original k-Clique instance. It is clear to see that any pair of horizontal (respectively vertical) canonical paths are both edge and vertex disjoint in D_{int}. On the other hand, for a fixed $i, j \in [k]$, the canonical paths $a_i \rightsquigarrow b_i$ and $c_j \rightsquigarrow d_j$ must intersect at some black vertex in the graph. That is that they are not vertex disjoint, but are edge-disjoint.

For our reduction from k-Clique to work, we need to modify D_{int} by applying a splitting operation (Definition 9) to its vertices so that there only exists edge disjoint paths $a_i \rightsquigarrow b_i$ and $c_j \rightsquigarrow d_j$ in D_{edge} if, for their intersection point $\mathbf{w}_{i,j}^{q,\ell}$, in D_{int}, $(q, \ell) \in S_{i,j}$ (Eq. 1). Chitnis [8] used a splitting technique to achieve this for the EDGE-DISJOINT-PATHS problem, and our splitting operation ensures that all shortest terminal paths in D_{int} as such in D_{edge} after the splitting operation.

Definition 8 (four neighbors of each grid vertex in D_{int}). For each vertex w in D_{int} (Fig. 1), we define $\text{west}(w)$, $\text{east}(w)$, $\text{north}(w)$ and $\text{south}(w)$ to be the vertices directly to the left, right, above and below w (as seen by the reader), respectively.

Definition 9 (informal description of the splitting operation which is used to obtain D_{edge} from D_{int}). Recall the definition Eq. 1. For each $i, j \in [k]$ and each $q, \ell \in [N]$:

- if $(q, \ell) \notin S_{i,j}$, we one-split (Fig. 2) the vertex $\mathbf{w}_{i,j}^{q,\ell}$, replacing it with 3 vertices each of cost 1 and arrange the edges such that the paths $\text{west}(w) \rightsquigarrow \text{east}(w)$ and $\text{south}(w) \rightsquigarrow \text{north}(w)$ cannot be edge-disjoint.
- Otherwise, if $(q, \ell) \in S_{i,j}$, we two-split (Fig. 3) the vertex $\mathbf{w}_{i,j}^{q,\ell}$, replacing it with 4 vertices each of cost 1 and arrange the edges such that the paths $\text{west}(w) \rightsquigarrow \text{east}(w)$ and $\text{south}(w) \rightsquigarrow \text{north}(w)$ can be edge-disjoint. .

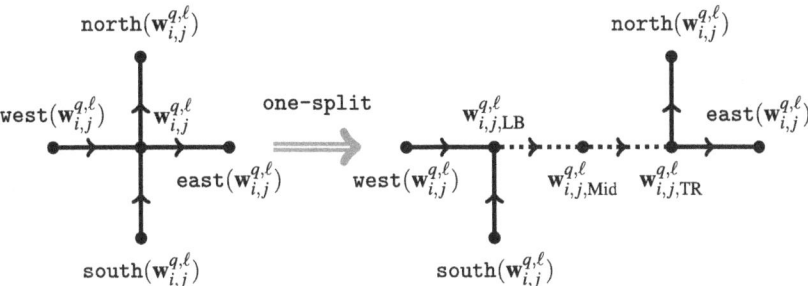

Fig. 2. The one-split operation for the vertex $w_{i,j}{}^{q,\ell}$ when $(q,\ell) \notin S_{i,j}$. The idea behind this splitting is that the horizontal path $\text{west}(w_{i,j}^{q,\ell}) \to w_{i,j}^{q,\ell} \to \text{east}(w_{i,j}^{q,\ell})$ and vertical path $\text{south}(w_{i,j}^{q,\ell}) \to w_{i,j}^{q,\ell} \to \text{north}(w_{i,j}^{q,\ell})$ are no longer edge-disjoint after the one-split operation as they must share the path $w_{i,j,\text{LB}}^{q,\ell} \to w_{i,j,\text{Mid}}^{q,\ell} \to w_{i,j,\text{TR}}^{q,\ell}$.

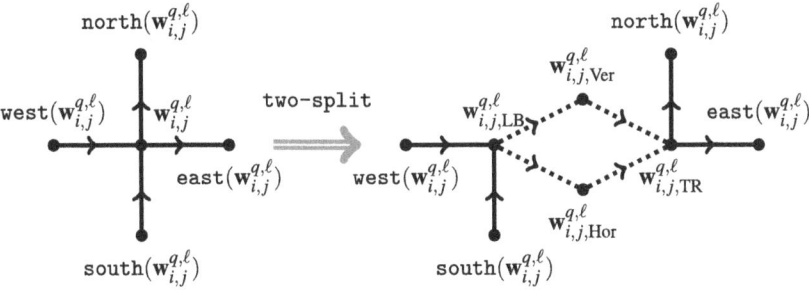

Fig. 3. The two-split operation for the vertex $w_{i,j}{}^{q,\ell}$ when $(q,\ell) \in S_{i,j}$. The idea behind this splitting is that the horizontal path $\text{west}(w_{i,j}^{q,\ell}) \to w_{i,j}^{q,\ell} \to \text{east}(w_{i,j}^{q,\ell})$ and vertical path $\text{south}(w_{i,j}^{q,\ell}) \to w_{i,j}^{q,\ell} \to \text{north}(w_{i,j}^{q,\ell})$ are still edge-disjoint after the two-split operation if we replace them with the paths $\text{west}(w_{i,j}^{q,\ell}) \to w_{i,j,\text{LB}}^{q,\ell} \to w_{i,j,\text{Hor}}^{q,\ell} \to w_{i,j,\text{TR}}^{q,\ell} \to \text{east}(w_{i,j}^{q,\ell})$ and $\text{south}(w_{i,j}^{q,\ell}) \to w_{i,j,\text{LB}}^{q,\ell} \to w_{i,j,\text{Ver}}^{q,\ell} \to w_{i,j,\text{TR}}^{q,\ell} \to \text{north}(w_{i,j}^{q,\ell})$ respectively.

This splitting operation modifies vertices in D_{int} such that we one-split if $(q,\ell) \notin S_{i,j}$, meaning that no two paths can pass through the point of splitting without violating edge-disjointness. Otherwise, we two-split if $(q,\ell) \in S_{i,j}$, meaning that two perpendicular paths can pass through the point of splitting without violating edge-disjointness. Observe that, regardless of whether we one-split or two-split, every path in D_{edge} will visit 3 vertices in place of 1, for each of its vertices that were split from D_{int}. Finally, we are now ready to define the instance of Directed-$2k$-EDSP that we have built, starting from an instance G of k-Clique.

Definition 10 (defining the $2k$-EDSP instance). The instance $(D_{\text{edge}}, \mathcal{T})$ of Directed-$2k$-EDSP is obtained by applying the splitting operation (Definition 9) to every black vertex in D_{int} and assigning a cost of 1 to every newly introduced vertex.

Characterizing Shortest Paths Between Terminal Pairs in D_{edge}. We now define canonical paths in D_{edge}, adapting the definition of canonical paths (Definition 6) in D_{int}. Vertical canonical paths can be defined analogously.

Definition 11 *(Informal - horizontal (resp. vertical) canonical paths in D_{edge}). Any horizontal (resp., vertical) canonical path in D_{int} (Definition 6) can be modified such that it visits exactly the vertices $\text{west}(w)$, $\text{east}(w)$ and w_{Mid}/w_{Hor} (resp., $\text{west}(w)$, $\text{east}(w)$ and w_{Mid}/w_{Ver}) that replaced any given black vertex from D_{int}'s canonical path.*

Observation 12 Any canonical path in D_{int} has a corresponding canonical path in D_{edge} which visits thrice as many black vertices.

Definition 13 and Lemma 14 analyze the structure of shortest horizontal paths between terminal pairs in D_{edge}. Observation 12 can be combined with Lemma 7 in order to obtain Lemma 14, which we omit here.

Definition 13 *(Informal - Image of a horizontal canonical path from D_{int} in D_{edge}). Any horizontal (resp. vertical) canonical path in D_{int} (Definition 6) can be modified such that it visits exactly 3 vertices that replaced any given black vertex from D_{int}'s canonical path.*

Note that every image of a horizontal canonical path in D_{edge} visits the same number of vertices and thus has the same cost.

Lemma 14 (**Horizontal Canonical Paths $\Leftrightarrow$ Shortest Horizontal Path in D_{edge}**). *Let $j \in [k]$. The horizontal canonical paths in D_{edge} satisfy the following two properties:*

(i) Every horizontal canonical path from $c_j \rightsquigarrow d_j$ is a shortest $c_j \rightsquigarrow d_j$ path in D_{edge}.
(ii) If P is a shortest $c_j \rightsquigarrow d_j$ path in D_{edge}, then P must be an image of a horizontal canonical $c_j \rightsquigarrow d_j$ path from D_{int}.

4.4 Using D_{edge} to Sketch Proof of Theorem 1(I) and Corollary 2(I)

By analysing the locations of split vertices in D_{edge}, we are able to show that our reduction is both *complete* and *sound*. For completeness, that is to say we show that given a satisfiable instance of k-*Clique*, we can construct a satisfiable instance of $2k$-EDSP in polynomial time. For soundness, on the other hand, we get that if at least $(\frac{1}{2} + \varepsilon)$-fraction of the $2k$ pairs from instance $(D_{\text{edge}}, \mathcal{T})$ of Directed-$2k$-EDSP can be satisfied, then graph G has a clique of size $2\varepsilon \cdot k$.

We now briefly describe how we obtain the proofs for Theorem 1(I) and Corollary 2(I).

Proof (**Proof sketch of Theorem** 1(I))**.** We have provided a polynomial-time parameterized reduction from an instance G of k-*Clique* to an instance $\mathcal{I}$ of $2k$-EDSP whose graph has size $n = O(N^2 k^2)$. Assume for contradiction that an algorithm $\mathbb{A}_{\text{EDSP}}$ exits which runs in time $f(k) \cdot n^{\zeta k}$ for some computable function f and $\zeta > 0$ that can distinguish between the two cases in Theorem 1(I). Our arguments for completeness and soundness then imply the existence of an algorithm for k-*Clique* which contradicts Theorem 4, by using our reduction to convert G to $\mathcal{I}$ and then applying $\mathbb{A}_{\text{EDSP}}$ to $\mathcal{I}$.

By setting $\varepsilon = \frac{1}{2}$ in the soundness part of our reduction, we obtain the lower bound from Corollary 2(I).

5 Obtaining the Graph D_{vertex} from D_{int} via the Splitting Operation

A natural corollary to the EDGE-DISJOINT-SHORTEST-PATHS results described above is to ask if such a result is feasible for VERTEX-DISJOINT-SHORTEST-PATHS. As mentioned, D_{edge} is a modified version of the graph used by Chitnis in [8], and our edge splitting operation allows us to ensure that D_{vertex} is a sound and complete encoding of the original k-Clique instance.

Likewise, we aim to obtain an equivalent result for the VERTEX-DISJOINT-SHORTEST-PATHS version of the problem and are able to do this by applying a different splitting operation (Fig. 4) to D_{int}. Similar to Sect. 4.3 for EDGE-DISJOINT-SHORTEST-PATHS, the splitting operation for VERTEX-DISJOINT-SHORTEST-PATHS ensures that there only exists vertex disjoint paths $a_i \rightsquigarrow b_i$ and $c_j \rightsquigarrow d_j$ in D_{edge} if, for their intersection point $\mathbf{w}_{i,j}^{q,\ell}$, in D_{int}, $(q,\ell) \in S_{i,j}$ (Eq. 1). That is to say that we vertex-split (Fig. 4), when $(q,\ell) \in S_{i,j}$. When $(q,\ell) \notin S_{i,j}$, we retain the structure of $\mathbf{w}_{i,j}^{q,\ell}$ and, analogously to Observation 12, we further note that any canonical path in D_{int} has a corresponding canonical path in D_{vertex} of equivalent cost.

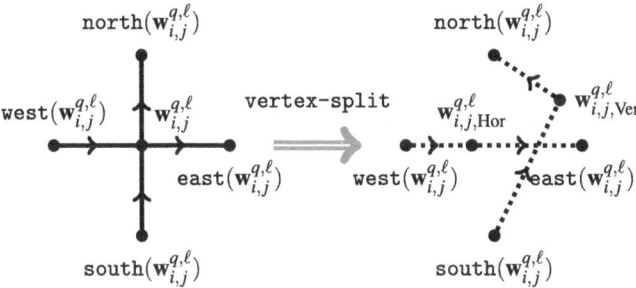

Fig. 4. The vertex-split operation for the vertex $\mathbf{w}_{i,j}^{q,\ell}$ when $(q,\ell) \in S_{i,j}$. The idea behind this is that the horizontal path west($\mathbf{w}_{i,j}^{q,\ell}$) $\rightarrow \mathbf{w}_{i,j}^{q,\ell} \rightarrow$ east($\mathbf{w}_{i,j}^{q,\ell}$) and the vertical path south($\mathbf{w}_{i,j}^{q,\ell}$) $\rightarrow \mathbf{w}_{i,j}^{q,\ell} \rightarrow$ north($\mathbf{w}_{i,j}^{q,\ell}$) become vertex-disjoint.

6 Using D_{vertex} to Sketch Proof of Lower Bounds for k-VDSP

Our lower bound results for the VERTEX-DISJOINT-SHORTEST-PATHS problem (Theorem 1(II) and Corollary 2(II)) can be obtained by analysing D_{vertex} in much the same way as D_{edge} is analysed when proving Theorem 1(I) and Corollary 2(I) in

Sect. 4.4. That is to say that by using our polynomial-time reduction's *completeness* and *soundness*, we show that an algorithm that matches the running time and exact/inapproximability guarantees in Theorem 1(II) or Corollary 2(II) would imply an algorithm for *k-Clique* which contradicts Theorem 4.

Cygan et al. [10] gave an FPT-time algorithm for the k-VERTEX-DISJOINT-PATHS problem which, with a planar splitting operation as in EDSP (Sect. 4.3), could be reduced-to from GRID-TILING-$\leq$. Since GRID-TILING-$\leq$ is known to be $W[1]$-hard[4], a planar splitting operation for k-VERTEX-DISJOINT-PATHS would imply that *FPT = W[1]*. Thus we do not expect that it is possible to obtain a planar splitting operation for the VDSP problem using our reduction, as has been shown for EDSP.

7 Modifications for Undirected EDSP and VDSP

By removing the directions of all edges in the constructions outlined in this paper, one can show that the reductions sketched in Sect. 4.4 and Sect. 6 also hold for the undirected EDSP and VDSP problems with two key modifications. Paths are more free to move around in undirected graphs since they can potentially backtrack or veer off, so our arguments for proving what constitutes as a shortest path must be more intricate.

Observe, also, that in an undirected graph, one would be able to use other terminal vertices as a shortcut. To give an example, imagine Fig. 1 to be undirected, a path between (e.g.) a_1 and b_1 could be shorter than the defined shortest path by using c_i for $1 \leq i \leq 3$ to jump from the 1st to the Nth row of $D_{1,i}$. We tackle this by modifying the weight of our terminal vertices to ensure that no shortest path could use them as a shortcut.

8 Conclusion and Open Problems

In this paper, we obtained approximate and exact lower bounds for all four variants of the k-DISJOINT-SHORTEST-PATHS problem. We leave open the following natural questions:

- Can we improve on the $(\frac{1}{2} + \varepsilon)$ factor of the FPT inapproximability results for EDGE-DISJOINT-SHORTEST-PATHS and VERTEX-DISJOINT-SHORTEST-PATHS on planar and 1-planar graphs respectively? Perhaps this could be achieved by modifying the reduction for the $o(k)$ factor lower bound on general graphs by Bentert et al. [3].
- Can we obtain inapproximability lower bounds also for the k-DISJOINT-PATHS problem on directed graphs, possibly on graph classes such as DAGs or planar graphs for which FPT or XP algorithms are known [10,14]?
- Corollary 2(II) gives W[1] hardness and, under ETH, an $f(k) \cdot n^{o(k)}$ lower bound for VERTEX-DISJOINT-SHORTEST-PATHS on directed 1-planar graphs. Can we get equivalent lower bounds for planar graphs, which would show Bérczi and

---
[4] We refer the interested reader to [8] for a formal definition of the GRID-TILING-$\leq$ problem and lower bounds for DISJOINT-PATHS.

Kobayashi's $n^{O(k)}$-time algorithm [5] to be tight? Alternatively, is an FPT algorithm for the problem possible by either adapting Cygan et al.'s $2^{2^{O(k^2)}} \cdot n^{O(1)}$-time algorithm for planar VERTEX-DISJOINT-PATHS [10], or through an entirely new technique.

References

1. Akhmedov, M.: Faster 2-disjoint-shortest-paths algorithm. In: Fernau, H. (ed.) CSR 2020. LNCS, vol. 12159, pp. 103–116. Springer, Cham (2020). https://doi.org/10.1007/978-3-030-50026-9_7
2. Amiri, S.A., Wargalla, J.: Disjoint shortest paths with congestion on DAGs. CoRR abs/2008.08368 (2020). https://arxiv.org/abs/2008.08368
3. Bentert, M., Fomin, F.V., Golovach, P.A.: Tight approximation and kernelization bounds for vertex-disjoint shortest paths, February 2024. https://doi.org/10.48550/arXiv.2402.15348
4. Bentert, M., Nichterlein, A., Renken, M., Zschoche, P.: Using a geometric lens to find k disjoint shortest paths. In: ICALP 2021, pp. 26:1–26:14 (2021). https://doi.org/10.4230/LIPIcs.ICALP.2021.26
5. Bérczi, K., Kobayashi, Y.: The directed disjoint shortest paths problem. In: ESA 2017, vol. 87, pp. 13:1–13:13 (2017). https://doi.org/10.4230/LIPIcs.ESA.2017.13
6. Chalermsook, P., et al.: From gap-exponential time hypothesis to fixed parameter tractable inapproximability: clique, dominating set, and more. SIAM J. Comput. **49**(4), 772–810 (2020). https://doi.org/10.1137/18M1166869
7. Chen, J., Huang, X., Kanj, I.A., Xia, G.: Strong computational lower bounds via parameterized complexity. J. Comput. Syst. Sci. **72**(8), 1346–1367 (2006). https://doi.org/10.1016/j.jcss.2006.04.007
8. Chitnis, R.: A tight lower bound for edge-disjoint paths on planar DAGs. SIAM J. Discret. Math. **37**(2), 556–572 (2023). https://doi.org/10.1137/21m1395089
9. Chitnis, R., Thomas, S., Wirth, A.: Lower bounds for approximate (& exact) k-disjoint-shortest-paths, August 2024. https://doi.org/10.48550/arXiv.2408.03933
10. Cygan, M., Marx, D., Pilipczuk, M., Pilipczuk, M.: The planar directed k-Vertex-Disjoint paths problem is fixed-parameter tractable. In: FOCS 2013, pp. 197–206 (2013). https://doi.org/10.1109/FOCS.2013.29
11. Dinur, I.: Mildly exponential reduction from gap 3SAT to polynomial-gap label-cover. Electron. Colloquium Comput. Complex, 128 (2016). https://eccc.weizmann.ac.il/report/2016/128
12. Downey, R.G., Fellows, M.R.: Fixed-parameter tractability and completeness II: on completeness for W[1]. Theoret. Comput. Sci. **141**, 109–131 (1995). https://doi.org/10.1016/0304-3975(94)00097-3
13. Eilam-Tzoreff, T.: The disjoint shortest paths problem. Discret. Appl. Math. **85**(2), 113–138 (1998). https://doi.org/10.1016/S0166-218X(97)00121-2
14. Fortune, S., Hopcroft, J.E., Wyllie, J.: The directed subgraph homeomorphism problem. Theoret. Comput. Sci. **10**, 111–121 (1980). https://doi.org/10.1016/0304-3975(80)90009-2
15. Frank, A.: Packing paths, circuits, and cuts - a survey. Paths, Flows, and VLSI-Layout (1990)
16. Gottschau, M., Kaiser, M., Waldmann, C.: The undirected two disjoint shortest paths problem. Oper. Res. Lett. **47**(1), 70–75 (2019). https://doi.org/10.1016/j.orl.2018.11.011
17. Impagliazzo, R., Paturi, R.: On the complexity of k-SAT. J. Comput. Syst. Sci. **62**(2), 367–375 (2001). https://doi.org/10.1006/jcss.2000.1727
18. Impagliazzo, R., Paturi, R., Zane, F.: Which problems have strongly exponential complexity? J. Comput. Syst. Sci. **63**(4), 512–530 (2001). https://doi.org/10.1006/jcss.2001.1774

19. Kobayashi, Y., Sako, R.: Two disjoint shortest paths problem with non-negative edge length. Oper. Res. Lett. **47**(1), 66–69 (2019). https://doi.org/10.1016/j.orl.2018.11.012
20. Lochet, W.: A polynomial time algorithm for the k-Disjoint shortest paths problem. In: SODA 2021, pp. 169–178 (2021). https://doi.org/10.1137/1.9781611976465.12
21. Manurangsi, P., Raghavendra, P.: A birthday repetition theorem and complexity of approximating dense CSPs. In: ICALP 2017, vol. 80, pp. 78:1–78:15 (2017). https://doi.org/10.4230/LIPIcs.ICALP.2017.78
22. Ogier, R., Rutenburg, V., Shacham, N.: Distributed algorithms for computing shortest pairs of disjoint paths. IEEE Trans. Inf. Theor. **39**(2), 443–455 (1993). https://doi.org/10.1109/18.212275
23. Robertson, N., Seymour, P.D.: An outline of a disjoint paths algorithm. In: Paths, Flows, and VLSI-Layout, pp. 267–292 (1990)
24. Srinivas, A., Modiano, E.: Finding minimum energy disjoint paths in wireless Ad-Hoc networks. Wirel. Netw. **11**(4), 401–417 (2005). https://doi.org/10.1007/s11276-005-1765-0

Approximating δ-Covering

Tim A. Hartmann[1] and Tom Janßen[2](✉)

[1] CISPA Helmholtz Center for Information Security, Saarbrücken, Germany
tim.hartmann@cispa.de
[2] Department of Computer Science, RWTH Aachen University, Aachen, Germany
janssen@algo.rwth-aachen.de

Abstract. δ-COVERING, for some covering range $\delta > 0$, is a continuous facility location problem on undirected graphs where all edges have unit length. The facilities may be positioned on the vertices as well as on the interior of the edges. The goal is to position as few facilities as possible such that every point on every edge has distance at most δ to one of these facilities. For large δ, the problem is similar to dominating set, which is hard to approximate, while for small δ, say close to 1, the problem is similar to vertex cover. In fact, as shown by Hartmann et al. [Math. Program. 22], δ-Covering for all unit-fractions δ is polynomial time solvable, while for all other values of δ the problem is NP-hard.

We study the approximability of δ-Covering for every covering range $\delta > 0$. For $\delta \geq 3/2$, the problem is log-APX-hard, and allows an $\mathcal{O}(\log n)$ approximation. For every $\delta < 3/2$, there is a constant factor approximation of a minimum δ-cover (and the problem is APX-hard when δ is not a unit-fraction). We further study the dependency of the approximation ratio on the covering range $\delta < 3/2$. By providing several polynomial time approximation algorithms and lower bounds under the Unique Games Conjecture, we narrow the possible approximation ratio, especially for δ close to the polynomial time solvable cases.

Keywords: Facility Location · Approximation Algorithms · Dominating Set · Vertex Cover

1 Introduction

We study the approximability of a continuous facility location problem. The input is a graph G whose edges have unit length. Let $P(G)$ be the continuum set of points on all the edges and vertices. The distance of two points $p, q \in P(G)$ is the length of a shortest path connecting p and q in the underlying metric space. For a rational $\delta > 0$, a subset of points $S \subseteq P(G)$ is a δ-*cover* if every point $p \in P(G)$ has distance $d(p,q) \leq \delta$ to some point $q \in S$. For a fixed $\delta > 0$, and given a graph G, our objective is to compute a minimum size δ-cover of

---

T. Janßen—funded by the Deutsche Forschungsgemeinschaft (DFG, German Research Foundation) – WO 1451/2-1.

G. We denote the minimum cardinality of a δ-cover S as δ-cover$(G) = |S|$. The decision-version of this problem is known as δ-COVERING.

The computational complexity of computing an optimal δ-cover is relatively well understood. δ-COVERING is polynomial time solvable for every covering range δ that is a unit-fraction, while it is NP-hard for all other δ, as shown by Hartmann et al. [13]. The same work also settles the parameterized complexity with the solution size as parameter. When $\delta < 3/2$, the problem is fixed-parameter tractable, while for $\delta \geq 3/2$, the problem is W[2]-hard. The approximability depending on δ has not yet been settled.

As the problem is polynomial time solvable when δ is a unit-fraction, we are particularly interested in the approximability when δ is near a unit-fraction. Also, as revealed by the parameterized complexity study, for large δ, the problem is similar to the dominating set problem – in the sense that the main focus is on covering the vertices, which is hard to approximate. On the other hand, for small δ, say close to 1, the problem is similar to the vertex cover problem – in the sense that the main focus is on covering the edges, which is well-known to allow a 2-approximation. Similarly, for the approximation, we expect some threshold for δ, till which the problem allows a constant factor approximation (like vertex cover), while beyond this threshold the problem is log-APX-hard (like dominating set).

Further Related Results. Facility Location in general is a wide research area. We refer to the books by Drezner [5] and Mirchandani & Francis [19]. Our model, where the metric space is defined by a graph, dates back to Dearing & Francis [3]. Several optimization goals have been considered, for example by works of Tamir [20, 21]. For δ-COVERING, Megiddo & Tamir [18] gave a polynomial time algorithm for trees, and show NP-hardness for $\delta = 2$.

A dual problem to δ-COVERING, is known as δ-DISPERSION. The task is to select a maximum number of points that have distance at least δ. Similarly to δ-COVERING, the problem is NP-hard for all δ that are not a unit fraction or twice a unit fraction, as shown by Grigoriev et al. [9]. The problem is fixed-parameter tractable in the solution size for $\delta \leq 2$, and W[1]-hard otherwise, as shown by Hartmann et al. [12]. The same work addresses the parameterized complexity with respect to various graph parameters. A similar task was studied when approximating the problem: Instead of fixing the minimum distance, the number of points to place is fixed, and the goal is to maximize the minimum pairwise distance between the placed points. For this task, Tamir [21] gave a $\frac{1}{2}$-approximation algorithm, while they showed that there is no ε-approximation for $\varepsilon > \frac{2}{3}$, unless P = NP.

Our Results. For $\delta \geq 3/2$, δ-COVERING is log-APX-hard, and allows an $\mathcal{O}(\log n)$ approximation. For every $\delta < 3/2$, the problem has a constant factor approximation and is APX-hard when δ is not a unit-fraction.

For a covering range $\delta < 3/2$, we give several polynomial time approximation algorithms and lower bounds under the Unique Games Conjecture (UGC), plotted in part in Figs. 1 and 3. Due to a result by Hartmann et al. [13] that lets us translate the value of δ, we may mostly focus on $\delta \geq 1/2$. One may expect that the approximation ratio approaches 1 if δ approaches a polynomial time solvable case such as $\delta = 1$ and $\delta = 1/2$.

- This is true for when δ approaches $1/2$ from above. We give families of upper and lower bounds which in the limit towards $1/2$ reach 1. That is, for $\delta \in [\frac{x+1}{2x+1}, \frac{x}{2x-1})$ for integer $x \geq 2$, we provide a $\frac{x+1}{x}$-approximation algorithm (Theorem 7). Our lower bounds under UGC are provided by three families of lower bounds that together cover every δ close to $1/2$. For example a lower bound of $1 + \frac{1}{2x+1}$ for covering range $\delta \in [\frac{x+1}{2x+1}, \frac{2x+1}{4x})$ for integer $x \geq 1$ (Theorem 2).
- On the other hand, this is not true when δ approaches 1 from below. We give an approximation lower bound under UGC of $6/5$ for $\delta \in [\frac{5}{6}, 1)$ (also Theorem 2). Further, we provide a 2-approximation algorithm for a superset of these δ (Theorem 6).

Further, for the interval $\delta \in [\frac{2}{3}, \frac{3}{4})$ we provide an upper and lower bound that only differ by a factor of $9/8$. Our upper bound is an algorithm that behaves uniformly for every $\delta \in [\frac{2}{3}, \frac{3}{4})$, while the minimum size of a δ-cover for such δ can vary by a factor of $9/8$ itself.

For covering range $\delta \in (1, \frac{3}{2})$, the situation is similar to $\delta \in (\frac{1}{2}, 1)$. We give a 2-approximation for this interval, while for $\delta < 5/4$ we reduce this factor to $5/3$ and for $\delta < 7/6$ we reduce this factor to $3/2$. The lower bounds are inherited from the case $\delta \in (\frac{1}{2}, \frac{3}{4})$.

For a covering range $\delta < \frac{1}{2}$, we inherit upper bounds from the case $\delta \in (\frac{1}{2}, 1)$, which we further improve for certain ranges of δ.

Organization of this Work. Preliminaries are in Sect. 2. Section 3 settles the APX-hardness and case $\delta \geq \frac{3}{2}$, while Sect. 4 provides hardness results under UGC. We give our upper bounds for $\delta > 1$, $\delta \in (\frac{1}{2}, 1)$ and $\delta < \frac{1}{2}$ in Sect. 5, Sect. 6, Sect. 7, respectively.

2 Preliminaries

All graphs we consider are simple and undirected. Every edge has unit length. We use $V(G)$ and $E(G)$ for the vertex respectively edge set of a graph G. Further, $N(v) := \{u \in V(G) \mid \{u,v\} \in E(G)\}$ and $N[v] := N(v) \cup \{v\}$. We use the word *vertex* in the graph-theoretic sense, while we use the word *point* for the elements of $P(G)$. We use the notation of [9]: For an edge $\{u, v\}$ and real $\lambda \in [0, 1]$, we denote by $p(u, v, \lambda)$ the point on edge $\{u, v\}$ that has distance λ to u. We have that $p(u, v, \lambda)$ coincides with $p(u, v, 1 - \lambda)$, and that $p(u, v, 0) = u$ and $p(u, v, 1) = v$. The *open ball* with radius r around a point p, denoted as $B^<(p, r)$, is the set of points with distance $< r$ from p. The *closed ball*, denoted as $B^{\leq}(p, r)$, is the set of points with distance $\leq r$ from p.

Lemma 1 (Hartmann et al. [13]). *Let δ be a unit fraction. Then a minimum δ-cover of a graph G can be found in polynomial time.*

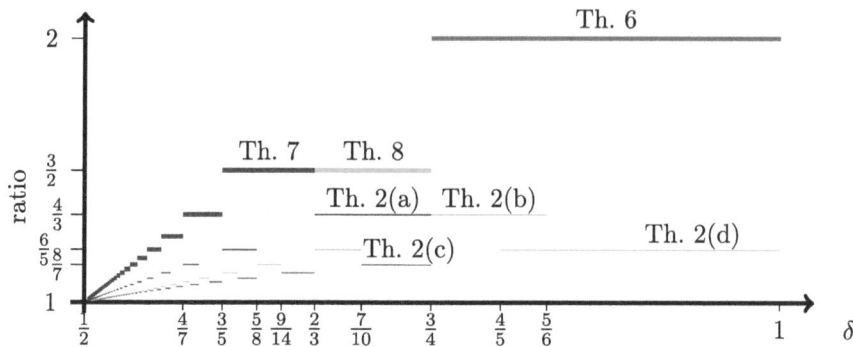

Fig. 1. Upper bounds (as bold lines) and lower bounds under UGC (as thin lines) on the approximation ratio of δ-COVERING plotted for $\delta \in (\frac{1}{2}, 1)$. The drawn intervals are half-open with the upper end excluded.

The proof in [13] uses the Edmonds-Gallai-Decomposition, see [6–8], to split G into three disjoint subgraphs G_0, G_1 and $G_{\geq 3}$. Graph G_0 has a perfect matching, G_1 is bipartite, and every component C of $G_{\geq 3}$ has size at least 3 and is *factor-critical*, that is for any vertex $u \in V(C)$ there is a maximum matching that misses exactly u. Let $c_{\geq 3}$ be the number of components of $G_{\geq 3}$. Since every component contains at least three vertices, we have $c_{\geq 3} \leq \frac{1}{3}|V(G)|$. Then the computed minimum 1-cover S^* has the size bound

$$|S^*| = \nu(G_0) + \nu(G_{\geq 3}) + c_{\geq 3} + \mathrm{vc}(G_1) \leq \tfrac{1}{2}(|V(G)| + c_{\geq 3}) \leq \tfrac{2}{3}|V(G)|, \quad (1)$$

where ν denotes the size of a maximum matching. Further, we use the following property. A subset $S \subseteq P(G)$ is *neat* if

- (N1) for every $\{u,v\} \in E(G)$, if $|S \cap P(G[\{u,v\}])| \geq 2$, then $S \cap P(G[\{u,v\}]) = \{u,v\}$.

Lemma 2. $(\star)$[1] *Let $\delta \geq \frac{1}{2}$ and G be a graph. There is a minimum δ-cover of G that is neat.*

To obtain an α-approximation of δ-cover(G), it suffices to compute an α-approximation for every connected component of G. Further, δ-COVERING is polynomial time solvable on trees, as shown by Megiddo et al. [18]. Hence, it suffices to consider connected non-tree graphs as input. Regarding an introduction to approximation preserving reductions, see [2,11].

3 APX-Hardness

This section derives the APX-hardness results of δ-COVERING. For every $\delta < \frac{3}{2}$ that is not a unit fraction, δ-COVERING is APX-hard. (We recall that the problem is polynomial time solvable when δ is a unit-fraction.) On the other hand, for $\delta \geq \frac{3}{2}$, δ-COVERING turns out to be log-APX-hard.

---
[1] The full proofs of theorems and lemmas marked with a $(\star)$ can be found in [11].

Theorem 1. ($\star$) δ-COVERING *is APX-hard for every non unit-fraction* δ.

Our proof establishes APX-hardness for all δ that are not unit-fractions in two steps. First, we show hardness for $\delta \geq \frac{3}{2}$ by several reductions from DOMINATING SET on 3-regular graphs. Then we obtain APX-hardness for the remaining $\delta < \frac{3}{2}$, by applying the following two self-reductions:

Definition 1 (Subdivision Reduction). *This reduction takes* $x \in \mathbb{N}^+$ *as a parameter. We define* $f_x^S(G) = G_x$. *For an* $(x\delta)$-*cover of* G_x, *we define* $g_x^S(G, S') = S$ *as the set containing*

- *every point in* $S' \cap V(G)$, *and*
- *for every point* $p' \in S' \setminus V(G)$, *the point* $p = p(u, v, \frac{\lambda}{x}) \in S$ *where* u, v *are the two closest vertices in* $V(G_x) \cap V(G)$ *and* λ *is the distance from* u *to* p' *in* G_x.

Graph G has a minimum δ-cover of size k, if and only if G_x has a minimum $(x\delta)$-cover of size k, as shown by Hartmann et al. [13]. It easily follows that the Subdivision Reduction is an L-reduction with $\alpha = \beta = 1$.

Definition 2 (Translation Reduction). *We define* $f^T(G) = G$. *For a* $\frac{\delta}{2\delta+1}$-*cover* S' *of* G, *we define* $g^T(G, S') = S \subseteq P(G)$ *as follows: For every edge* $\{u, v\} \in E(G)$,

- *if* $|S' \cap P(G[\{u,v\}])| \leq 1$, *then* S *contains no point from* e; *and*
- *if* $S' \cap P(G[\{u,v\}]) = \{p'_1, \ldots, p'_k\}$ *for a* $k \geq 2$ *then* S *contains the* $k-1$ *points* $p_1, \ldots, p_{k-1}$ *where* $p_i = p(u, v, \lambda_i)$ *and* $\lambda_1 = \mu \cdot (2\delta + 1)$ *and* $\lambda_{i+1} = \lambda_i + 2\delta$ *for* $i \in \{2, \ldots, k-1\}$ *and* $\mu = \min_{i \in [k]} d(p_i, u)$.

Intuitively, a $\frac{\delta}{2\delta+1}$-cover compared to a δ-cover must contain exactly one more point on every edge. Indeed, the minimum δ-cover of G has size k, if and only if the minimum $\frac{\delta}{2\delta+1}$-cover of G has size $k+|E(G)|$, as shown by Hartmann et al. [13]. It easily follows that the Translation Reduction (f^T, g^T) is an L-reduction for graphs where δ-cover$(G) = c \cdot |E|$ for some constant $c > 0$ with $\alpha = 1/c$ and $\beta = 1$.

On the other hand, we give an $\mathcal{O}(\log n)$-approximation algorithm by using a discretization argument and then solving a SET COVER-instance using [1,14,17].

Lemma 3. ($\star$) *There is an* $\mathcal{O}(\log n)$-*approximation algorithm for* δ-COVERING.

4 Hardness Under Unique Game Conjecture

This section derives lower bound on the approximation ratio for δ-COVERING, assuming the Unique Games Conjecture, see [15]. We reduce from VERTEX COVER, which is UG-hard to approximate within any factor better than 2, as shown by Khot et al. [16]. In particular, assuming the Unique Games Conjecture, it is NP-hard to distinguish graphs that have a weighted vertex cover of size $\frac{1}{2} + \varepsilon_1$ and $1 - \varepsilon_2$, where the total weight of all vertices is 1. Further, Khot et al. note that their construction can be converted to an unweighted graph by the same technique as in [4], which essentially duplicates each vertex to model the weights by the number of copies. Therefore it is UG-hard to decide whether a graph with n vertices has a vertex cover of size $\frac{n}{2} + \varepsilon_1$ or no vertex cover of size $n - \varepsilon_2$.

We give several sets of reductions showing UG-hardness for different ranges of δ, as also shown in Fig. 1. The core construction, given a vertex cover instance consisting of a graph G, adds a gadget to every vertex $u \in V(G)$, see Fig. 2. For each gadget, we may assume that a minimum δ-cover contains points from this gadget that cover a maximum area of points in the original graph G, some ball of radius $\ell < 1/2$ around each vertex $u \in V(G)$. For $\delta \geq 1 - \ell$, then a vertex cover of G (as points in $P(G)$) covers all the remaining points $P(G)$.

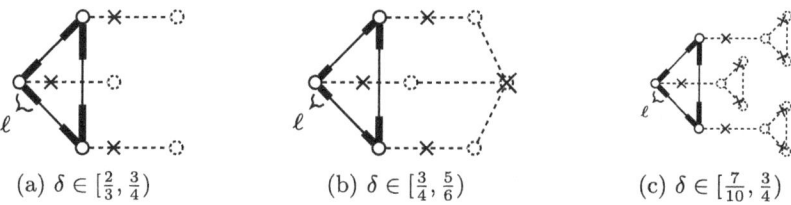

(a) $\delta \in [\frac{2}{3}, \frac{3}{4})$ (b) $\delta \in [\frac{3}{4}, \frac{5}{6})$ (c) $\delta \in [\frac{7}{10}, \frac{3}{4})$

Fig. 2. Constructions (a), (b) and (c) for Theorem 2 for $x = 1$ and input graph K_3. The dashed edges and vertices are the added gadgets, and the crosses mark the optimal placement of points on them. The thick edge segments are covered by those points. By the choice of δ, for each construction $1 - \delta \leq \ell < \frac{1}{2}$. Thus a vertex cover of the original graph covers the remaining edge segments.

Theorem 2. ($\star$) *For every $x \in \mathbb{N}^+$ and $\varepsilon > 0$, it is UG-hard to approximate δ-COVERING*

(a) *within $1 + \frac{1}{2x+1}$, for any $\delta \in [\frac{x+1}{2x+1}, \frac{2x+1}{4x})$.*
(b) *within $1 + \frac{1}{2x+1+\varepsilon}$, for any $\delta \in [\frac{x+2}{2x+2}, \frac{2x+3}{4x+2})$.*
(c) *within $1 + \frac{1}{2x+5}$, for any $\delta \in [\frac{2x+5}{4x+6}, \frac{x+2}{2x+2})$.*
(d) *within $\frac{6}{5}$, for any $\delta \in [\frac{4}{5}, 1)$.*

We note that every lower bound for a $\delta \in (\frac{1}{2}, \frac{3}{4})$ also holds for $2\delta \in (1, \frac{3}{2})$, by applying Subdivision Reduction (see Definition 1) for $x = 2$.

5 Approximation Algorithm for $1 < \delta < \frac{3}{2}$

This section derives the approximation algorithms for $\delta > 1$. We do so by constructively bounding the size of minimum δ-cover S_δ in the size of a 1-cover by some factor α. Then a polynomial time computable 1-cover constitutes an

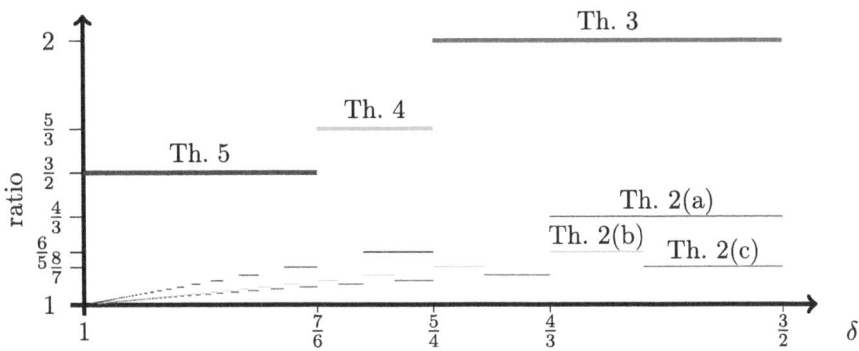

Fig. 3. Upper bounds (as bold lines) and lower bounds under UGC (as thin lines) on the approximation ratio of δ-COVERING plotted for $\delta \in (1, 3/2)$. The drawn intervals are half-open with the upper end excluded. Applying the Subdivision Reduction to Theorem 2 yields the lower bounds.

α-approximation. For $\delta < 3/2$, we construct a 1-cover of size at most twice that of a minimum δ-cover, while for $\delta < 5/4$ we reduce this factor to $5/3$ and for $\delta < 7/6$ we further reduce this factor to $3/2$.

For $\delta \geq 9/8$, we show that these approximation ratios are best possible with this approach, as we give explicit examples where a 1-cover and a δ-cover differ asymptotically by these factors.

To show that 1-cover$(G) \leq 2 \cdot \delta$-cover(G), we construct a 1-cover S_1 from a δ-cover S_δ, by considering each point $p \in S_\delta$ individually.

- If $p \in S_\delta \cap V(G)$, we add p to S_1.
- If p is on the interior of an edge $\{u,v\}$, we add u and v to S_1.

It is not hard to see, that S_1 is a 1-cover, and obviously $|S_1| \leq 2|S_\delta|$.

Theorem 3. ($\star$) *For every $\delta < \frac{3}{2}$ and graph G, 1-cover$(G) \leq 2 \cdot \delta$-cover(G). Further, for $\delta \geq \frac{5}{4}$, this bound is asymptotically tight.*

The proof for 1-cover$(G) \leq \frac{5}{3} \cdot \delta$-cover$(G)$ for $\delta < \frac{5}{4}$ works similar, but the 1-cover S_1 is constructed in two parts. First, we construct a partial 1-cover S_1' by considering each point $p \in S_\delta$:

- If $p \in S_\delta \cap V(G)$, then $p \in S_1'$.
- If $p = p(u,v,\lambda)$ and $\lambda < \frac{1}{2}$, then $u \in S_1'$ and otherwise $v \in S_1'$

We then let $S_1 = S_1' \cup S_1''$, where S_1'' is a polynomial time computable minimum 1-cover of the graph induced by the edges not 1-covered by S_1'. The ratio then follows from size bound of a minimum 1-cover given by Eq. (1).

Theorem 4. ($\star$) *For every graph G and $\delta < \frac{5}{4}$, 1-cover$(G) \leq \frac{5}{3} \cdot \delta$-cover$(G)$. Further, for $\delta \geq \frac{7}{6}$, this bound is asymptotically tight.*

For $\delta < \frac{7}{6}$, we refine this approach in several ways. We construct the partial 1-cover S'_1 more sensitive to S_δ. We analyze the remaining edges more carefully, and finally obtain a good approximation by considering two different alternatives.

Theorem 5. ($\star$) *For every graph G and $\delta < \frac{7}{6}$, 1-cover$(G) \leq \frac{3}{2} \cdot \delta$-cover$(G)$. Further, for $\delta \geq \frac{9}{8}$, this bound is asymptotically tight.*

Proof. Let a point $p(u, v, \lambda)$ be *u-close* if $\lambda \leq \frac{1}{6}$ and *centric* if $\frac{1}{6} < \lambda < \frac{5}{6}$. Conveniently, let an edge be *u-close* and be a *center edge* if it contains a u-close point or a centric point, respectively. We say that δ-cover S of G is *nice* if it is neat (i.e., satisfies (N1)) and

- (A1) no two center edges are adjacent.

Claim. ($\star$) For every graph G and $\delta \in [1, \frac{3}{2})$, a nice minimum δ-cover S_δ exists.

Now consider a minimum δ-cover S_δ of G that is nice. We construct a 1-cover S_1 of G from S_δ in two steps. First, we construct a partial 1-cover S'_1:

- For every vertex $u \in V(G)$, if S_δ contains a u-close point, we add u to S'_1. In particular, this includes all points on vertices.
- For every center edge $\{u, v\} \in E(G)$ with $N(v) \setminus \{u\} \subseteq S'_1$, we add u to S'_1, unless v was already added to S'_1 by this rule.

Let G' be the graph induced by the edges $\{u, v\} \in E(G)$ where $u, v \notin S'_1$. Note that by construction of S'_1, there are no edges in $E(G')$ that are u-close for some vertex $u \in V(G')$, and the center edges in $E(G')$ form a matching of G'. Then each edge $\{u, v\} \in E(G')$ satisfies at least one of the following properties. By symmetry, assume that u has shorter or equal distance to a point in S_δ than v.

(a) The edge $\{u, v\}$ is a center edge.
(b) Vertices u and v are incident to center edges in $E(G')$.
(c) Vertex u is incident to a center edge in $E(G')$ and v is adjacent to a vertex $w \in V(G)$ with a w-close point in $S_\delta \cap \{p(w, v, \lambda) \mid \lambda \in (0, \frac{1}{6}]\}$.
(d) Vertex u is incident to a center edge in $E(G')$ and v is adjacent to a vertex $w \in V(G)$ with a w-close point in $S_\delta \setminus \{p(w, v, \lambda) \mid \lambda \in (0, \frac{1}{6}]\}$.

We classify each edge $\{u, v\} \in E(G')$ as type-(a), -(b), -(c) or -(d) depending on the first property $\{u, v\}$ satisfies. Let f be the mapping that assigns type-(c) and type-(d) edges their unique incident center edge. Since $\delta < \frac{7}{6}$, for every type-(c) edge $\{u, v\}$, the center edge $f(\{u, v\})$ contains a point in distance $< \frac{1}{2}$ to the common vertex of $\{u, v\}$ and $f(\{u, v\})$. Further, for every type-(d) edge $\{u, v\}$, the center edge $f(\{u, v\})$ contains a point in distance $< \frac{1}{3}$ to the common vertex of $\{u, v\}$ and $f(\{u, v\})$.

If $E(G')$ contains a type-(d) edge $\{u, v\}$, we add the common vertex of $\{u, v\}$ and $f(\{u, v\})$ to S'_1, w.l.o.g., let this be u. Then we delete u from G' and reclassify the type-(b) edges adjacent to the other endpoint of $f(\{u, v\})$ as type-(e). After this step, $E(G')$ contains one less type-(d) edge. We apply this modification exhaustively. Finally, we delete all isolated vertices from G'.

Now with the only edge types of G' being (a), (b), (c) and (e), we proceed with the second part of the construction of S_1. For that, we consider each connected component C of G' separately. Let $V_a(C)$ denote the vertices of C that are adjacent to a center edge. Further, let $V_{ce}(C) = V(C) \backslash V_a(C)$ denote the vertices of C adjacent to a type-(c) or type-(e) edge but not to a center edge. Note that both endpoints of a type-(b) edge are in $V_a(C)$. For a component C of G', let C^* be the component of G defined by $G[V(C) \cup \{v \mid v \notin V(C), u \in V(C) \wedge \{u, v\} \in E(G) \wedge |P(G[\{u,v\}]) \cap S_\delta| \geq 1\}]$. Informally, each component gets enlarged by the edges containing a point δ-covering parts of the type-(c) and type-(e) edges. For every vertex in $V_{ce}(C)$, for every component C of G', there is a unique point in S_δ, by definition of the type-(c) and type-(e) edges. Further, this point is contained in $P(G[C^*])$. Then for any two components C_1, C_2 of G', the corresponding components C_1^* and C_2^* are disjoint. Additionally, for every component C, there are $\frac{1}{2}|V_a(C)|$ unique points in S_δ, that are different from the previous points, as the center edges form a matching in G'. Then, for each component C of G', we compute a 1-cover S_1^C depending on the size of $V_a(C)$:

1. If $|V_a(C)| < 2 \cdot |V_{ce}(C)|$, we set $S_1^C = V_a(C)$.
2. If $|V_a(C)| \geq 2 \cdot |V_{ce}(C)|$, we use Lemma 1 to output an optimal 1-cover of C as S_1^C.

Finally, we set $S_1 = S_1' \cup \bigcup_{C \in \mathcal{C}(G')} S_1^C$ where $\mathcal{C}(G')$ are the connected components of G'. By the arguments above, we have $|S_\delta \cap P(G[C^*])| = \frac{1}{2}|V_a(C)| + |V_{ce}(C)|$.

In case 1, we have $|S_1 \cap P(G[C^*])| = |V_a(C)| + |V_{ce}(C)|$. Further, $\frac{1}{2}|V_a(C)| < |V_{ce}(C)| \Leftrightarrow 2 \cdot (|V_a(C)| + |V_{ce}(C)|) < 3 \cdot (\frac{1}{2}|V_a(C)| + |V_{ce}(C)|) \Leftrightarrow |S_1(C^*)| < \frac{3}{2}|S_\delta(C^*)|$.

In case 2, we have $|S_1 \cap P(G[C^*])| \leq \frac{2}{3}|V_a(C)| + \frac{5}{3}|V_{ce}(C)|$. Further, $|V_{ce}(C)| \leq \frac{1}{2}|V_a(C)| \Leftrightarrow 4|V_a(C)| + 10|V_{ce}(C)| \leq \frac{9}{2}|V_a(C)| + 9|V_{ce}(C)| \Leftrightarrow |S_1(C^*)| \leq \frac{3}{2}|S_\delta(C^*)|$.

Thus in both cases, the number of points S_1 places on C^* is at most $\frac{3}{2}$ times the number of points S_δ places on C^*. Since the components C^* for all $C \in \mathcal{C}(G)$ are mutually disjoint and S_1 places the same number of points as S_δ in the parts of G that are not part of a component C^*, we get an overall ratio of $\frac{3}{2}$.

6 Approximation Algorithms for $\frac{1}{2} < \delta < 1$

This section derives our approximation algorithms for $\delta \in (\frac{1}{2}, 1)$. We begin with a bound for the whole interval. Since a minimum $\frac{1}{2}$-cover for a connected non-tree graph G simply is $V(G)$, the following gives a 2-approximation of a δ-cover of G for $\delta \in (\frac{1}{2}, 1)$.

Theorem 6. $\delta\text{-cover}(G) \geq \frac{1}{2}|V(G)|$ *for every graph G and $\delta < 1$.*

Proof. Dual to δ-cover(G) is δ-disp(G), which is the maximum size subset of points $I \subseteq P(G)$ such that no two points in I have distance less than δ. It is easy to see that δ-cover$(G) \geq$ 1-cover(G) for $\delta < 1$. Tamir [21] showed that for every graph G and $\delta, \varepsilon > 0$, it holds that δ-cover$(G) \geq (2\delta + \varepsilon)$-disp$(G)$. Thus also δ-cover$(G) \geq$ 2-disp(G) for $\delta < 1$. Further, Hartmann [10] observed that 1-cover(G)+2-disp$(G) = |V(G)|$. It therefore follows that δ-cover$(G) \geq \frac{1}{2}|V(G)|$.

6.1 General Approximation for $\frac{1}{2} < \delta < \frac{2}{3}$

Our family of approximation algorithms for $\delta \in (\frac{1}{2}, \frac{2}{3})$, rely on bounding the size of a δ-cover linearly in $|V(G)|$. Intuitively, such a δ-cover S can be smaller than $V(G)$ on a long path Q by spacing the points S far apart such that eventually Q contains an edge e where $S \cap P(G[e]) = \emptyset$.

Consider a nice minimum δ-cover S of a graph G. We say that a point $p(u, v, \lambda) \in P(G)$ with $\lambda \in [0, 1 - \delta)$ is u-close. We call a δ-cover S *humble* if it is neat (i.e., satisfies (N1)) and

- (N2) there is no cycle C in G where each edge contains a point from S in its interior,
- (N3) for every vertex $u \in V(G)$, set S contains at most one u-close point.

Lemma 4. ($\star$) *For $\delta \in (\frac{1}{2}, \frac{2}{3})$, there is a humble minimum δ-cover S of G.*

For an integer $x \geq 2$ and $\delta \in [\frac{x+1}{2x+1}, \frac{x}{2x-1})$, we give an approximation of a δ-cover by bounding its size in $|V(G)|$, which is the size of a $\frac{1}{2}$-cover for connected non-tree graphs G. E.g., for $x = 2$ and $\delta \in [\frac{3}{5}, \frac{2}{3})$, this gives a $\frac{3}{2}$-approximation.

Lemma 5. ($\star$) *Let $x \geq 2$ be integer and $\delta \in [\frac{x+1}{2x+1}, \frac{x}{2x-1})$. Then $|V(G)| \leq \frac{x+1}{x} \cdot \delta$-cover$(G)$, for connected graphs G with $|E(G)| \geq x$. This bound is asymptotically tight.*

Proof. Let S_δ be a humble minimum δ-cover of G. Let E' be the set of edges e where S contains a point in the interior of e. Then E' induces a forest in G by property (N2). Consider a component C of $G[E']$. We claim that $|V(C)| \leq \frac{x+1}{x}|S_\delta \cap P(G[C])|$. This then also implies $|V(G)| \leq \frac{x+1}{x}|S_\delta|$ since every point $p \in S_\delta$ is either a vertex or is contained in $P(G[C])$ of a component C of $G[E']$.

By property (N1) we have $|S_\delta \cap P(G[C])| = |E(C)|$. Let $\beta(C)$ be the set of edges of G incident to C and incident to $V(G) \setminus C$. In case $\beta(C) = \emptyset$, then $G[C]$ is the whole graph G. We have that $|S_\delta| = |E(C)| = |E(G)| \geq x$ and $|V(C)| \leq x + 1$. That means $|V(G)| \leq \frac{x+1}{x}|S_\delta|$.

In case $\beta(C) \neq \emptyset$, we show that $|V(C)| \geq x + 1$. We have $|E(C)| \geq x$, such that similarly to before, $V(C)$ has size at most $\frac{x+1}{x}|S_\delta \cap P(G[C])|$. We fix an edge $\{u_0, u_1\} \in \beta(C)$ where $u_1 \in V(C)$. Inductively, we construct a path $Q = (u_0, u_1, \ldots, u_{x'}, u_{x'+1})$ of some length $x' + 1 \geq 2$. In step i, for increasing $i \geq 0$, we identify a point p_i having distance $1 - \lambda_i$ to u_{i+1}. For step $i = 0$, we use that there is a unique u_0-close point p_0, which is not in the interior of the edge $\{u_0, u_1\}$. We define $-\lambda_0$ as the distance of p_0 to u_0. Consequently, p_0 has distance $1 - \lambda_0$ to u_1. We have $-\lambda_0 \leq \delta - 1 \leq 1 - \delta$, as $\delta \leq 1$. We proceed with step $i = 1$. Step $i \geq 1$ is defined as follows.

- If there is a u_i-close point $p_i = p(u_i, u_{i+1}, \lambda_i)$, let $\{u_i, u_{i+1}\}$ be the edge containing the unique u_i-close point p_i, which must be distinct from the previous edge $\{u_{i-1}, u_i\}$. They define vertex u_{i+1} and edge position λ_i. Then proceed with step $i + 1$.
- Otherwise, there is an incident edge $\{u_i, u_{i+1}\}$ where S contains a point $p(u_i, u_{i+1}, \lambda_i)$ with $\lambda_i \geq 1 - \delta$, which exists as otherwise vertex u_i is not covered. They define vertex u_{i+1} and edge position λ_i. Again $\{u_i, u_{i+1}\}$ is different from $\{u_{i-1}, u_i\}$. In this case, our procedure terminates. Let $Q = (u_0, u_1, \ldots, u_{x'}, u_{x'+1})$ be the resulting path.

We claim that our procedure terminates and that $u_1, u_2, \ldots, u_{x'+1}$ are distinct vertices in $V(C)$, such that $|V(C)| \geq x' + 1$. Further, we claim that $\lambda_i \leq i(2\delta - 1)$ for $i \in \{0, \ldots, x'\}$. As $\delta < \frac{x}{2x-1}$, we have $x' > x - 1$. This then implies $|V(C)| \leq \frac{x+1}{x}|S_\delta \cap P(G[C])|$.

To compute a $\frac{x+1}{x}$-approximation of a δ-cover for instances with $|E(G)| < x$, we may use a brute-force algorithm as x can be considered a constant.

Theorem 7. Let $x \geq 2$ be integer and $\delta \in [\frac{x+1}{2x+1}, \frac{x}{2x-1})$. Then δ-cover(G) allows a polynomial time $\frac{x+1}{x}$-approximation algorithm.

6.2 3/2-Approximation Algorithm for $\delta \in [\frac{2}{3}, \frac{3}{4})$

For δ in the interval $[\frac{2}{3}, 1)$, Theorem 3 (and similarly the idea of Lemma 5) only give a 2-approximation. In case $\delta < \frac{3}{4}$, we improve this upper bound to 3/2. As the lower bound proof of Theorem 2(a) reveals, the neighbors of the leaves hide a vertex cover instance. Our algorithm computes an approximate vertex cover for these vertices. Then our analysis makes use of a carefully chosen partition of the input graph.

We define levels of the input graph G as follows. Let L_0 be the vertices of degree one of G, and $L_i = \{u \in V(G) \mid d(u, u_0) = i, u_0 \in L_0\}$, for $i \in \{1, 2\}$, possibly $L_1 \cap L_2 \neq \emptyset$. For $i, j \in \{0, 1, 2\}$, with $i \neq j$, let $E_{i,j}$ be set of tuples (u_i, u_j) where $\{u_i, u_j\} \in E(G)$ and $u_i \in L_i$ and $u_j \in L_j$. For $i \in \{0, 1, 2\}$, let $E_{i,i}$ be the set of edges $E(G) \cap (L_i \times L_i)$.

Our approximation algorithm proceeds as follows.

1. Let L be the set of points $p(u_0, u_1, \frac{2}{3})$ for every leaf-edge $(u_0, u_1) \in E_{0,1}$.
2. We compute a 2-approximation X of the vertex cover of $G[E_{1,1}]$.
3. Let $W = V(G) \setminus (L_0 \cup L_1)$ and output $S = L \cup X \cup W$.

Since a 2-approximation of a vertex cover can be computed in polynomial time, the above algorithm runs in polynomial time. Further, we observe that the output S is a $\frac{2}{3}$-cover, and hence also a δ-cover for $\delta \in [\frac{2}{3}, \frac{3}{4})$. Indeed, the points

$L \cup W$ $\frac{2}{3}$-cover every point on every edge $\{u_1, u_1'\} \in E(G) \setminus E_{1,1}$ as well as every point in $B^{\leq}(u, \frac{1}{3})$ for every $u \in L_1$. The remaining points are the points on every edge $\{u_1, u_1'\} \in E_{1,1}$ with distance $\leq \frac{2}{3}$ to u_1 and to u_1', which are covered by the vertex cover X of $G[E_{1,1}]$.

To bound the approximation ratio, we compare the output S with a humble minimum (i.e., satisfying (N1)–(N3)) δ-cover $S^\star$ that also satisfies the following properties:

- (B1) $p(u_1, u_1', \frac{1}{2}) \notin S^\star$ for every edge $\{u_1, u_1'\} \in E_{1,1}$.
- (B2) Every point $p(u_1, u_2, \lambda) \in S^\star$ with $(u_1, u_2) \in E_{1,2}$ and $\lambda \geq 1 - \delta$, has $\lambda = 1$.
- (B3) $X^{\star\star} := \{u_1 \in L_1 \mid S^\star \cap P_{u_1} \neq \emptyset\}$ is a vertex cover of the graph $G[E_{1,1}]$, where $P_{u_1} := B^{\leq}(u_1, \delta - \frac{1}{2}) \cup (B^<(u_1, \frac{1}{2}) \cap P(G[E_{1,1}]))$ for $u_1 \in L_1$.

Claim. ($\star$) There is a humble minimum δ-cover $S^\star$ that satisfies (B1)–(B3).

Our goal is to identify disjoint sets $P_{V^\star}, P_{\mathcal{C}_{\geq 2}}, P_L \subseteq P(G)$ such that $S \subseteq P_{V^\star} \cup P_{\mathcal{C}_{\geq 2}} \cup P_L$ and $|S \cap P| \leq \frac{3}{2}|S^\star \cap P|$ for each set $P \in \{P_{V^\star}, P_{\mathcal{C}_{\geq 2}}, P_L\}$, which we do below in (I), (II) and (III). Then also $|S| \leq \frac{3}{2}|S^\star|$, and hence our algorithms outputs a $\frac{3}{2}$-approximation. Let $V^\star := S^\star \cap W$ and let $P_{V^\star} := B^<(V^\star, 1)$.

(I) We claim that $|S \cap P_{V^\star}| \leq |S^\star \cap P_{V^\star}|$. Any point $p \in S^\star$ contained in the interior of an edge incident to $V^\star \subseteq S^\star$ contradicts property (N1). In other words, $S^\star \cap P_{V^\star} = V^\star$. Since also S satisfies property (N1) and $V^\star \subseteq W \subseteq S$, analogously to $S^\star$, we have $S \cap P_{V^\star} = V^\star$. Then, trivially, $|S \cap P_{V^\star}| \leq |S^\star \cap P_{V^\star}|$.

Let $E_\emptyset^\star$ be the set of edges $\{u, v\} \in E(G)$ where $S^\star \cap P(G[\{u,v\}]) = \emptyset$, which particularly means that $u, v \notin S^\star$. Let $G^\star$ be the graph resulting from G after removing vertices $V^\star \cup L_0 \cup L_1$ from $V(G)$ with their incident edges and edges $E_\emptyset^\star$ from $E(G)$. Let $\mathcal{C} \subseteq 2^{V(G)}$ be the components of $G^\star$. Let $P_{\mathcal{C}_{\geq 2}}$ be the union of $P(G[C])$ over all components $C \in \mathcal{C}$ where C contains at least 2 edges.

(II) We claim that $|S \cap P_{\mathcal{C}_{\geq 2}}| \leq \frac{3}{2}|S^\star \cap P_{\mathcal{C}_{\geq 2}}|$. For every component $C \in \mathcal{C}_{\geq 2}$, set S contains $|V(C)| \leq |E(C)| + 1$ vertices while $S^\star$ contains $|E(C)| \geq 2$. Hence $|S \cap P(G[C])| \leq \frac{3}{2}|S^\star \cap P(G[C])|$. Since components $C \in \mathcal{C}$ are disjoint, we conclude that $|S \cap P_{\mathcal{C}_{\geq 2}}| \leq \frac{3}{2}|S^\star \cap P_{\mathcal{C}_{\geq 2}}|$.

To show (III), we first consider $\mathcal{C}$ more closely. It is easy to see that no component $C \in \mathcal{C}$ consists solely of single vertex u. Indeed, the point of $S^\star$ closest to the vertex u is either a neighbor $v \in N(u)$ or a point $p(v, u, \lambda)$ for a neighbor $v \in L_1$ with $\lambda < 1 - \delta$ by (B2), which in both cases do not δ-cover u. Let $\mathcal{C}_1 \subseteq \mathcal{C}$ be the subset of components that consist of exactly 1 edge. Let $X^\star := S^\star \cap P_1$ where $P_1 := \bigcup_{u_1 \in L_1} P_{u_1}$, and P_{u_1} defined as in (B3). By definition, $|X^{\star\star}| \leq |X^\star|$.

Claim. There is an injective mapping $f : \mathcal{C}_1 \to X^\star$ such that every component $C \in \mathcal{C}_1$ consists of an edge $\{u, v\}$ incident to the vertex $f(\{u, v\})$.

(III) Let $P_L := P(G[E_{0,1}]) \cup \bigcup_{u_1 \in L_1} P_{u_1} \cup P_{\mathcal{C}_1}$ where $P_{\mathcal{C}_1}$ is the union of $P(G[\{u, v\}])$ for every edge $\{u, v\}$ forming a component in $\mathcal{C}_1$. We claim that $|S \cap P_L| \leq \frac{3}{2}|S^\star \cap P_L|$. Let $L^\star := S^\star \cap B^{\leq}(L_0, \delta)$, which has $|L^\star| = |L_0| = |L|$.

Let $X_C^\star := S^\star \cap \bigcup_{C \in \mathcal{C}_1} P(G[C])$. We note that $|X_C^\star| \leq |\mathcal{C}_1|$ since $S^\star$ contains at most one vertex of every $P(G[C])$, by property (N1). Since further $f : \mathcal{C}_1 \to X^\star$ is injective, we have $|X_C^\star| \leq |\mathcal{C}_1| \leq |X^\star|$. The approximation ratio restricted to P_L then is

$$\frac{|S \cap P_L|}{|S^\star \cap P_L|} = \frac{|L \cup X \cup \bigcup_{C \in \mathcal{C}_1} V(C)|}{|L^\star \cup X^\star \cup X_C^\star|} \leq \frac{|L| + |X| + 2|X^\star|}{|L| + 2|X^\star|}.$$

We recall that $|X^{\star\star}| \leq |X^\star|$. The set X, of size at most $|L_1| \leq |L|$, is a 2-approximation of $X^{\star\star}$, and hence $|X| \leq \min\{|L|, 2|X^\star|\}$. In case $|X^\star| \leq \frac{1}{2}|L|$, the above ratio is at most $(2|L|+4|X^\star|)/(|L|+2|X^\star|)$, which is at most $\frac{3}{2}$. Otherwise, in case $|X^\star| > \frac{1}{2}|L|$, the above ration is at most $(2|L| + 2|X^\star|)/(|L| + 2|X^\star|)$, which also is at most $\frac{3}{2}$.

Sets $P_{V^\star}, P_{\mathcal{C}_{\geq 2}}, P_L$ are disjoint by their definition. It remains to show that $S \subseteq P_{V^\star} \cup P_{\mathcal{C}_{\geq 2}} \cup P_L$. Every point $p \in S \setminus V(G)$, has form $p(u_0, u_1, \frac{2}{3})$ for some edge $(u_0, u_1) \in E_{0,1}$, and hence $p \in P(G[E_{0,1}]) \subseteq P_L$. Every point $p \in S \cap V(G)$ is either contained in $V^\star$ or in $V(G^\star)$. In the former case, $p \in P_{V^\star}$. In the latter case $p \in P_{\mathcal{C}_1} \subseteq P_L \cup P_{\mathcal{C}_{\geq 2}}$. Hence:

Theorem 8. *For every $\delta \in [\frac{2}{3}, \frac{3}{4})$, there is a $\frac{3}{2}$-approximation algorithm for δ-COVERING.*

There is a gap to the lower bound under UGC of $\frac{4}{3}$, as shown Theorem 2(a). However, our algorithm as a *uniform* algorithm for $\delta \in [\frac{3}{2}, \frac{3}{4})$, is reasonably close to the lower bound in the following sense. The upper and lower bound only differ by a factor of $\frac{3}{2}/\frac{4}{3} = \frac{9}{8}$. Hence any improvement of the approximation factor for such a uniform algorithm implies a better approximation guarantee for the class of graphs $\mathcal{G}$ where the minimum size of a δ-cover can be apart by a factor of asymptotically $\frac{9}{8}$. An example is a path of length $24x$ for $x \geq 1$. A minimum $\frac{3}{4}$-cover has size $18(x-1)$ while a minimum $(\frac{3}{4} - \varepsilon)$-cover has size $16(x-1)+1$ for a small enough $\varepsilon > 0$.

7 Approximation Algorithms for $\delta < \frac{1}{2}$

By the Subdivision Reduction (Definition 1), an upper bound for $\delta > \frac{1}{2}$ implies the same upper bound for $\frac{\delta}{c}$ for $c \in \mathbb{N}^+$. This can be further improved by using the Translation Reduction (Definition 2). Let $\Delta'(G)$ denote the average vertex degree of G.

Lemma 6. $(\star)$ *For every $k \in \mathbb{N}^+$, $\delta \in (\frac{1}{2k+2}, \frac{1}{2k+1})$ and graph G, $\frac{1}{2k+2}$-$\mathrm{cover}(G) \leq (1 + \frac{1}{k\Delta'(G)+1}) \cdot \delta$-$\mathrm{cover}(G)$.*

To derive bounds for the remaining intervals, $(\frac{1}{3}, \frac{1}{2}), (\frac{1}{5}, \frac{1}{4})$ and so on, we recall the Eq. (1): $|S^*| = \nu(G_0) + \nu(G_{\geq 3}) + c_{\geq 3} + \mathrm{vc}(G_1) \leq \frac{1}{2}(|V(G)| + c_{\geq 3}) \leq \frac{2}{3}|V(G)|$.

Further, we use that no component of $G_{\geq 3}$ is a tree, and thus $|E(G)| \geq |V(G)| - 1 + c_{\geq 3}$.

Lemma 7. ($\star$) *Let* $k \in \mathbb{N}^+$, $\delta \in (\frac{1}{2k+1}, \frac{1}{2k})$. *For every* $\varepsilon > 0$, *there is an* n_0 *such that for every graph* G *on at least* n_0 *vertices,* $\frac{1}{2k+1}$-$\mathrm{cover}(G) \leq \min\{1 + \frac{4}{3k\Delta'(G)}, 1 + \frac{1}{2k} + \varepsilon\} \cdot \delta$-$\mathrm{cover}(G)$.

As there are only constant many graphs on less than n_0 vertices, we obtain an approximation of δ-cover with approximation ratio $\min\{1 + \frac{4}{3k\Delta'(G)}, 1 + \frac{1}{2k} + \varepsilon\}$ for any $\varepsilon > 0$.

References

1. Chvátal, V.: A greedy heuristic for the set-covering problem. Math. Oper. Res. **4**(3), 233–235 (1979). https://doi.org/10.1287/moor.4.3.233
2. Crescenzi, P.: A short guide to approximation preserving reductions. In: Proceedings of the Twelfth Annual IEEE Conference on Computational Complexity, Ulm, Germany, 24–27 June 1997, pp. 262–273. IEEE Computer Society (1997). https://doi.org/10.1109/CCC.1997.612321
3. Dearing, P.M., Francis, R.L.: A minimax location problem on a network. Transp. Sci. **8**(4), 333–343 (1974)
4. Dinur, I., Safra, S.: On the hardness of approximating vertex cover. Ann. Math. **162**(1), 439–485 (2005). https://doi.org/10.4007/annals.2005.162.439
5. Drezner, E.: Facility location: a survey of applications and methods. J. Oper. Res. Soc. **47**(11), 1421–1421 (1996)
6. Edmonds, J.: Paths, trees, and flowers. Can. J. Math. **17**, 449–467 (1965)
7. Gallai, T.: Kritische Graphen II. A Magyar Tudományos Akadémia Matematikai Kutató Intézetének Közleményei **8**, 373–395 (1963)
8. Gallai, T.: Maximale Systeme unabhängiger Kanten. A Magyar Tudományos Akadémia Matematikai Kutató Intézetének Közleményei **9**, 401–413 (1964)
9. Grigoriev, A., Hartmann, T.A., Lendl, S., Woeginger, G.J.: Dispersing obnoxious facilities on a graph. In: Niedermeier, R., Paul, C. (eds.) 36th International Symposium on Theoretical Aspects of Computer Science, STACS 2019. LIPIcs, 13–16 March 2019, Berlin, Germany, vol. 126, pp. 33:1–33:11. Schloss Dagstuhl - Leibniz-Zentrum für Informatik (2019). https://doi.org/10.4230/LIPIcs.STACS.2019.33
10. Hartmann, T.A.: Facility location on graphs. Dissertation, RWTH Aachen University, Aachen (2022). https://doi.org/10.18154/RWTH-2023-01837
11. Hartmann, T.A., Janßen, T.: Approximating δ-Covering (2024). https://arxiv.org/abs/2408.04517
12. Hartmann, T.A., Lendl, S.: Dispersing obnoxious facilities on graphs by rounding distances. In: Szeider, S., Ganian, R., Silva, A. (eds.) 47th International Symposium on Mathematical Foundations of Computer Science, MFCS 2022. LIPIcs, 22–26 August 2022, Vienna, Austria, vol. 241, pp. 55:1–55:14. Schloss Dagstuhl - Leibniz-Zentrum für Informatik (2022). https://doi.org/10.4230/LIPIcs.MFCS.2022.55
13. Hartmann, T.A., Lendl, S., Woeginger, G.J.: Continuous facility location on graphs. Math. Program. **192**(1), 207–227 (2022). https://doi.org/10.1007/s10107-021-01646-x
14. Johnson, D.S.: Approximation algorithms for combinatorial problems. In: Aho, A.V., et al. (eds.) Proceedings of the 5th Annual ACM Symposium on Theory of Computing, 30 April–2 May 1973, Austin, Texas, USA, pp. 38–49. ACM (1973). https://doi.org/10.1145/800125.804034

15. Khot, S.: On the power of unique 2-prover 1-round games. In: Reif, J.H. (ed.) Proceedings on 34th Annual ACM Symposium on Theory of Computing, 19–21 May 2002, Montréal, Québec, Canada, pp. 767–775. ACM (2002). https://doi.org/10.1145/509907.510017
16. Khot, S., Regev, O.: Vertex cover might be hard to approximate to within 2-epsilon. J. Comput. Syst. Sci. **74**(3), 335–349 (2008). https://doi.org/10.1016/j.jcss.2007.06.019
17. Lovász, L.: On the ratio of optimal integral and fractional covers. Discret. Math. **13**(4), 383–390 (1975). https://doi.org/10.1016/0012-365X(75)90058-8
18. Megiddo, N., Tamir, A.: New results on the complexity of p-center problems. SIAM J. Comput. **12**(4), 751–758 (1983). https://doi.org/10.1137/0212051
19. Mirchandani, P.B., Francis, R.L.: Discrete Location Theory (1990)
20. Tamir, A.: On the solution value of the continuous p-center location problem on a graph. Math. Oper. Res. **12**(2), 340–349 (1987). https://doi.org/10.1287/moor.12.2.340
21. Tamir, A.: Obnoxious facility location on graphs. SIAM J. Discret. Math. **4**(4), 550–567 (1991). https://doi.org/10.1137/0404048

Fast Approximation Algorithms for Euclidean Minimum Weight Perfect Matching

Stefan Hougardy and Karolina Tammemaa(✉)

Research Institute for Discrete Mathematics and Hausdorff Center for Mathematics, University of Bonn, Bonn, Germany
hougardy@or.uni-bonn.de, karolina.tammemaa@gmail.com

Abstract. We study the problem of finding a Euclidean minimum weight perfect matching for n points in the plane. It is known that a deterministic approximation algorithm for this problems must have at least $\Omega(n \log n)$ runtime. We propose such an algorithm for the Euclidean minimum weight perfect matching problem with runtime $O(n \log n)$ and show that it has approximation ratio $O(n^{0.2995})$. This improves the so far best known approximation ratio of $n/2$. We also develop an $O(n \log n)$ algorithm for the Euclidean minimum weight perfect matching problem in higher dimensions and show it has approximation ratio $O(n^{0.599})$ in all fixed dimensions.

Keywords: Euclidean matching · Approximation algorithms

1 Introduction

A *perfect matching* in a graph G is a subset of edges such that each vertex of G is incident to exactly one edge in the subset. When each edge e of G has a real weight w_e, then the *minimum weight perfect matching problem* is to find a perfect matching M that minimizes the weight $\sum_{e \in M} w_e$. In case where the vertices are points in the Euclidean plane and we have a complete graph where each edge e has weight w_e equal to the Euclidean distance between its two endpoints, then we call it *Euclidean minimum weight perfect matching problem*.

The Euclidean minimum weight perfect matching problem can be solved in polynomial time by applying Edmonds' blossom algorithm [3,4] to the complete graph where the edge weights are the Euclidean distances between the edge endpoints. Gabow [5] and Lawler [7] have shown how Edmonds' algorithm can be implemented to achieve the running time $O(n^3)$ on graphs with n vertices. This implies an $O(n^3)$ running time for the Euclidean minimum weight perfect matching problem on point sets of size n. By exploiting the geometry of the problem Vaidja [10] developed an algorithm for the Euclidean minimum weight perfect matching problem with running time $O(n^{\frac{5}{2}} \log n)$. In 1998, Varadarajan [13] improved on this result by presenting a $O(n^{\frac{3}{2}} \log^5(n))$ algorithm that uses geometric divide and conquer. This is the fastest exact algorithm currently known for the Euclidean minimum weight perfect matching problem.

Faster approximation algorithms for the Euclidean minimum weight perfect matching problem are known. In [11] Vaidja presented a $1+\varepsilon$-approximation algorithm with runtime $O(n^{\frac{3}{2}}\log^{\frac{5}{2}} n(1/\varepsilon^3)\sqrt{\alpha(n,n)})$, where α is the inverse Ackermann function and $\varepsilon \leq 1$. Even faster approximation algorithms are known when we allow randomization. Arora [1] presented a $1+\varepsilon$ Monte-Carlo approximation algorithm with runtime $O(n \log^{O(1/\varepsilon)} n)$, where $\varepsilon < 1$. Varadarajan and Agarwal [14] improved on this, presenting a $1+\varepsilon$ approximation algorithm with runtime $O(n/\varepsilon^3 \log^6 n)$, where $\varepsilon > 0$. Rao and Smith [8] gave a constant factor $\frac{3}{2}e^{8\sqrt{2}} \approx 122\,905.81$ Monte-Carlo approximation algorithm with runtime $O(n \log n)$. In the same paper, they also propose a deterministic $\frac{n}{2}$-approximation algorithm with runtime $O(n \log n)$.

Das and Smid have shown [2] that any deterministic approximation algorithm for the Euclidean minimum weight perfect matching problem needs to have a runtime of at least $\Omega(n \log n)$. It is therefore natural to ask: *What approximation ratio can be achieved in $O(n \log n)$ by a deterministic approximation algorithm for the Euclidean minimum weight perfect matching problem for n points in the plane?* In this paper we will significantly improve the approximation ratio of $n/2$ due to Rao and Smith [8] and show:

Theorem 1. *For n points in $\mathbb{R}^2$ there exists a deterministic $O(n^{0.2995})$-approximation algorithm for the Euclidean minimum weight perfect matching problem with runtime $O(n \log n)$.*

Our algorithm is based on the idea of clustering the given points in the Euclidean plane into components of even cardinality. We show that there is a way to compute these components such that for all but one of the components we can find a perfect matching with small enough weight. We then apply our algorithm iteratively on this single remaining component. After sufficiently many iterations the number of points contained in the remaining component is small enough to apply an exact algorithm. Note that in the Euclidean minimum weight perfect matching problem we do not have the property that the value of an optimum solution is at least as large as the optimum value for any even cardinality subset of the given points. Crucial for the analysis of our approach is therefore that we can show that the size of the remaining component decreases faster than the value of an optimum solution for the remaining component increases. The algorithm of Rao and Smith [8] can be extended to higher (fixed) dimension resulting in an approximation ratio of n in runtime $O(n \log n)$. Our approach also extends to higher (fixed) dimensions and we obtain the following result:

Theorem 2. *For any fixed dimension d there exists a deterministic $O(n^{0.599})$-approximation algorithm for the Euclidean minimum weight perfect matching problem in $\mathbb{R}^d$ with runtime $O(n \log n)$.*

The paper is organized as follows. In Sect. 2 we introduce the nearest neighbor graph and study its relation to minimum spanning trees and matchings. A basic ingredient to our main algorithm is a subroutine we call the EVEN-COMPONENT-ALGORITHM. We explain this subroutine and prove some basic facts about it in Sect. 3. This subroutine was also used in the *Even Forest Heuristic* of Rao and Smith [8] that deterministically achieves in $O(n \log n)$ the so far best approximation ratio of $n/2$. We will apply the EVEN-COMPONENT-ALGORITHM as a subroutine within our NODE-REDUCTION-ALGORITHM. The idea of the NODE-REDUCTION-ALGORITHM is to partition a point set into some subsets of even cardinality and a single remaining set of points such that two properties hold. First, the even subsets should allow to compute a short perfect matching within linear time. Second, the remaining set of points should be sufficiently small. In Sect. 4 we will present our NODE-REDUCTION-ALGORITHM and will derive bounds for its runtime and the size of the remaining subset. The next idea is to iterate the NODE-REDUCTION-ALGORITHM on the remaining subset until it becomes so small that we can apply an exact minimum weight perfect matching algorithm to it. We call the resulting algorithm the ITERATED-NODE-REDUCTION-ALGORITHM and analyze its runtime and approximation ratio in Sect. 5 and also prove there Theorem 1. Finally, in Sect. 6 we extend the ITERATED-NODE-REDUCTION-ALGORITHM to higher (fixed) dimension and prove Theorem 2.

2 Preliminaries

A crucial ingredient to our algorithm is the so called *nearest neighbor graph*. For a given point set in R^d we first fix a random total ordering on all points, which we use to break ties when creating the nearest neighbor graph to avoid getting cycles. We compute for each point all other points that have the smallest possible distance to this point. Among all these points we select, as its nearest neighbor, a point that is minimal with respect to the total ordering. Now we get the nearest neighbor graph by taking as vertices all points and adding an undirected edge between each point and its nearest neighbor. We will denote the nearest neighbor graph for a point set $V \subseteq \mathbb{R}^d$ by $NN(V)$. Immediately from the definition we get that the nearest neighbor graph is a forest. It is well known that the nearest neighbor graph for a point set in $\mathbb{R}^d$ is a subgraph of a Euclidean minimum spanning tree for this point set. For $k \geq 2$ the nearest neighbor graph can be generalized to the k-nearest neighbor graph. To obtain this graph choose for each point its k nearest neighbors (ties broken arbitrarily) and connect them with an edge.

Shamos and Hoey [9] have shown that the nearest neighbor graph and a Euclidean minimum spanning tree for a point set of cardinality n in $\mathbb{R}^2$ can be computed in $O(n \log n)$. For the nearest neighbor graph this result also holds in higher (fixed) dimension as was shown by Vaidya [12]. The algorithm of Vaidya [12] even allows to compute the k-nearest neighbor graph for point sets in $\mathbb{R}^d$ in $O(n \log n)$ as long as k and d are fixed.

We call a connected component of a graph an *odd connected component* if it has odd cardinality. Similarly we define an *even connected component*. We will denote by $\ell(e)$ the Euclidean length of an edge e and for a set E of edges we define $\ell(E) := \sum_{e \in E} \ell(e)$. For a point set V we denote a Euclidean minimum weight matching for this point set by $MWPM(V)$. Clearly, the point set V must have even cardinality for a perfect matching to exist. Throughout this paper by log we mean the logarithm with base 2.

There is a simple connection between the length of the nearest neighbor graph for a point set V and a Euclidean minimum weight perfect matching for this point set:

Lemma 1. *For a point set $V \subseteq \mathbb{R}^d$ we have $\ell(NN(V)) \leq 2 \cdot \ell(MWPM(V))$.*

Proof. In a Euclidean minimum weight perfect matching for V each point $v \in V$ is incident to an edge that is at least as long as the distance to a nearest neighbor of v. Thus, if we assign to each point $v \in V$ the length of the edge that it is incident to in a Euclidean minimum weight perfect matching of V and compare it to the distance of a nearest neighbor then we get $\ell(NN(V)) \leq 2 \cdot \ell(MWPM(V))$.

3 The Even Component Algorithm

The currently best deterministic $O(n \log n)$ approximation algorithm for the Euclidean minimum weight perfect matching problem is the Even Forest Heuristic due to Rao and Smith [8]. It achieves an approximation ratio of $n/2$ and tight examples achieving this approximation ratio are known [8]. The idea of the Even Forest Heuristic is to compute first a minimum spanning tree and then to remove all so called *even edges* from the tree. An edge is called even, if removing this edge results in two connected components of even cardinality. This way one gets a forest where all connected components of the forest have even cardinality. Within each component of the forest they then compute a Hamiltonian cycle by first doubling all edges of the tree and then short-cutting a Eulerian cycle. They then obtain a matching from the Hamiltonian cycle by taking every second edge. We also make use of this second part of the Even Forest Heuristic and call it the EVEN-COMPONENT-ALGORITHM (see Algorithm 1). In line 5 of this algorithm we shortcut the edges of a Eulerian cycle. By this we mean that we iteratively replace for three consecutive vertices x, y, z the edges xy and yz by the edge xz if the vertex y has degree more than two. Notably, Rao and Smith applied this algorithm to a random forest derived from minimum spanning tree, we will apply it for a random forest derived from nearest neighbor graph.

For completeness we briefly restate the following two results and their proofs from [8]:

Lemma 2 ([8]). *The* EVEN-COMPONENT-ALGORITHM *applied to an even connected component returns a matching with length at most the length of all edges in the even component.*

Proof. By doubling all edges in a component we double the total edge length. The Eulerian cycle computed in line 4 has therefore exactly twice the length of the edges in the connected component. Shortcutting this cycle in line 5 cannot make

---
Algorithm 1. EVEN-COMPONENT-ALGORITHM

Input: a forest F where all connected components have even cardinality
Output: a perfect matching M
1: $M := \emptyset$
2: **for** each connected component of F **do**
3:   double the edges in the component
4:   compute a Eulerian cycle in the component
5:   shortcut the edges in the Eulerian cycle to get a Hamiltonian cycle
6:   add the shorter of the two perfect matchings inside the Hamiltonian cycle to M.
7: **end for**
---

it longer. The Hamiltonian cycle therefore has length at most twice the length of all edges in the connected component. One of the two perfect matchings into which we can decompose the Hamiltonian cycle has length at most half of the length of the Hamiltonian cycle. Therefore, the length of the matching computed by the EVEN-COMPONENT-ALGORITHM is upper bounded by the length of all edges in the connected component.

Lemma 3 ([8]). *The* EVEN-COMPONENT-ALGORITHM *has linear runtime.*

Proof. We can use depth first search to compute in linear time the connected components and double the edges in all components. A Eulerian cycle can be computed in linear time using for example Hierholzer's algorithm [6]. Short-cutting can easily be done in linear time by running along the Eulerian cycle. We find the smaller of the two matchings by selecting every second edge of the Hamiltonian cycle.

4 The Node Reduction Algorithm

A central part of our algorithm is a subroutine we call the NODE-REDUCTION-ALGORITHM. This algorithm gets as input a point set $V \subseteq \mathbb{R}^2$ of even cardinality and returns a subset $W \subseteq V$ and a perfect matching M for $V \setminus W$. The idea of this algorithm is to first compute the nearest neighbor graph $NN(V)$ of V. If $NN(V)$ has more than $|V|/4$ odd connected components, then we add some more edges to reduce this number. From each of the remaining odd connected components we remove one leaf vertex and put it into W. We are left with a set of even connected components and apply the EVEN-COMPONENT-ALGORITHM to each of these. Algorithm 2 shows the pseudo code of the NODE-REDUCTION-ALGORITHM. In the following two lemmas we analyze the performance and runtime of this algorithm.

Lemma 4. *For a point set $V \subseteq \mathbb{R}^2$ let W be the point set and M be the matching returned by the* NODE-REDUCTION-ALGORITHM.

(a) If $NN(V)$ has at most $|V|/4$ odd connected components then we have:
$|W| \leq |V|/4$, $\ell(M) \leq 2 \cdot \ell(MWPM(V))$, *and* $\ell(MWPM(W)) \leq 3 \cdot \ell(MWPM(V))$.

(b) If $NN(V)$ has more than $|V|/4$ odd connected components then we have:
$|W| \leq |V|/6$, $\ell(M) \leq 4 \cdot \ell(MWPM(V))$, and $\ell(MWPM(W)) \leq 5 \cdot \ell(MWPM(V))$.

Algorithm 2. NODE-REDUCTION-ALGORITHM

Input: a set $V \subseteq \mathbb{R}^2$ of even cardinality
Output: $W \subseteq V$ and a perfect matching M for $V \setminus W$
1: $G := NN(V)$
2: **if** the number of odd connected components of G is $> |V|/4$ **then**
3:    $S := \emptyset$
4:    compute a minimum spanning tree T of V
5:    **for** each odd connected component of G **do**
6:       add to S a shortest edge from T going out of this component
7:    **end for**
8:    add the edges of S to G
9: **end if**
10: $W := \emptyset$
11: **for** each odd connected component of G **do**
12:    choose one leaf node in the component and add it to W
13: **end for**
14: $M := $ EVEN-COMPONENT-ALGORITHM$(G[V \setminus W])$

Proof. We first prove statement (a). In this case $NN(V)$ has at most $|V|/4$ odd connected components and lines 3–8 of the NODE-REDUCTION-ALGORITHM are not executed. By lines 11–13 the set W contains exactly one node from each odd connected component of $NN(V)$. We therefore get $|W| \leq |V|/4$.

As M is obtained by the EVEN-COMPONENT-ALGORITHM applied to a subset of the edges of $NN(V)$ we have $\ell(M) \leq \ell(NN(V))$ by Lemma 2. From Lemma 1 we get $\ell(M) \leq \ell(NN(V)) \leq 2 \cdot \ell(MWPM(V))$.

Let H be the graph obtained from $NN(V)$ by adding the edges of $MWPM(V)$. All connected components in H have even cardinality as by definition H contains a perfect matching. This implies that each connected component of H contains an even number of odd connected components from $NN(V)$. As W contains exactly one point from each odd connected component of $NN(V)$ this implies that each connected component of H contains an even number of vertices of W. Within each connected component of H we can therefore pair all vertices from W and connect each pair by a path. By taking the symmetric difference of all these paths we get a set of edge disjoint paths such that each vertex of W is an endpoint of exactly one such path. Thus, the total length of all these paths is an upper bound for $\ell(MWPM(W))$. As the total length of all these edge disjoint paths is bounded by the total length of all edges in H we get by using Lemma 1:
$\ell(MWPM(W)) \leq \ell(E(H)) \leq \ell(NN(V)) + \ell(MWPM(V)) \leq 3 \cdot \ell(MWPM(V))$.
(Part of this argument would easily follow from the theory of T-joins which we avoid to introduce here.)

We now prove statement (b). In this case $NN(V)$ has more than $|V|/4$ odd connected components and in lines 3–8 of the NODE-REDUCTION-ALGORITHM we add the edges of S to G. Each edge of S that we add to G joins an odd component of $NN(V)$ to some other component of $NN(V)$. Each odd connected component of G must contain an odd number of odd connected components of $NN(V)$. Let n_1 denote the number of odd connected components in G that contain exactly one odd connected component from $NN(V)$ and let n_3 be the number of all other odd connected components of G. Let k denote the number of odd connected components of $NN(V)$. Then we have $n_1 + 3n_3 \leq k$. For each odd connected component counted by n_1 there is exactly one even connected component in $NN(V)$ to which the edge from S connects it. As each even connected component contains at least two points and as each odd connected component contains at least three points we get $2n_1 \leq |V| - 3k$. By adding these two inequalities we get $3n_1 + 3n_3 \leq |V| - 2k$. Now $k > |V|/4$ implies $n_1 + n_3 \leq (|V| - 2k)/3 \leq |V|/6$.

Next we claim that the edges in the set S computed in lines 3–7 of the NODE-REDUCTION-ALGORITHM have total length at most $2 \cdot \ell(MWPM(V))$. For each odd connected component of G there must exist an edge e in $MWPM(V)$ that leaves this component. By the cut property of minimum spanning trees the tree T contains an edge that leaves the component and has at most the length of e. As each edge in $MWPM(V)$ can connect at most two odd connected components we get the upper bound $\ell(S) \leq 2 \cdot \ell(MWPM(V))$. In line 14 of the NODE-REDUCTION-ALGORITHM the matching M is computed by the EVEN-COMPONENT-ALGORITHM on a subset of the edges in $NN(V) \cup S$. Lemma 2 therefore implies $\ell(M) \leq \ell(NN(V)) + \ell(S)$. Now Lemma 1 together with the above bound on $l(S)$ yields $\ell(M) \leq \ell(NN(V)) + \ell(S) \leq 2 \cdot \ell(MWPM(V)) + 2 \cdot \ell(MWPM(V)) = 4 \cdot \ell(MWPM(V))$.

Let H be the graph obtained from $NN(V)$ by adding the edges of S and $MWPM(V)$. We can now follow the same arguments as in case (a) to derive that $\ell(MWPM(W)) \leq \ell(E(H)) \leq \ell(NN(V)) + \ell(S) + \ell(MWPM(V)) \leq 5 \cdot \ell(MWPM(V))$.

Lemma 5. *The* NODE-REDUCTION-ALGORITHM *(Algorithm 2) on input V with $|V| = n$ has runtime $O(n \log n)$.*

Proof. In line 1 of the algorithm the nearest neighbor graph $NN(V)$ can be computed in $O(n \log n)$ as was proved by Shamos and Hoey [9]. As the nearest neighbor graph has a linear number of edges we can use depth first search to compute its connected components and their parity in $O(n)$. The minimum spanning tree T for V in line 4 of the algorithm can be computed in $O(n \log n)$ by using the algorithm of Shamos and Hoey [9]. To compute the set S in lines 3–7 of the algorithm simply run through all edges of T and store for each component the shortest edge leaving that component. Choosing a leaf node in a connected component which is a tree can be done in time proportional to the size of the connected component, thus the total runtime of lines 10–13 is $O(n)$. Finally, by Lemma 3 the runtime of the EVEN-COMPONENT-ALGORITHM is linear in the size of the input graph. Therefore, line 14 requires $O(n)$ runtime. Summing up all these time complexities gives us a time complexity $O(n \log n)$.

5 Iterating the NODE-REDUCTION-ALGORITHM

The NODE-REDUCTION-ALGORITHM (Algorithm 2) on input $V \subseteq \mathbb{R}^2$ returns a set $W \subseteq V$ and a perfect matching on $V \setminus W$. The idea now is to iterate the NODE-REDUCTION-ALGORITHM on the set W of unmatched vertices. By Lemma 4 we know that after each iteration the set W shrinks by at least a factor of four. Therefore, after a logarithmic number of iterations the set W will be empty. However, we can do a bit better by stopping as soon as the set W is small enough to compute a Euclidean minimum weight perfect matching on W in $O(n \log n)$ time. We call the resulting algorithm the ITERATED-NODE-REDUCTION-ALGORITHM, see Algorithm 3. In line 3 of this algorithm we apply the NODE-REDUCTION-ALGORITHM to the point set V_i. This gives us a matching we denote by M_i and a set of unmatched points which we denote by V_{i+1}.

Algorithm 3. ITERATED-NODE-REDUCTION-ALGORITHM

Input: a set $V \subseteq \mathbb{R}^2$ of even cardinality, $\epsilon > 0$
Output: a perfect matching M for V
1: $V_1 := V$, $i := 1$
2: **while** $|V_i| > |V|^{2/3-\epsilon}$ **do**
3:    $V_{i+1}, M_i \leftarrow$ NODE-REDUCTION-ALGORITHM(V_i)
4:    $i := i + 1$
5: **end while**
6: $M := MWPM(V_i) \cup M_1 \cup M_2 \cup \ldots$

Clearly, the ITERATED-NODE-REDUCTION-ALGORITHM returns a perfect matching on the input set V. The next lemma states the runtime of the ITERATED-NODE-REDUCTION-ALGORITHM.

Lemma 6. *The* ITERATED-NODE-REDUCTION-ALGORITHM *(Algorithm 3) on input V with $|V| = n$ has runtime $O(n \log n)$.*

Proof. We have $|V_1| = |V|$ and for $i \geq 2$ Lemma 4 implies $|V_i| < |V|/4^{i-1}$. By Lemma 5 the runtime of the NODE-REDUCTION-ALGORITHM on the set $|V_i|$ is $O(|V_i| \log |V_i|)$. The total runtime for lines 2–5 of the ITERATED-NODE-REDUCTION-ALGORITHM is therefore bounded by

$$O\left(\sum_{i=1}^{\infty} |V_i| \log |V_i|\right) \leq O\left(\log |V| \cdot \sum_{i=1}^{\infty} \frac{|V|}{4^{i-1}}\right) = O(|V| \log |V|).$$

In line 6 of the ITERATED-NODE-REDUCTION-ALGORITHM we can use the algorithm of Varadarajan [13] to compute a Euclidean minimum weight perfect matching on V_i. For a point set of size s Varadarajan's algorithm has runtime $O(s^{\frac{3}{2}} \log^5(s))$. As the set V_i in line 6 of the Iterated-Node-Reduction-Algorithm has size at most $|V|^{2/3-\epsilon}$ we get a runtime of

$$O\left((|V|^{2/3-\epsilon})^{\frac{3}{2}} \log^5(|V|^{2/3-\epsilon})\right) = O\left(\frac{|V|}{|V|^{\frac{3\epsilon}{2}}} \log^5(|V|^{2/3-\epsilon})\right) = O(|V|).$$

In total the Iterated-Node-Reduction-Algorithm has runtime $O(n \log n)$.

We now analyze the approximation ratio of the Iterated-Node-Reduction-Algorithm. For this we will bound the length of the matching $MWPM(V_i)$ computed in line 6 of the algorithm and the total length of all matchings M_i computed in all iterations of the algorithm. We start with a bound on $\ell(MWPM(V_i))$.

Lemma 7. *For input V and ϵ the length of the matching $MWPM(V_i)$ computed in line 6 of the* Iterated-Node-Reduction-Algorithm *can be bounded by*

$$15 \cdot |V|^{(1/3+\epsilon) \cdot \frac{\log 5}{\log 6}} \cdot \ell(MWPM(V)).$$

Proof. The Node-Reduction-Algorithm has two cases as was shown in Lemma 4. In an iteration of case (a) we have $|V_{i+1}| \leq |V_i|/4$ and $\ell(MWPM(V_{i+1})) \leq 3 \cdot \ell(MWPM(V_i))$. In case (b) we have $|V_{i+1}| \leq |V_i|/6$ and $\ell(MWPM(V_{i+1})) \leq 5 \cdot \ell(MWPM(V_i))$. Let a and b be the number of iterations of the Iterated-Node-Reduction-Algorithm where case (a) respectively case (b) occurs. By the condition of the while-loop we have

$$\frac{|V|}{4^a \cdot 6^b} \leq |V|^{2/3-\epsilon} \quad \text{or equivalently} \quad 4^a \cdot 6^b \geq |V|^{1/3+\epsilon}$$

and the length of the Euclidean minimum weight perfect matching computed by the exact algorithm can be bounded by

$$3^a \cdot 5^b \cdot \ell(MWPM(V)).$$

The condition of the while loop implies that we can choose a' and b' with $a-1 < a' \leq a$ and $b-1 < b' \leq b$ such that

$$4^{a'} \cdot 6^{b'} = 6^{a' \cdot \frac{\log 4}{\log 6} + b'} = |V|^{1/3+\epsilon} \tag{1}$$

and we have

$$3^a \cdot 5^b \cdot \ell(MWPM(V)) \leq 3^{a'+1} \cdot 5^{b'+1} \cdot \ell(MWPM(V)) = 15 \cdot 5^{a' \frac{\log 3}{\log 5} + b'} \cdot \ell(MWPM(V)).$$

We have

$$0.774 \approx \frac{\log 4}{\log 6} > \frac{\log 3}{\log 5} \approx 0.683.$$

Therefore, the expression $a' \frac{\log 3}{\log 5} + b'$ attains its maximum under the constraint (1) at $a' = 0$. Thus we have $6^{b'} = |V|^{1/3+\epsilon}$ and get as an upper bound for the length of the perfect matching computed by the exact algorithm:

$$15 \cdot 5^{b'} \cdot \ell(MWPM(V)) = 15 \cdot 6^{b' \cdot \frac{\log 5}{\log 6}} \cdot \ell(MWPM(V)) = 15 \cdot |V|^{(1/3+\epsilon) \cdot \frac{\log 5}{\log 6}} \cdot \ell(MWPM(V)).$$

We now prove a bound on the total length of all matchings M_i computed in line 3 of the ITERATED-NODE-REDUCTION-ALGORITHM.

Lemma 8. *For input V and ϵ let q denote the number of iterations made by the* NODE-REDUCTION-ALGORITHM. *Then we have*

$$\sum_{i=1}^{q} \ell(M_i) \leq 2 \cdot x_q \cdot \prod_{i=1}^{q-1} y_i \cdot \ell(MWPM(V))$$

were M_i is the matching computed in line 3 of the algorithm and x_i and y_i are defined as follows: if iteration i of the ITERATED-NODE-REDUCTION-ALGORITHM *is of case (a) of Lemma 4 then $y_i = 3$ and $x_i = 2$, and if it is of case (b) then $y_i = 5$ and $x_i = 4$.*

Proof. By Lemma 4 and the definition of x_i and y_i we have $\ell(M_i) \leq x_i \cdot \ell(MWPM(V_i))$ for all $i = 1, \ldots, q$. Moreover, we have $\ell(MWPM(V_i)) \leq y_{i-1} \cdot \ell(MWPM(V_{i-1}))$ and therefore we get for all $i = 1, \ldots, q$:

$$\ell(M_i) \leq x_i \cdot \prod_{j=1}^{i-1} y_j \cdot \ell(MWPM(V)). \quad (2)$$

We now prove the statement of the lemma by induction on q. For $q = 1$ we have by Lemma 4: $\ell(M_1) \leq x_1 \cdot \ell(MWPM(V))$. Now let us assume $q > 1$ and that the statement holds for $q - 1$. Using (2) we get

$$\sum_{i=1}^{q} \ell(M_i) = \ell(M_q) + \sum_{i=1}^{q-1} \ell(M_i)$$

$$\leq \ell(M_q) + 2 \cdot x_{q-1} \cdot \prod_{i=1}^{q-2} y_i \cdot \ell(MWPM(V))$$

$$\leq x_q \cdot \prod_{j=1}^{q-1} y_j \cdot \ell(MWPM(V)) + 2 \cdot x_{q-1} \cdot \prod_{i=1}^{q-2} y_i \cdot \ell(MWPM(V))$$

$$= (x_q \cdot y_{q-1} + 2 \cdot x_{q-1}) \cdot \prod_{i=1}^{q-2} y_i \cdot \ell(MWPM(V)).$$

Now we have $x_q \cdot y_{q-1} + 2 \cdot x_{q-1} \leq x_q \cdot y_{q-1} + 2 \cdot y_{q-1} \leq 2 \cdot x_q \cdot y_{q-1}$ which proves the lemma.

Combining Lemmas 7 and 8 we can now state the approximation ratio for the ITERATED-NODE-REDUCTION-ALGORITHM.

Lemma 9. *For input V and ϵ the* ITERATED-NODE-REDUCTION-ALGORITHM *has approximation ratio $O\left(|V|^{(1/3+\epsilon) \cdot \frac{\log 5}{\log 6}}\right)$.*

Proof. Similar as in the proof of Lemma 7 let a and b be the number of iterations of the ITERATED-NODE-REDUCTION-ALGORITHM where case (a) respectively case (b) of Lemma 4 occurs. By Lemma 8 we have

$$\sum_{i=1}^{q} \ell(M_i) \leq 2 \cdot x_q \cdot \prod_{i=1}^{q-1} y_i \cdot \ell(MWPM(V)) \leq 2 \cdot 3^a \cdot 5^b \cdot \ell(MWPM(V)).$$

We already know from the proof of Lemma 7 that this expression attains its maximum under the constraint (1) for $a = 0$ and similar as in the proof of Lemma 7 this results in the upper bound $\sum_{i=1}^{q} \ell(M_i) = O\left(|V|^{(1/3+\epsilon) \cdot \frac{\log 5}{\log 6}} \cdot \ell(MWPM(V))\right)$. Together with Lemma 7 this implies that the length of the perfect matching returned by the ITERATED-NODE-REDUCTION-ALGORITHM is bounded by $O(|V|^{(1/3+\epsilon) \cdot \frac{\log 5}{\log 6}} \cdot \ell(MWPM(V)))$.

We can now prove our main result on approximating Euclidean minimum weight perfect matchings for point sets in $\mathbb{R}^2$

Proof. We claim that our algorithm ITERATED-NODE-REDUCTION-ALGORITHM has the desired properties. From Lemma 6 we know that the runtime of this algorithm is $O(n \log n)$. By Lemma 9 its approximation ratio is $O\left(|V|^{(1/3+\epsilon) \cdot \frac{\log 5}{\log 6}}\right)$. We have $\frac{1}{3}\frac{\log 5}{\log 6} < 0.29942$ and therefore by choosing ϵ sufficiently small we get an approximation ratio of $O(n^{0.2995})$.

6 Extension to Higher Dimensions

We now want to extend our result for the 2-dimensional case to higher dimensions. For a fixed dimension $d > 2$ we can use essentially the same approach as in two dimensions but need to adjust two things. First, for $d > 2$ no $O(n \log n)$ algorithm is known that computes a Euclidean minimum spanning tree in $\mathbb{R}^d$. Instead we will use Vaidya's [12] $O(n \log n)$ algorithm to compute a 7-nearest neighbor graph in $\mathbb{R}^d$ for fixed d. Secondly, the fastest known exact algorithm to compute a Euclidean minimum weight perfect matching in $\mathbb{R}^d$ is the $O(n^3)$ implementation of Edmonds' algorithm due to Gabow [5] and Lawler [7]. Using these two changes we get:

Theorem 2. *For any fixed dimension d there exists a deterministic $O(n^{0.599})$-approximation algorithm for the Euclidean minimum weight perfect matching problem in $\mathbb{R}^d$ with runtime $O(n \log n)$.*

Proof. We will use essentially the same approach as for the 2-dimensional case but need to apply the following changes. In line 4 of the NODE-REDUCTION-ALGORITHM (Algorithm 2) instead of a minimum spanning tree we compute the 7-nearest neighbor graph using Vaidya's algorithm [12]. Moreover, we restrict the for-loop in line 5 of the algorithm to odd connected components of size at most 7. With these modifications Lemma 4 still holds. The reason for this is that the 7-nearest neighbor graph satisfies the cut property for all odd components of size at most 7. Moreover, an odd component of size at least 9 can be replaced in the arguments by three smaller odd components, making the bounds of the lemma not worse. Second, in line 2 of the ITERATED-NODE-REDUCTION-ALGORITHM we have to replace the bound $|V|^{2/3-\epsilon}$ by $|V|^{1/3}$. In Lemma 7 we have to replace $|V|^{1/3+\epsilon}$ by $|V|^{2/3}$. If we apply all these changes then the approximation ratio in Lemma 9 changes to $O\left(|V|^{\frac{2}{3}\frac{\log 5}{\log 6}}\right)$. As we have $\frac{2}{3}\frac{\log 5}{\log 6} \approx 0.5988$ we get the claimed approximation ratio $O(n^{0.599})$.

Remark. Theorem 2 also holds if instead of Euclidean metric, we have some other L_p, $p = 1, 2, \ldots, \infty$ metric. Running time stays the same, as we can do all the computational steps, including the computationally expensive steps like finding the 7-nearest neighbor graph with the same runtime according to [12]. The approximation ratio also stays the same, as when finding different bounds, we only use the triangle inequality which also holds in L_p metric spaces.

Acknowledgement. This study was supported by the Deutsche Forschungsgemeinschaft (DFG, German Research Foundation) under Germany's Excellence Strategy - EXC-2047/1 - 390685813

Disclosure of Interests. The authors have no competing interests to declare that are relevant to the content of this article.

References

1. Arora, S.: Polynomial time approximation schemes for Euclidean traveling salesman and other geometric problems. J. ACM **45**, 753–782 (1998). https://doi.org/10.1145/290179.290180
2. Das, G., Smid, M.: A lower bound for approximating the geometric minimum weight matching. Inf. Process. Lett. **74**, 253–255 (2000)
3. Edmonds, J.: Maximum matching and a polyhedron with 0,1-vertices. J. Res. Nat. Bur. Stand. B **69B**, 125–130 (1965)
4. Edmonds, J.: Paths, trees, and flowers. Can. J. Math. **17**, 449–467 (1965)
5. Gabow, H.: An Efficient Implementation of Edmonds' Maximum Matching Algorithm. Digital Systems Laboratory, Stanford University (1972)
6. Hierholzer, C.: Ueber die Möglichkeit, einen Linienzug ohne Wiederholung und ohne Unterbrechung zu umfahren. Math. Ann. **6**, 30–32 (1873)
7. Lawler, E.: Combinatorial Optimization: Networks and Matroids. Holt, Rinehart (1976)
8. Rao, S., Smith, W.: Improved approximation schemes for geometrical graphs via "spanners" and "banyans" (1998)

9. Shamos, M., Hoey, D.: Closest-point problems. In: 16th Annual Symposium on Foundations of Computer Science, FOCS 1975, pp. 151–162 (1975)
10. Vaidya, P.: Geometry helps in matching. SIAM J. Comput. **18**, 1201–1225 (1989)
11. Vaidya, P.: Approximate minimum weight matching on points in k-dimensional space. Algorithmica **4**, 569–583 (1989)
12. Vaidya, P.: An $O(n \log n)$ algorithm for the all-nearest-neighbors problem. Discrete Comput. Geom. **4**, 101–115 (1989)
13. Varadarajan, K.: A divide-and-conquer algorithm for min-cost perfect matching in the plane. In: Proceedings 39th Annual Symposium on Foundations of Computer Science, pp. 320–329 (1998)
14. Varadarajan, K., Agarwal, P.: Approximation algorithms for bipartite and non-bipartite matching in the plane. In: Proceedings of the Tenth Annual ACM-SIAM Symposium On Discrete Algorithms, pp. 805–814 (1999)

Approximation Algorithms for k-Scenario Matching

Danny Blom, Dylan Hyatt-Denesik[✉], Afrouz Jabal Amelia, and Bart Smeulders

Eindhoven University of Technology, 5612 AZ Eindhoven, The Netherlands
{d.v.p.hyatt-denesik,a.jabal.ameli,b.smeulders}@tue.nl

Abstract. Matching theory is among the most fundamental graph optimization problems. Several different variants of this problem have been introduced and studied. Maximum matching, capacitated matching, and perfect matchings are all well studied examples of problems in matching theory. Here we motivate and study a generalization of the maximum weighted matching problem, so-called k-*scenario matching*. In k-scenario matching we are given a graph $G = (V, E)$, in which each edge belongs to at least one of the k subsets $E_1, ..., E_k$ (scenarios), and each scenario is equipped with probability p_i for each $1 \le i \le k$ (hence $\sum_{i=1}^{k} p_i = 1$). We are provided with budget, some integer B, and our goal is to find a subset $\hat{E} \subseteq E$ of size at most B, that maximizes the expected value of the maximum cardinality matchings on all the possible k scenarios (i.e. $\sum_{i=1}^{k} p_i |M_i|$ where M_i is a maximum matching on edges $\hat{E} \cap E_i$.) This problem is motivated by applications such as kidney exchange, where compatibility between donors and patients is uncertain. Tests must be performed to ascertain whether two vertices can be matched, but due to the time and cost involved, only a limited number of tests are possible.

In this paper we show that this problem is NP-hard even in the simple case where we have two scenarios, and the input graph is both bipartite and sub-cubic. We also study the approximability of this problem in several cases from the positive direction. For the general case of this problem, we find a 2.314-approximation by a natural greedy algorithm. We then present two approximation algorithms for the case where there are two scenarios. When $p_1 = p_2 = \frac{1}{2}$, our first algorithm achieves a $\frac{11}{9}$-approximation. When $p_1 \ne p_2$, we show that this problem admits a $\frac{5}{4}$-approximation. For the general case of this problem, we find a 2.314-approximation by a natural greedy algorithm.

Keywords: Approximation Algorithms · Matching Theory · Combinatorial Optimization

1 Introduction

In this paper, we consider a stochastic variant of the matching problem, which we call k-scenario matching. Given is a graph $G = (V, E)$, a maximum number of tests B, and a set of scenarios S. Each scenario $s \in S$ occurs with probability p_s (with $\sum_{s \in S} p_s = 1$) and corresponds to an edge set $E_s \subseteq E$. The goal is to select a set of edges $\hat{E} \subseteq E$, with $|\hat{E}| \leq B$, such that $v(\hat{E}) = \sum_{s \in S} p_s |M(\hat{E} \cap E_s)|$ is maximized, with $M(E)$ being a maximum size matching on the edge set E. Figure 1 shows an example.

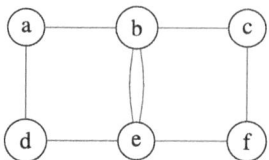

Fig. 1. An instance of k-scenario matching. $|S| = 2, p_1 = p_2 = 0.5$. E_1 is depicted with red edges and E_2 with blue. $B = 3$. For each scenario, there is an optimal matching of size 3, with 6 distinct edges in total. Due to the limited budget, the optimal solution is $\hat{E}^* = \{ad, be, cf\}$, selecting an edge contained in the maximum matching of neither scenario. The expected maximum matching size $\sum_{i \in S} p_s |M(\hat{E}^* \cap E_i)| = 2$

This problem is motivated by kidney exchange, where a matching between patient-donors pairs must be found, but the compatibility of patients and donors is uncertain. Transplants are only allowed if an additional test specific to the patient-donor combination confirms compatibility. If recipients are revealed to be incompatible with donors with particular genetic characteristics, then edges to all such donors are removed. Moreover, due to time and budgetary constraints, there is a limit to the number of these additional tests that can be made [5,6,10].

One can view this as a k-scenario matching instance where the scenarios capture the correlation of compatibility between recipients and groups of genetically similar donors, and the edges we select for our matchings correspond to the tests we make, limited by our budget. Other possible applications are labour markets, where evaluation of candidates is time consuming [4]. We discuss two streams of literature related to our problem.

First, Stochastic Matching problems, which have received significant attention, are closely related in their applications. However, this existing literature assumes independent edge realizations, as opposed to the given set of scenarios in our model. These scenarios allow for correlation in edge realizations, which may be relevant due to underlying mechanisms. For example, incompatibility in kidney exchange has a genetic component, and thus there is a positive correlation for edges between a single patient and multiple genetically similar donors. We now give a brief overview of the literature on Stochastic Matching. [1–3,5] study a matching problem where edge realizations are independent and there is no global budget. Their goal is to pick $\hat{E}$ such that $\mathbb{E}(M(\hat{E})) \geq (1 - \epsilon)M(E)$, and $\hat{E}$ has limited maximum degree. Maehara and Yamaguchi [12] generalize this to packing problems with random variables in the objective. For matching problems they guarantee $\mathbb{E}(M(\hat{E})) \geq (\frac{1}{2} - \epsilon)M(E)$, but with a global limit on the total number of tests. Smeulders et al. [15] use integer programming techniques to maximize the expectation of $M(\hat{E})$ given a budget B. While they assume independent realizations of the edges, they generate a set of scenarios that are included in the integer program, which is solved using Benders' decomposition.

A second related stream of literature is on the Flexibility Design Problem (FDP), first described by [11]. As in our k-Scenario Matching problem, a set of edges, subject to a budget constraint, must be selected such that the expected value of a lower-level problem is optimized. While for FDP this lower-level problem is a bipartite transportation problem, we consider matching on general graphs. A series of heuristics have been proposed for this problem [7,8,14,16,17]. Recently, [9] provided the first worst-case guarantees and proposed a LP-based approximation algorithm achieving a $\frac{2e^2}{e^2-1}$ approximation.

1.1 Our Results

We provide several approximations for the k-Scenario Matching Problem. For our first algorithm, provided in Sect. 2, gives an approximation for the general k-Scenario Matching problem. Interestingly, the approximation ratio we find for this problem does not depend on k, the number of scenarios, but is a function of the budget B, which converges to a constant.

Theorem 1. *There is a polynomial time $\frac{2}{1-(\frac{B-2}{B})^B} \leq \frac{2e^2}{e^2-1} \approx 2.314$ - approximation algorithm for the k-Scenario Matching problem.*

The second approximation we show in Sect. 3 is for the case when $k = 2$ and the scenarios have equal probability.

Theorem 2. *There is a polynomial time $\frac{11}{9}$-approximation algorithm for the 2-Scenario Matching problem when $p_1 = p_2 = \frac{1}{2}$.*

In Sect. 4 we provide an approximation algorithm for 2-Scenario Matching when the instances do not have equal probability.

Theorem 3. *There is a polynomial time $\frac{5}{4}$-approximation algorithm for the 2-Scenario Matching problem.*

To complement these approximations, we also provide the following result on hardness, which is proven in the full version of this paper.

Theorem 4. *The k-Scenario Matching problem is NP-hard even when $k = 2$, $p_1 = p_2 = \frac{1}{2}$, and the input graph G is bipartite and sub-cubic.*

1.2 Preliminaries

Definition 1. MAXIMUM w-WEIGHTED B-BUDGETED MATCHING PROBLEM.
Given: *A graph $G = (V, E)$ and weight function $w \colon E \to \mathbb{R}$ and a non-negative integer budget B.*
Goal: *Find a subset $E' \subseteq E$, with $|E'| \leq B$, such that E' is a matching and $\sum_{e \in E'} w(e)$ is maximized.*

For a set of edges E, we denote by $M_B^w(E)$ a maximum w-weighted B-budgeted matching on E. In case of uniform weights on the edges, we omit the w-superscript. The following lemma is a consequence of Theorem 1 of [13]. We provide a proof of this claim in the full version of this paper. Note that this lemma will be so ubiquitous throughout this paper that we will omit explicit reference to it,

Lemma 1 ([13]). *Let B be a nonnegative integer, $G = (V, E)$ with n vertices, and $w \colon E \to \mathbb{R}$ be a weight function. The maximum w-weighted B-budgeted matching problem can be solved in polynomial time.*

The following lemma describes a set of instances in which we can find an optimal solution in polynomial time. We provide a proof in the full version of this paper.

Lemma 2. *Let $M_1 := M_B(E_1)$, $M_2 := M_B(E_2)$ and $M_p := M_B(E_1 \cap E_2)$. Suppose scenarios G_1 and G_2 have probabilities $p_2 \leq p_1$. If at least one of the following conditions hold, then we can verify that condition, and find an optimal solution for the 2-scenario matching in polynomial time:*

1. *$|M_1| + |M_2| \leq B$;*
2. *$B \in O(\frac{1}{\varepsilon})$, for some constant $\varepsilon > 0$;*
3. *$|M_p| \geq B$, and ;*
4. *$p_1 < 1 - \delta$ for some constant $\delta > 0$ and $\nu(OPT) < \frac{1}{\varepsilon}$, for some optimal solution OPT, and for some constant $\varepsilon > 0$.*

2 Constant Approximation for k-Scenario Matching

In this section, we will prove Theorem 1, and see a constant sized approximation for the k-Scenario Matching Problem. We are given k scenarios $E_1, \ldots, E_k$, where each scenario has probability $p_i \in [0, 1]$ and $\sum_{i=1}^{k} p_i = 1$.

We briefly describe our approximation algorithm for this problem, Algorithm 1. We greedily select edges, based on the sum of probabilities of the scenarios for which the edge is present. To account for previously selected edges, we keep track of "relevant" scenarios for each edge. When an edge is selected, we remove all scenarios relevant for that selected edge from the relevant scenarios of all of its adjacent edges. We will initially define a function $w_0 : E \to \mathbb{R}$, where $w_0(e)$ is the sum of scenario probabilities that contain edge e ($w_0(e) = \sum_{s \in S : e \in E_s} p_s$). We iteratively add edges until we have a maximal solution, where in each iteration, we select an edge $f_i \in \operatorname{argmax}\{w_i(f_i) | f_i \in E \setminus \{f_1, \ldots, f_{i-1}\}\}$, breaking ties arbitrarily. To account for previously selected edges in $w_i(e)$, we define sets $E_{s,i}$, with $E_{s,0} = E_s$. If $f_i \in E_{s,i}$, we remove f_i and each edge adjacent to f_i from $E_{s,i}$ to construct $E_{s,i+1}$. Then, we define $w_{i+1}(f') := w_i(f') - \sum_{\{j | f, f' \in E_j\}} p_j$. We repeat the argument until our solution has B many edges. We present this algorithm in full detail in Algorithm 1.

Define $F_0 = \emptyset$ and F_i as the set of edges $\{f_0, \ldots, f_{i-1}\}$. We now consider instances I_i for each iteration, with scenario edge sets $E_{s,i} \subseteq E_S$ as defined in

Approximation Algorithms for k-Scenario Matching   93

Algorithm 1: Constant Approximation (outline)

---
1 *Input:* $G = (V, E)$, edge sets $E_s \subseteq E, \forall s \in S$, budget B
2 **if** $B < 3$, *or* $|E| \leq B$ **then**
3 $\quad$ Return optimal solution by enumeration
4 Fix arbitrary means of breaking ties.
5 $F \leftarrow \emptyset$
6 For all $s \in S$, $E_{s,0} = E_s$
7 **for** $i = 0, \ldots, B-1$ **do**
8 $\quad$ $w_i(f) \leftarrow \sum_{\{s | f \in E_{s,i}\}} p_s$,
9 $\quad$ $f_i \leftarrow \text{argmax}\{w_i(f) | f \in E\}$
10 $\quad$ $F \leftarrow F \cup f_i$
11 $\quad$ $E \leftarrow E \setminus \{f_i\}$
12 $\quad$ **for** *each* $s \in S$ **do**
13 $\quad\quad$ $E_{s,i+1} \leftarrow E_{s,i} \setminus \{f_i\}$
14 $\quad\quad$ **if** $f_i \in E_{s,i}$ **then**
15 $\quad\quad\quad$ **for** *each* $f' \in E_{s,i}$, *that shares an endpoint with* f_i **do**
16 $\quad\quad\quad\quad$ $E_{s,i+1} \leftarrow E_{s,i+1} \setminus \{f'\}$

17 **Return** F

---

the algorithm. The budget in I_i is equal to B. We denote by OPT_i the optimal solution of the instance I_i and let $\nu_i(OPT_i)$ denote its value. Note that I_0 is the original instance, as such OPT_0 is the optimal solution overall. Denote by r_i the maximum value of $w_i(f)$ in iteration i.

Claim. $r_i \geq \frac{\nu_i(OPT_i)}{B}$

Proof. Let f_i be the edge selected in iteration i. By definition r_i is the value $w(f_i)$ takes in iteration i, and by definition we see $r_i = w_i(f_i) = \max_{f' \in E} w_i(f') \geq \max_{f' \in OPT_i} w_i(f') \geq \frac{1}{B} \sum_{f' \in OPT_i} w_i(f') = \frac{\nu_i(OPT_i)}{B}$.

Claim. $\nu_i(OPT_i) \geq \nu_{i-1}(OPT_{i-1}) - 2r_{i-1} \geq \nu_0(OPT_0) - \sum_{j=0}^{i-1} 2r_j$.

Proof. First, note that by construction of the sets $E_{s,i}$, we have $M(E' \cap E_{s,i}) = M(E' \cap E_{s,i-1})$ if $f_{i-1} \notin E_{s,i-1}$ and $M(E' \cap E_{s,i}) \geq M(E' \cap E_{s,i-1}) - 2$ if $f_{i-1} \in E_{s,i}$ This is true, since $E_{s,i} = E_{s,i-1}$ if $f_{i-1} \notin E_{s,i}$, and otherwise only f_{i-1} and it's adjacent edges are removed, so the size of the matching can reduce by at most two. Since OPT_{i-1} is a feasible solution for I_i, we have.

$$\nu_i(OPT_i) \geq \sum_{s \in S: f_{i-1} \notin E_{s,i-1}} p_s M(OPT_{i-1} \cap E_{s,i}) + \sum_{s \in S: f_{i-1} \in E_{s,i-1}} p_s M(OPT_{i-1} \cap E_{s,i})$$

$$\geq \sum_{s \in S: f_{i-1} \notin E_{s,i-1}} p_s M(OPT_{i-1} \cap E_{s,i-1}) + \sum_{s \in S: f_{i-1} \in E_{s,i-1}} p_s (M(OPT_{i-1} \cap E_{s,i-1}) - 2)$$

$$= \nu_{i-1}(OPT_{i-1}) - \sum_{s \in S: f_{i-1} \in E_{s,i-1}} 2p_s = \nu_{i-1}(OPT_{i-1}) - 2r_{i-1}.$$

To see the second inequality, we apply induction. For $i = 0$, the claim is clear. Assume the claim holds for i. To see it holds for $i + 1$, we apply the inductive hypothesis

$$\nu_{i+1}(OPT_{i+1}) \geq \nu_i(OPT_i) - 2r_i \geq \nu_0(OPT_0) - 2\sum_{j=0}^{i-1} r_j - 2r_i = \nu_0(OPT_0) - 2\sum_{j=0}^{i} r_j,$$

and the claim holds.

We now have the necessary ingredients to prove Theorem 1.

Proof (Proof of Theorem 1). The for loop on line 7 runs B many times, with the subloops at line 12 and line 15 running $|S|$ and at most $|E|$ times respectively. Thus, the algorithm terminates in polynomial time. By construction, we have $\nu_0(ALG) \geq \sum_{i=0}^{B-1} r_i$. We wish to show that $\sum_{j=0}^{i} r_j \geq \frac{\nu_0(OPT_0)}{B} \sum_{j=0}^{i} \left(\frac{B-2}{B}\right)^j$. We prove this claim by induction. Assume the case holds for i, we show the case for $i + 1$.

$$\sum_{j=0}^{i+1} r_j = \sum_{j=0}^{i} r_j + r_{i+1} \geq \sum_{j=0}^{i} r_j + \frac{\nu_{i+1}(OPT_{i+1})}{B} \geq \sum_{j=0}^{i} r_j + \frac{\nu_0(OPT_0) - \sum_{j=0}^{i} 2r_j}{B}$$

$$= \frac{B-2}{B} \sum_{j=0}^{i} r_j + \frac{\nu_0(OPT_0)}{B} \geq \frac{(B-2)\nu_0(OPT_0)}{B^2} \sum_{j=0}^{i} \left(\frac{B-2}{B}\right)^j + \frac{\nu_0(OPT_0)}{B}$$

$$= \frac{\nu_0(OPT_0)}{B} \left(\sum_{j=0}^{i} \left(\frac{B-2}{B}\right)^{j+1} + 1\right) = \frac{\nu_0(OPT_0)}{B} \sum_{j=0}^{i} \left(\frac{B-2}{B}\right)^j.$$

The first inequality above follows from Claim 17, the second inequality follows from Claim 17, and finally, the third inequality follows by applying the induction hypothesis. Applying this inequality, we can see

$$\nu(ALG) \geq \sum_{i=0}^{B-1} r_i \geq \frac{\nu(OPT)}{B} \sum_{j=0}^{B-1} \left(\frac{B-2}{B}\right)^j = \frac{\nu(OPT)}{B} \times \left(\frac{1 - \left(\frac{B-2}{B}\right)^B}{1 - \frac{B-2}{B}}\right)$$

$$= \nu(OPT) \left(\frac{1 - \left(\frac{B-2}{B}\right)^B}{2}\right) \geq \nu(OPT)\left(\frac{1}{2} - \frac{1}{2e^2}\right).$$

Where the last inequality above follows since $\left(\frac{B-2}{B}\right)^B$ is an increasing function in B for $B > 0$, and we have $\left(\frac{B-2}{B}\right)^B = \left(1 - \frac{2}{B}\right)^B \leq \left(e^{\frac{-2}{B}}\right)^B = \frac{1}{e^2}$, since $1 + x \leq e^x$ for every $x \in \mathbb{R}$.

3 Constant Approximation for Uniform 2-Scenario Matching

In this section, we consider the case with 2 scenarios, where each scenario has an equal probability ($p_1 = p_2 = 0.5$). Since each scenario has equal probability, we therefore abuse notation slightly in this section and consider the unweighted objective $\nu(\hat{E}) := 2\sum_{i=1}^{2} \frac{1}{2}M_B(\hat{E} \cap E_i)$ for simplicity (observe that in this instance, $\nu(\hat{E})$ is simply twice the usual cost). Throughout this section, we consider a fixed optimal solution $OPT \subseteq E$, and fixed matchings $M_i^* = M_B(OPT \cap E_i)$. We define OPT_1 $(OPT_2) := M_1^* \backslash M_2^*$ $(M_2^* \backslash M_1^*)$, i.e. tested edges that are only in the maximum matching in a single scenario. We define $OPT_3 := M_1^* \cap M_2^*$. It is not hard to see that $\nu(OPT) = 2|OPT_3| + |OPT_1| + |OPT_2|$. Using Lemma 2, we assume that any given instance has $|M_B(E_1)| + |M_B(E_2)| > B$, $B \notin O(1)$, and $|M_B(E_1 \cap E_2)| < B$.

We now describe two natural algorithms which will be essential in finding the desired $\frac{11}{9}$-approximation algorithm for Uniform 2-Scenario Matching. The third, and most complex, algorithm (Algorithm 4) is introduced in Sect. 3.1. Algorithm 2 computes a weighted matching, A_i, on the edge set of scenario E_i, with double the weight for edges appearing in both.

Algorithm 2: Weighted Budget for Scenario i.

1 *Input:* $G = (V, E)$, edge sets $E_i \subseteq E, i \in \{1, 2\}$, budget B
2 Define weights $w^i : E_i \to \{1, 2\}$, $w^i(e) = 2$ for $e \in E_1 \cap E_2$ and $w^i(e) = 1$ otherwise
3 Compute $A_i := M_B^{w^i}(E_i)$
4 **Return** (A_i)

Lemma 3. *For $i = 1, 2$, let A_i be the matching returned by Algorithm 2. We have $\nu(A_i) \geq |OPT_i| + 2|OPT_3|$*

Proof. $OPT_i \cup OPT_3$ is a feasible, B-Budgeted matching on G_i, therefore $|OPT_i| + 2|OPT_3| = \nu(OPT_i \cup OPT_3) \leq \nu(A_i)$.

Algorithm 3: Budget Based matching.

1 *Input:* $G = (V, E)$, edge sets $E_i \subseteq E, i \in \{1, 2\}$, budget B
2 Compute $M_3 := M_B(E_1)$
3 Compute $N_3 := M_{B-|M_3|}(E_2)$
4 **Return** (A_3)

Algorithm 3 computes a maximum matching for the edge set of each scenario, first a from one scenario with budget B, then from the other scenario with the remaining budget.

Lemma 4. *We have $\nu(A_3) \geq |OPT_1| + |OPT_2| + |OPT_3|$*

Proof. We have that $\nu(A_3) = |M_B(A_3 \cap E_1)| + |M_B(A_3 \cap E_2)| \geq |M_3| + |N_3| = B \geq |OPT_1| + |OPT_2| + |OPT_3|$. We remark that the second equality follows as we assume condition 1 of Lemma 2 does not hold.

3.1 11/9-Approximation

In this section, we find an improved approximation for the case with 2 scenarios with equal probability. In principle, the philosophy behind Algorithm 3 is to buy edges from maximum matchings with the goal of simply spending our budget. In this section we will show that one can take a more careful approach while following this philosophy in order to find a set of algorithms, the best of which finds the desired approximation factor.

Before we state our algorithm, we need the following useful lemma, which we prove in the full version of this paper. We note here that the proof of this lemma is non-trivial, requiring a reduction to a novel combinatorial problem, and may be of independent interest.

Lemma 5. *Consider matchings $M_1 := M_B(E_1)$, and $M_2 := M_B(E_2)$ (these sets might have non-empty intersection) and M be any other matching. Given a number $i \leq |M|$ we can find in polynomial time a size i subset $F_i \subseteq M$ such that there are x_1^i edges in M_1 and x_2^i edges in M_2 that F_i do not share an endpoint with, where:*

$$|M_1| - x_1^i + |M_2| - x_2^i \leq \max\{3i+1, \frac{|M_1|+|M_2|}{|M|}i + 1\},$$

and for every $0 \leq i \leq |M|-1$, we have: $x_1^i + x_2^i \geq x_1^{i+1} + x_2^{i+1} \geq x_1^i + x_2^i - 4$.

With Lemma 5, we are ready to state our improving algorithm. In summary, we apply Lemma 5 on M_1, M_2, and $M_p = M(E_1 \cap E_2)$ for each $i = 0, \ldots, |M_p|$, computing x_1^i, x_2^i and subset $F_i \subseteq M_p$. If there is no value of i for which $i + x_1^i + x_2^i < B$, then in particular $|M_p| + x_1^{|M_P|} + x_2^{|M_P|} \geq B$, so we buy M_p for our solution and a subset of $M_1^{|M_P|} \cup M_2^{|M_P|}$ of size $(B - |M_p|)$ (Note that by Lemma 2, condition 3, we assume $|M_p| < B$). We will show in Lemma 6 that this solution is enough to find a $\frac{6}{5}$-approximation. If there is some value of i such that $i + x_1^i + x_2^i < B$, then we pick the minimum value of i such that $i + x_1^i + x_2^i < B$ and return $(F_i \cup M_1^i \cup M_2^i)$.

Recall, we consider a fixed optimal solution $OPT \subseteq E$, and fixed matchings $M_i^* = M_B(OPT \cap E_i)$. We define OPT_1 (OPT_2) $:= M_1^* \backslash M_2^*$ ($M_2^* \backslash M_1^*$) and $OPT_3 := M_1^* \cap M_2^*$. Denote by A^* the solution provided by Algorithm 4. Recall that by employing Algorithm 2 for $i = 1, 2$ we find solutions A_1 and A_2, and by employing Algorithm 3, we find solution A_3. The following lemma, proven in the full version of this paper, will allow us to take the best solution from Algorithm 2, 3, and 4 to find a $\frac{6}{5}$-approximation.

Lemma 6. *If the condition on line 7 of Algorithm 4 holds, then $\max\{A_1, A_2, A_3, A^*\} \geq \frac{5}{6}\nu(OPT)$.*

With Lemma 6, we assume that the condition on line 7 of Algorithm 4 does not hold. That is, we assume that there exists an $i \in \{0, \ldots, B\}$ such that $i + x_1^i + x_2^i < B$. Going forward, we will let i denote the minimum value of i

Algorithm 4: Improved Budget algorithm for 2-Scenario Matching.

1 *Input:* $G = (V, E)$, edge sets $E_i \subseteq E, i \in \{1, 2\}$, budget B
2 Compute $M_1 := M_B(E_1)$, $M_2 := M_B(E_2)$, and $M_p := M(E_1 \cap E_2)$
3 **for** $i = 0, \ldots, |M_p|$ **do**
4   Apply Lemma 5 with i and M_1 and M_2 to find F_i
5   Define x_1^i and x_2^i as in the statement of Lemma 5
6   Let M_1^i, and M_2^i be subsets of M_1 and M_2 that do not share an endpoint with F_i
7 **if** $|M_p| + x_1^{|M_p|} + x_2^{|M_p|} \geq B$ **then**
8   Take a subset Q of $M_1^{|M_p|} \cup M_2^{|M_p|}$ size $B - |M_p|$
9   **Return** $(M_p \cup Q)$
10 **else**
11   Pick minimum i such that $i + x_1^i + x_2^i < B$
12   **Return** $(F_i \cup M_1^i \cup M_2^i)$

such that $i + x_1^i + x_2^i < B$ and for simplicity we define $x_i := x_1^i + x_2^i$. With these definitions, we can immediately observe the following claim regarding the size of the output.

Claim. Let $(F \cup M_1^i \cup M_2^i)$ be the solution computed by Algorithm 4. Then we have $\nu(A^*) := \nu(F_i \cup M_1^i \cup M_2^i) = 2i + x_i$

Consider $M_1 = M_B(E_1)$, $M_2 = M_B(E_2)$, and $M_p = M_B(E_1 \cap E_2)$ as in Algorithm 4. We define $\alpha := 4 - \frac{|M_1| + |M_2|}{|M_p|}$. We define $\gamma \in \mathbb{R}$ such that $(1 + \gamma)\nu(OPT) = |M_1| + |M_2|$. Obviously, $\gamma \geq 0$ or else we have that $|M_1| + |M_2| < \nu(OPT)$, which is a contradiction.

Lemma 7. *The following inequalities hold*

1. $(1 + \gamma)\nu(OPT) \leq \max\{(4 - \alpha)i, 3i\} + 1 + x_i;$
2. $B - 3 \leq i + x_i.$

Proof. **1.** Follows from the definitions of α and γ. That is, $(1 + \gamma)\nu(OPT) = |M_1| + |M_2|$. Now using Lemma 5 we have $|M_1| + |M_2| - x_i \leq \max\{(4 - \alpha), 3\}i + 1$ and hence the claim.

2. As $x_1^0 + x_2^0 = |M_1| + |M_2|$ and we assume $|M_1| + |M_2| \leq B$ does not hold (Lemma 2) then $x_1^0 + x_2^0 > B$. This implies that $i \neq 0$. Furthermore as i is the smallest index such that $i + x_1^i + x_2^i < B$ then $i - 1 + x_1^{i-1} + x_2^{i-1} \geq B$. Therefore, using Lemma 5 we have $i + x_1^i + x_2^i \geq i + x_1^{i-1} + x_2^{i-1} - 4 \geq B - 3$.

Let A_i be the output of Algorithm 2 for $i \in \{1, 2\}$. Now as the output of the final algorithm we select the solution among A_1, A_2 and A^* with the maximum revenue. Note that by Lemma 3, we have $\max\{\nu(A_1), \nu(A_2)\} \geq 2|OPT_3| + \frac{1}{2}(|OPT_1| + |OPT_2|)$. For notational simplicity, we denote by $|OPT_{12}| := |OPT_1| + |OPT_2|$.

Recall, we assume by Lemma 2 that $\nu(OPT) > \frac{1}{\varepsilon}$, and to simplify some notation, we define $\varepsilon' = \frac{49}{33}\varepsilon$. Thus, $\nu(OPT) > \frac{49}{33\varepsilon'}$. We can prove the following lemma which will provide us our main approximation factor for this section. Its proof can be found in the full version of this paper.

Lemma 8. $\max\{\nu(A_1), \nu(A_2), \nu(A^*)\} \geq \frac{9\nu(OPT)}{11} - \varepsilon'$.

We now have the ingredients necessary to prove Theorem 2.

Proof (Proof of Theorem 2). Let $M_1 = M_B(E_1)$, $M_2 = M_B(E_2)$ and $M_3 = M_B(E_1 \cap E_2)$. We first apply Lemma 2 to find an optimal solution if any of the following holds: (1) $|M_1| + |M_2| \leq B$; (2) $B \in O(\frac{1}{\varepsilon})$, for some constant $\varepsilon > 0$; (3) $|M_3| \geq B$, and; (4) $p_1 < 1 - \delta$ for some constant $\delta > 0$ and $\nu(OPT) < \frac{1}{\varepsilon}$, for some optimal solution OPT, and for some constant $\varepsilon > 0$. We assume that none of the above hold.

Denote by A_i the output of Algorithm 2 for $i = 1, 2$, denote by A_3 the output of Algorithm 3, and denote the solution returned by Algorithm 4 by A^*. If the condition of line 7 in Algorithm 4 holds, then by Lemma 6, we can achieve a $\frac{6}{5}$-approximation in polynomial time using $\max\{\nu(A^*), \nu(A_1), \nu(A_3)\}$. If the condition of line 7 in Algorithm 4 does not hold, then by Lemma 8, $\max\{\nu(A^*), \nu(A_1), \nu(A_2)\}$ is a $\frac{11}{9} + \varepsilon$-approximation.

4 Constant Approximation for Non-uniform 2-Scenario Matching

In this section, we consider the case with 2 scenarios, where scenario G_1 has probability $p_1 = r \in [0.5, 1]$, and scenario G_2 has probability $p_2 = 1 - p_1$. We will provide a $\frac{5}{4}$-approximation algorithm for this case. Recall that Sect. 3.1 is the case when $r = 0.5$, thus, the algorithm provided in this section provides a weaker approximation than that in Sect. 3.1. Indeed, one can generalize this algorithm to achieve a good approximation when r is very close to $\frac{1}{2}$. For further details on this special case, see the full version of this paper.

Recall, we consider a fixed optimal solution $OPT \subseteq E$, and fixed matchings $M_i^* = M_B(OPT \cap E_i)$. We define OPT_1 (OPT_2) := $M_1^* \setminus M_2^*$ ($M_2^* \setminus M_1^*$) and $OPT_3 := M_1^* \cap M_2^*$. We have $\nu(OPT) = \nu(OPT_1) + \nu(OPT_2) + \nu(OPT_3) = r|OPT_1| + (1-r)|OPT_2| + |OPT_3|$.

In this section, the algorithm we present is dependent on the value of r. For $r \in [\frac{2}{3}, 1]$, we apply Algorithms 2 and 3, noting that these algorithms do not rely on scenario probabilities, and show in the analysis that the best among these provides a $\frac{5}{4} - \varepsilon$-approximation. For $r \in [\frac{1}{2}, \frac{2}{3}]$, we apply an algorithm that is similar to the one found in Sect. 3.1, where we apply Algorithm 2 for $i = 1, 2$, and then a version of Algorithm 4 that does not utilize Lemma 5.

The following lemmas show bounds on the solution revenue when we apply Algorithms 2 and 3 to compute solutions for 2-scenario matching when the scenario probability is non-uniform.

Lemma 9. *Algorithm 2 finds a solution A_i such that $\nu(A_i) \geq \nu(OPT_3) + \nu(OPT_i)$ when run on scenario i, for $i = 1, 2$.*

Proof. Let $OPT_i := M(OPT \cap E_i)$ for $i \in \{1,2\}$ and $OPT_3 := (OPT \cap E_1 \cap E_2)$. We have $|OPT_i \cup OPT_3| \leq B$ and $\nu(OPT_i \cup OPT_3) = \nu(OPT_i) + \nu(OPT_3)$. Therefore $OPT_i \cup OPT_3$ is a feasible solution such $OPT_i \cup OPT_3 \subseteq E_i$. Thus for the solution S obtained by algorithm $\nu(S) \geq \nu(OPT_i) + \nu(OPT_3)$

The following lemma is proven in the full version of this paper.

Lemma 10. *Consider an instance of weighted 2-scenario matching. Algorithm 3 finds a solution A_3 with $\nu(A_3) \geq \nu(OPT_1) + \nu(OPT_2) + r|OPT_3| = \nu(OPT_1) + \nu(OPT_2) + r \cdot \nu(OPT_3)$ in polynomial time.*

Let us start with the case where $r \geq \frac{2}{3}$. Using Lemmas 9 and 10 we will find the previously referred to $\frac{5}{4}$-approximation. Denote by A_i for $i = 1,2$ the solution found by applying Algorithm 2, and by Lemma 9 we see that $\nu(A_i) \geq \nu(OPT_i) + \nu(OPT_3)$. Denote by A_3 the solution found by applying Algorithm 3, and by Lemma 10 we see that $\nu(A_3) \geq \nu(OPT_1) + \nu(OPT_2) + r \cdot \nu(OPT_3)$.

Lemma 11. *If $r \in [\frac{2}{3}, 1]$, then $\max\{\nu(A_1), \nu(A_2), \nu(A_3)\} \geq \frac{4}{5}\nu(OPT)$.*

Proof. $\max\{\nu(A_1), \nu(A_2), \nu(A_3)\}$ is at least

$$\geq \frac{1}{3-2r}\nu(A_3) + \frac{1 - \frac{1}{3-2r}}{2}\nu(A_1) + \frac{1 - \frac{1}{3-2r}}{2}\nu(A_2)$$

$$\geq \frac{2-r}{3-2r}\left(\nu(OPT_1) + \nu(OPT_2) + \nu(OPT_3)\right)$$

$$= \frac{2-r}{3-2r}\nu(OPT) = \left(\frac{1}{2} + \frac{1}{6-4r}\right)\nu(OPT) \geq \frac{4}{5}\nu(OPT).$$

With Lemma 11, we only need to consider the case where $r \in [\frac{1}{2}, \frac{2}{3}]$. As discussed previously, the core of this case is to provide an algorithm that is similar in spirit to Algorithm 4. We summarize Algorithm 5 here. We begin by computing $M_1 := M_B(E_1 \setminus E_2)$ and $M_2 := M_B(E_2 \setminus E_1)$, and $M_p := M_B(E_1 \cap E_2)$. We begin with $M_1 \cup M_2$ as our solution. We then add edges of M_p to our solution, one by one, removing edges of $M_1 \cup M_2$ that are adjacent to the added edges. We repeat this process until either some iteration i where the number of remaining edges from $M_1 \cup M_2$ plus the edges added from M_p are less than B, and return that as our solution. We also handle the case where this does not happen, and show that we can easily find a $\frac{6}{5}$-approximation.

We note that this algorithm differs from Algorithm 4 in the construction of sets F_i. While the analysis in Sect. 3.1 requires the specific construction of F_i as described in Lemma 5, this is not used in the analysis here. Therefore, we explicitly include the possibility of constructing F_i through the addition of edges in an arbitrary order.

Define by A^* solution provided by Algorithm 5. The following lemma is proven in the full version of this paper, whose proof is completely analogous to the proof of Lemma 6.

Algorithm 5: Computing Solution for Weighted 2-Scenario.

1 *Input:* $G = (V, E)$, edge sets $E_i \subseteq E, i \in \{1, 2\}$, budget B
2 Compute $M_1 := M_B(E_1 \setminus E_2)$, $M_2 := M_B(E_2 \setminus E_1)$, and $M_p := M_B(E_1 \cap E_2)$
3 Initialize $N_1^0 \leftarrow M_1$, $N_2^0 \leftarrow M_2$, $F_0 = \emptyset$, $n_1^0 = |M_r|$, $n_2^0 = |M_b|$
4 Fix an arbitrary ordering on edges of M_P as $f_1, f_2, \ldots$.
5 for $i = 1, \ldots, |M_p|$ do
6 $\quad F_i \leftarrow F_{i-1} \cup \{f_i\}$
7 $\quad$ Let N_1^i and N_2^i be edges of N_1^{i-1} and N_2^{i-1} that do not share an endpoint with f_i
8 $\quad n_1^i \leftarrow |N_1^i|$
9 $\quad n_2^i \leftarrow |N_2^i|$
10 if $|M_P| + n_1^{|M_P|} + n_2^{|M_P|} \geq B$ then
11 $\quad$ Let $Q \subseteq N_1^{|M_P|} \cup N_2^{|M_P|}$ be a subset of size $B - |M_P|$
12 $\quad$ **Return** $|M_P| \cup Q$
13 else
14 $\quad$ Pick minimum i such that $n_1^i + n_2^i + i < B$
15 $\quad$ **Return** $(F_i \cup N_1^i \cup N_2^i)$

Lemma 12. *If the condition on line 10 of Algorithm 5 holds, then* $\max\{A_1, A_2, A_3, A^*\} \geq \frac{5}{6}\nu(OPT)$.

With Lemma 12, it remains to show that we can find a $\frac{5}{4}$-approximation when the condition on line 10 of Algorithm 5 does not hold. Therefore, we will assume that there exists an $i \in \{1, \ldots, |M_P|\}$ such that $i + n_1^i + n_2^i < B$. We find $i \in \{1, \ldots, |M_P|\}$ to be the smallest such value satisfying $i + n_1^i + n_2^i < B$.

If $i = 1$ then as we are assuming that the first condition on Lemma 2 does not hold, then we have $n_1^0 + n_2^0 = |M_B| + |M_R| > B$. Else if $i \geq 2$ then $i - 1 + n_1^{i-1} + n_2^{i-1} \geq B$. Furthermore, each time we add an edge from M_p, we remove at most 2 edges from M_1 (and 2 from M_2), since M_1 (M_2) is a matching. Therefore, $n_1^{i-1} \leq n_1^i + 2$ and $n_2^{i-1} \leq n_2^i + 2$, we can see that $B - 3 \leq i - 1 + n_1^{i-1} + n_2^{i-1} - 3 \leq i + n_1^i + n_2^i < B$.

To simplify notation, let $n_1 := n_1^i$, $n_2 := n_2^i$, and $x_i = rn_1 + (1 - r)n_2$. Let (F_i, N_1^i, N_2^i) be the solution computed by Algorithm 5 (assuming condition on line 10 does not hold), therefore we have $A^* = F_i \cup N_1^i \cup N_2^i$. With these definitions, it is not hard to see the following claim.

Claim. $\nu(A^*) = x_i + i$.

The algorithm for this section is to compute Algorithm 2 for $i = 1, 2$, which we denote by A_1 and A_2, and compute A^* using Algorithm 5. We then return the solution that maximizes $\nu(A_1), \nu(A_2)$, and $\nu(A^*)$. The goal of the next lemma is to show that if OPT_1 is very small in comparison to OPT, then we can achieve a $\frac{4}{5}$-approximation with Algorithm 2 on scenario 2.

Lemma 13. *If* $|OPT_1| \leq \frac{|OPT|}{5r}$, *then* $\nu(A_2) \geq \frac{4}{5}\nu(OPT)$.

Proof. By definition we have $\nu(OPT_1) \leq r\frac{\nu(OPT)}{5r} = \frac{\nu(OPT)}{5}$. Therefore, by Lemma 9, we see that Algorithm 2 returns a solution with revenue at least $\nu(OPT_2) + \nu(OPT_3) \geq \nu(OPT)\left(1 - \frac{1}{5}\right) = \frac{4}{5}\nu(OPT)$.

Therefore, it remains to find an approximation when $|OPT_1| > \frac{\nu(OPT)}{5r}$, which we assume without loss of generality. The following lemma is proven in the full version of this paper.

Lemma 14. $x_i + i \geq \nu(OPT)\left(1 - \frac{3r+1}{15r^2}\right) + B\frac{1-r}{3r} - 3(1-r)$.

Note that $\nu(OPT) = \nu(OPT_1) + \nu(OPT_2) + \nu(OPT_3)$, and $B \geq |OPT_1| + |OPT_2| + |OPT_3| = \frac{\nu(OPT_1)}{r} + \frac{\nu(OPT_2)}{1-r} + \nu(OPT_3)$. Thus, we have $\nu(A^*) + 3(1-r) = x_i + i + 3(1-r)$ is at least

$$\geq \nu(OPT_1)\left(1 - \frac{3r+1}{15r^2} + \frac{1-r}{3r^2}\right) + \nu(OPT_2)\left(1 - \frac{3r+1}{15r^2} + \frac{1-r}{3r(1-r)}\right)$$
$$+ \nu(OPT_3)\left(1 - \frac{3r+1}{15r^2} + \frac{1-r}{3r}\right)$$
$$= \nu(OPT_1)\left(1 - \frac{3r+1}{15r^2} + \frac{1-r}{3r^2}\right) + \nu(OPT_2)\left(1 - \frac{3r+1}{15r^2} + \frac{1}{3r}\right) + \nu(OPT_3)\left(\frac{2}{3} + \frac{2r-1}{15r^2}\right)$$
$$= \nu(OPT_1)\left(1 + (-4)\frac{2r-1}{15r^2}\right) + \nu(OPT_2)\left(1 + \frac{2r-1}{15r^2}\right) + \nu(OPT_3)\left(\frac{2}{3} + \frac{2r-1}{15r^2}\right).$$

The following useful claim is proven in the full version of this paper.

Claim. For $\frac{1}{2} \leq r \leq \frac{2}{3}$, we have: $0 \leq \frac{2r-1}{15r^2} \leq \frac{1}{20}$

We are now ready to show a bound on $\max\{\nu(A_1), \nu(A_2), \nu(A^*)\}$ using Lemma 14.

Lemma 15. *We have:*

$$\max\{\nu(A_1), \nu(A_2), \nu(A^*)\} \geq \frac{4}{5}\nu(OPT) - \frac{3}{10}.$$

Proof. For notational simplicity, we denote $x := \frac{2r-1}{15r^2}$. By Claim 15, we have, $0 \leq x \leq \frac{1}{20}$. Thus, $\frac{3}{5(1-3x)}$, $1 - \frac{4}{5(1-3x)}$, $\frac{1}{5(1-3x)}$, and $1 - 4x$ are all positive. Moreover, we have $\frac{3}{5(1-3x)} + 1 - \frac{4}{5(1-3x)} + \frac{1}{5(1-3x)} = 1$. Hence:

$$\max\{\nu(A_1), \nu(A_2), \nu(A^*)\} \geq \frac{\nu(A_1)}{5(1-3x)} + \frac{\nu(A_2)(1-15x)}{5(1-3x)} + \frac{3\nu(A^*)}{5(1-3x)}.$$

By applying Lemma 9, and Lemma 14 we have that $\max\{\nu(A_1), \nu(A_2), \nu(A^*)\}$ is at least

$$\geq \frac{\nu(OPT_1) + \nu(OPT_3)}{5(1-3x)} + \frac{5(1-3x) - 4}{5(1-3x)}(\nu(OPT_2) + \nu(OPT_3))$$
$$+ \frac{3(\nu(OPT_1)(1-4x) + \nu(OPT_2)(1+x) + \nu(OPT_3)(\frac{2}{3}+x)) - 3(1-r)}{5(1-3x)}$$
$$= \frac{4}{5}(\nu(OPT_1) + \nu(OPT_2) + \nu(OPT_3)) - \frac{3(1-r)}{5(1-3x)}$$
$$= \frac{4}{5}\nu(OPT) - \frac{3(1-r)}{5(1-3x)} > \frac{4}{5}\nu(OPT) - \frac{3}{10}.$$

Proof (Proof of Theorem 3). Let $M_1 = M_B(E_1)$, $M_2 = M_B(E_2)$ and $M_3 = M_B(E_1 \cap E_2)$. We first apply Lemma 2 to find an optimal solution if any of the following holds: (1) $|M_1| + |M_2| \leq B$; (2) $B \in O(\frac{1}{\varepsilon})$, for some constant $\varepsilon > 0$; (3) $|M_p| \geq B$, and ; (4) $p_1 < 1 - \delta$ for some constant $\delta > 0$ and $\nu(OPT) < \frac{1}{\varepsilon}$, for some optimal solution OPT, and for some constant $\varepsilon > 0$. We assume that none of these cases hold. Denote by A_i the output of Algorithm 2 for $i = 1, 2$, and denote by A_3 the output of Algorithm 3. Denote the solution returned by Algorithm 5 by A^*, If the condition of line 10 in Algorithm 5 holds, then by Lemma 12, we can achieve a $\frac{6}{5}$-approximation in polynomial time using A_1, A_3, and A^*. Therefore, we assume that the condition of line 10 in Algorithm 5 does not hold.

Recall we fix an optimal solution OPT, and defined OPT_i $i = 1, 2$ as the edges of scenario E_i that provide revenue only from scenario E_i, and OPT_3 as the edges that provide revenue from both scenarios. By Lemma 13, if $|OPT_1| \leq \frac{|OPT|}{5r}$, then $\nu(A_2)$ is a $\frac{5}{4}$-approximation algorithm.

So we assume that $|OPT_1| > \frac{|OPT|}{5r}$. By Lemma 15, the solution A^*, A_1, or A_2 that maximizes $\max\{\nu(A^*), \nu(A_1), \nu(A_2)\}$ is a $\frac{5}{4} + \varepsilon$-approximation, since we assume that $\nu(OPT) > \frac{1}{\varepsilon}$.

References

1. Assadi, S., Khanna, S., Li, Y.: The stochastic matching problem with (very) few queries. In: Proceedings of the 2016 ACM Conference on Economics and Computation, pp. 43–60. ACM (2016)
2. Assadi, S., Khanna, S., Li, Y.: The stochastic matching problem: beating half with a non-adaptive algorithm. In: Proceedings of the 2017 ACM Conference on Economics and Computation, pp. 99–116 (2017)
3. Behnezhad, S., Derakhshan, M., Hajiaghayi, M.T.: Stochastic matching with few queries: (1-ε) approximation. In: Proceedings of the 52nd Annual ACM SIGACT Symposium on Theory of Computing, pp. 1111–1124 (2020)
4. Behnezhad, S., Reyhani, N.: Almost optimal stochastic weighted matching with few queries. In: Proceedings of the 2018 ACM Conference on Economics and Computation, pp. 235–249 (2018)
5. Blum, A., Dickerson, J.P., Haghtalab, N., Procaccia, A.D., Sandholm, T., Sharma, A.: Ignorance is almost bliss: near-optimal stochastic matching with few queries. In: Proceedings of the Sixteenth ACM Conference on Economics and Computation, pp. 325–342 (2015)
6. Carvalho, M., Klimentova, X., Glorie, K., Viana, A., Constantino, M.: Robust models for the kidney exchange problem. INFORMS J. Comput. **33**(3), 861–881 (2021). https://doi.org/10.1287/ijoc.2020.0986
7. Chan, T.C., Letourneau, D., Potter, B.G.: Sparse flexible design: a machine learning approach. Flex. Serv. Manuf. J. **34**(4), 1066–1116 (2022)
8. Chou, M.C., Chua, G.A., Teo, C.P., Zheng, H.: Process flexibility revisited: the graph expander and its applications. Oper. Res. **59**(5), 1090–1105 (2011)
9. DeValve, L., Pekeč, S., Wei, Y.: Approximate submodularity in network design problems. Oper. Res. **71**(4), 1021–1039 (2023)

10. Dickerson, J.P., Procaccia, A.D., Sandholm, T.: Failure-aware kidney exchange. In: Proceedings of the Fourteenth ACM Conference on Electronic Commerce, pp. 323–340 (2013)
11. Jordan, W.C., Graves, S.C.: Principles on the benefits of manufacturing process flexibility. Manage. Sci. **41**(4), 577–594 (1995)
12. Maehara, T., Yamaguchi, Y.: Stochastic packing integer programs with few queries. Math. Program. **182**, 141–174 (2020)
13. Plesnik, J.: Constrained weighted matchings and edge coverings in graphs. Discret. Appl. Math. **92**(2–3), 229–241 (1999)
14. Simchi-Levi, D., Wei, Y.: Worst-case analysis of process flexibility designs. Oper. Res. **63**(1), 166–185 (2015)
15. Smeulders, B., Bartier, V., Crama, Y., Spieksma, F.C.: Recourse in kidney exchange programs. INFORMS J. Comput. **34**(2), 1191–1206 (2022)
16. Wang, S., Wang, X., Zhang, J.: Robust optimization approach to process flexibility designs with contribution margin differentials. Manuf. Serv. Oper. Manage. **24**(1), 632–646 (2022)
17. Yan, Z., Gao, S.Y., Teo, C.P.: On the design of sparse but efficient structures in operations. Manage. Sci. **64**(7), 3421–3445 (2018)

Tight Approximation Bounds on a Simple Algorithm for Minimum Average Search Time in Trees

Svein Høgemo[✉]

University of Bergen, Bergen, Norway
svein.hogemo@uib.no

Abstract. The graph invariant EPT-sum has cropped up in several unrelated fields in later years: As an objective function for hierarchical clustering, as a more fine-grained version of the classical edge ranking problem, and, specifically when the input is a vertex-weighted tree, as a measure of average/expected search length in a partially ordered set. The EPT-sum of a graph G is defined as the minimum sum of the depth of every leaf in an edge partition tree (EPT), a rooted tree where leaves correspond to vertices in G and internal nodes correspond to edges in G. A simple algorithm that approximates EPT-sum on trees is given by recursively choosing the most balanced edge in the input tree G to build an EPT of G. Due to its fast runtime, this balanced cut algorithm can be used in practice, and has earlier been analysed to give a 1.62-approximation on trees. In this paper, we show that the balanced cut algorithm gives a 1.5-approximation of EPT-sum on trees, which amounts to a tight analysis and answers a question posed by Cicalese et al. in 2014.

1 Introduction

Searching in ordered structures is a basic problem in computer science and has seen a lot of attention since the dawn of the field [11,19]. Some special cases are well understood, e.g. the best search strategy in a totally ordered set is a binary search tree, and the problem of optimal search in totally ordered sets given a probability distribution on the elements was treated by Knuth [14] and Hu and Tucker [13] in 1971. Also, if the search space is every single subset of elements, then it is known that the optimum search strategy is given by a Huffman tree (see [13]). Apart from such specific subcases, little is known except that they are hard problems [7,18].

Searching for an element x in a *partially ordered set* $(A, \preceq)$ can be achieved by a series of queries of the type "is $x \preceq a$ (for some element a)?", and restricting the next query to the subset $\{a' \in A \mid a' \preceq a\}$ (or $\{a' \in A \mid a' \npreceq a\}$, depending on the answer to the query). The search finishes when the eligible subset of A contains only one element, which must be x.

A full version is available on arxiv: https://arxiv.org/abs/2402.05560.

© The Author(s), under exclusive license to Springer Nature Switzerland AG 2025
M. Bieńkowski and M. Englert (Eds.): WAOA 2024, LNCS 15269, pp. 104–118, 2025.
https://doi.org/10.1007/978-3-031-81396-2_8

An interesting subclass of partially ordered sets are "tree-like posets", those posets whose Hasse diagrams are rooted trees, or in other words those posets where for any two elements a, b, the sets $\{x \mid x \preceq a\}$ and $\{x \mid x \preceq b\}$ either are disjoint, or one is contained in the other. They generalize totally ordered sets (whose Hasse diagrams are paths), but are still much easier to work with than the general case of posets. The average performance of search strategies for tree-like posets have been studied by Laber and Molinaro in [15] and later by Cicalese et al. in [6,7]. For these posets, the optimum search strategy with respect to average search time can be shown to be equal to the *EPT-sum* of their Hasse diagrams, a graph invariant recently named in [12], but defined and used several times earlier in other circumstances.

1.1 EPT's and EPT-Sum

An *edge partition tree*, or EPT for short, of a graph G is a rooted tree where every leaf corresponds to a vertex in G, the root corresponds to an edge e in G, and each child of the root is itself the root of an EPT of a component of $G \setminus e$. The EPT-sum of G with respect to some EPT T is defined as the sum of the depth of each leaf in T; and the EPT-sum of G is defined as the minimum over all EPT's. Viewing the EPT as a certain search strategy in the graph G where, upon each edge e, one can query which of the components of $G \setminus e$ contains the wanted vertex, the EPT-sum of G with respect to T is the combined search time of locating every vertex in G with this search strategy. Giving a probability distribution to $V(G)$, the weighted EPT-sum is equal to the expected search time for a vertex pulled at random from the given distribution. When G is a tree, this search strategy is equivalent to the usual search strategy on the tree-like poset that has G as its Hasse diagram. The aforementioned edge rank problem, on the other hand, is interpreted as finding a search strategy with optimal *worst-case* performance on a tree-like poset.

Given a tree G, we are interested in the following algorithm (the *balanced cut algorithm*) to make an approximately optimal EPT of G (with respect to EPT-sum):

- Find a balanced cut in G, i.e. an edge e that minimizes the size of the biggest component in $G \setminus e$.
- Make e the root of T.
- Make EPT's of the components of $G \setminus e$ recursively, and make the roots of those trees children of e.

The balanced cut algorithm is attractive because of its runtime: Finding a balanced edge in a tree takes $O(n)$ time, so making a balanced EPT can trivially be done in $O(n^2)$ time. However, this runtime can be reduced to $O(n \log n)$, as shown in Theorem 1. This makes it practical for any potential applications of searching in trees.

For the case of vertex-weighted trees, Cicalese et al. [6,7] show a lower and upper bound on the performance of the balanced cut algorithm of 1.5 and φ, respectively (where φ is the golden ratio $\frac{\sqrt{5}+1}{2}$, approximately equal to 1.62). They ask what the actual performance of this algorithm is, and conjecture that it does provide a 1.5-approximation, matching the lower bound.

In this paper we will affirm this conjecture, by proving that the balanced cut algorithm indeed does give a 1.5-approximation of EPT-sum on vertex-weighted trees. Our proof proceeds by constructing a so-called *augmented tree*, built from an optimum EPT by subdividing half of its edges, and showing that it has EPT-sum at most 1.5 times the optimum. Thereafter we iteratively apply operations to this tree that are guaranteed to not increase the EPT-sum, eventually arriving at the EPT constructed by the balanced cut-algorithm. We believe that there might be some interest in the proof technique used here. Specifically, the trick of building an augmented tree that has cost 1.5 times the optimum has, to our knowledge, not been used before.

1.2 Applications of EPT-Sum to Hierarchical Clustering

Employing edge-weighted graphs, a measure equal to EPT-sum was considered by Dasgupta in [10] as a suitable objective function for measuring the quality of hierarchical clusterings of similarity graphs (where large weights mean high similarity and vice versa). In this paper and a string of follow-ups [5,9,20], several attractive properties of EPT-sum as an objective function were highlighted, and approximation algorithms for EPT-sum were found. In light of this, a balanced edge of a tree is just a special case of the *sparsest cut* of an edge-weighted graph, a partition (A, B) of $V(G)$ that minimizes the ratio $(\sum_{e \in E(G[A,B])} x_e)/(|A| \cdot |B|)$ (where x_e is the weight of the edge e).

Charikar and Chatziafratis show that the balanced cut algorithm gives an 8-approximation of the EPT-sum of edge-weighted trees (and other graph classes for which an optimal sparsest cut can be found in polynomial time, like planar graphs, see [1] for more information). Whether this is the real approximation ratio is, however, unknown.

1.3 Related Parameters

There are also several relevant results on the related parameter of *VPT-sum*; the VPT-sum of a graph G with respect to a *vertex partition tree* (*VPT*) T is the sum of depth plus one of every node in the VPT. Optimizing VPT-sum is, on trees, equivalent to optimizing a different search model where one has access to an oracle which, for each vertex $v \in V(G)$, answers which of the components of $G \setminus v$ (or, potentially, $\{v\}$ itself) contains the vertex one is searching for. For this parameter, the corresponding strategy of recursively choosing the most balanced vertex (i.e. the *centroid*) to build a VPT, was recently shown by Berendsohn et al. [3] to give exactly a 2-approximation of VPT-sum – both for weighted and unweighted trees. In addition, there exists a PTAS for VPT-sum on trees [4].

EPT-sum and the related parameters that are treated in [12] are all NP-hard to calculate exactly on general, unweighted graphs. In contrast, on unweighted trees, the complexity of EPT-sum and VPT-sum are unknown. For vertex-weighted trees, the complexity of calculating VPT-sum is also, to the best of our knowledge, unknown, but it is NP-hard to calculate EPT-sum [6]. One should note that in the NP-hardness reductions employed in [6], the trees have exponential weights. In [12], an equivalence between EPT-sum on vertex-weighted trees and unweighted trees was demonstrated. So, if EPT-sum should turn out to be NP-hard on trees where every vertex has polynomial weight, then it is also NP-hard on unweighted trees.

1.4 Organization

The rest of the paper is organized as follows: in Sect. 2 we give basic definitions. We also show that there is an algorithm for making balanced EPT's that run in $O(n \log n)$ time. In Sect. 3 we prove the main result of this paper, namely the fact that a balanced EPT has EPT-sum at most 1.5 the optimum. Finally, in Sect. 4 we restate some of the most relevant results and open problems regarding computing and approximating EPT-sum on trees.

2 Preliminaries

2.1 Basic Notions

In general, we follow the notation established in [12]. The vertex set and edge set of a graph G is denoted $V(G)$ and $E(G)$, respectively. Given vertices $S \subseteq V(G)$, $G[S]$ signifies the induced subgraph of G on S. As both the input graphs and the output data structures are trees, we employ the following conventions to avoid confusion: An unrooted tree that is given as input is denoted G, while a rooted tree generated as output is denoted T, and its vertices are called *nodes*. For a node $x \in V(T)$, $T[x]$ signifies the rooted subtree of T rooted in x. The set of *leaves* in T is denoted $L(T)$, a node which is not a leaf is called an *internal node*. A *full binary tree* is a binary tree where every internal node has exactly two children.

The *depth* of a node x in a rooted tree T is equal to the distance from the root r to x in T. For example, r itself has depth 0. The *height* of T is the maximum depth out of any node in T. The nodes on the path from r to x (including both r and x) are the *ancestors* of x, and x is a *descendant* of all these nodes.

2.2 Edge Partition Trees

Given a connected graph G, an *edge partition tree*, or EPT for short, of G is a rooted tree with $|V(G)|$ leaves and $|E(G)|$ internal nodes, that can be defined inductively as follows:

– The EPT of a graph with a single vertex is a tree with a single node.

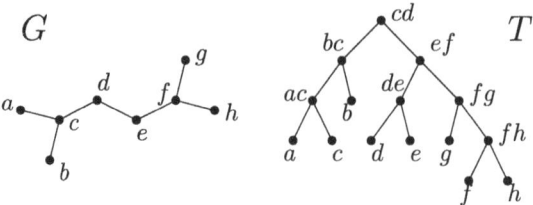

Fig. 1. An unweighted tree G and an EPT T of G. Adding up the depth of each leaf, one sees that EPT-sum$(G, T) = 25$. This is optimal for G; however, T is not balanced. Any EPT that has the most balanced edge de as root has an EPT-sum of 26, which is suboptimal.

– For a bigger graph G, the root r of any EPT T corresponds to an edge $e \in E(G)$, and r has at most two children c_1, c_2; one for each component of $G \setminus e$. $T[c_1]$ (and $T[c_2]$, if it exists) are, in turn, EPT's of the component(s) of $G \setminus e$ (Fig. 1).

An EPT is, as such, a binary tree. A graph G is a tree if and only if any (and every) EPT of G is a full binary tree (since in that case, every edge of G is a bridge). Since we focus on trees as input, we assume that EPT's are full binary trees.

There is a bijection from the vertex set of G to the leaves of T, and likewise from the edge set of G to the internal vertices of T. To simplify notation somewhat, we will not notate these bijections and rather say that each leaf of T *is* a vertex of G and each internal node of T *is* an edge of G. For an internal node $v \in T$, we define $G_T[v]$ as a shorthand for $G[L(T[v])]$, the (connected) subgraph of G induced by the rooted subtree under v.

2.3 EPT-Sum and Searching in Trees

The focus of this paper is on the graph measure called EPT-sum. We give a definition where the input graph G is equipped with vertex weights, but not edge weights:

Given a graph G with vertex weights $W = \{w_v \mid v \in V(G)\}$, and an EPT T of G, we define the EPT-sum of G with respect to T as follows:

Definition 1 (EPT-sum). EPT-sum(G, T) *is equal to*

$$\sum_{e \in E(G)} \sum_{v \in L(T[e])} w_v$$

The EPT-sum of G, EPT-sum(G) *is the minimum of* EPT-sum(G, T) *over all EPTs T.*

Definition 2 (EPT-sum, alternative def.). *We can easily verify through reordering of terms that EPT-sum can be equivalently defined as follows:*

$$\sum_{v \in V(G)} w_v \cdot |\text{ancestors of } v \text{ in } T|$$

It is easy to see that the number of ancestors of a node $v \in T$ is equal to its *depth*, i.e. its distance to the root. The formula $\sum_{v \in V(G)} (w_v \cdot \text{dist}_T(r, v))$, where r is the root of T, is therefore a natural choice to use. It is also this formulation that makes EPT-sum and similar measures attractive for measuring the performance of search trees.

We must note that the equivalence between EPT-sum and average search time only holds if the Hasse diagram actually is a rooted tree; otherwise these are two different problems. This is evident from the fact that the sparsest cut algorithm from [5] provides a $O(\sqrt{\log n})$-time approximation of EPT-sum in edge-weighted graphs, while it was proven in [6] that searching in posets (i.e. vertex-weighted DAG's) has no $o(\log n)$-approximation unless every problem in NP admits a quasipolynomial algorithm (see also [16]).

2.4 A Fast Balanced Cut Algorithm

Before moving on to the main result, we show that finding a balanced EPT can be done more quickly than the basic algorithm outlined in the introduction. This may well have been noted beforehand; however we have been unable to find such a result in the literature.

Theorem 1. *Given a tree G, one can compute a balanced EPT of G in time $O(n \log n)$.*

The proof is to be found in the full version. This result shows that the balanced cut algorithm is a practical way of generating an approximately optimal EPT of a tree.

3 Balanced EPT's Have EPT-Sum at Most 1.5 the Optimum

We will go through the proof with unweighted graphs in mind, but note that the proof is agnostic as to whether the vertices have weights (see Corollary 1). The proof therefore also works for vertex-weighted trees.

The first step is building the tree that has EPT-sum at most 1.5 times as high as the optimal tree.

Definition 3 (Augmented tree). *Given a full binary tree T, the* augmented tree *of T, denoted $\text{aug}(T)$, is constructed in the following manner: For any internal node $v \in V(T)$ with children c_l, c_r, choose one child (say, c_r) with the property that $|L(T[c_r])| \le |L(T[c_l])|$ (this is obviously true for at least one of c_l, c_r). Then, subdivide the edge vc_r once.*

If T is an EPT of a tree G, then $aug(T)$ is not an EPT of G, since it has more internal nodes than there are edges in G. Nonetheless, we define EPT-sum($aug(T)$) to be the sum of depths of leaves in $aug(T)$.

Lemma 1. *For any full binary tree T, $\text{EPT-sum}(aug(T)) \leq \frac{3}{2}\text{EPT-sum}(T)$.*

Proof. The proof goes by induction. For the base case, we observe that a full binary tree with one leaf has EPT-sum equal to 0, and no edges to augment.

For the inductive step, assume that the lemma holds for any full binary tree with at most $n - 1 \geq 1$ leaves, and let T be an arbitrary full binary tree with n leaves, and $aug(T)$ the augmented tree of T. Furthermore, let r be the root of T, and c_l, c_r the children of r. Note that c_l and c_r must exist, by the assumption that T is a full binary tree with at least two leaves. Also, note that $aug(T)[c_l]$ (resp. $aug(T)[c_r]$) is the augmented tree $aug(T[c_l])$ (resp. $aug(T[c_r])$).

W.l.o.g. assume that c_r is the child of r such that the edge rc_r is subdivided in $aug(T)$. Then we know that $|L(T[c_r])| \leq \frac{n}{2}$. Also, both $T[c_l]$ and $T[c_r]$ have at most $n - 1$ leaves.

By Definition 1,

$$\text{EPT-sum}(T) = \text{EPT-sum}(T[c_l]) + \text{EPT-sum}(T[c_r]) + n,$$

while

$$\text{EPT-sum}(aug(T)) = \text{EPT-sum}(aug(T[c_l])) + \text{EPT-sum}(aug(T[c_r])) + n + |L(T[c_r])|,$$

which we have seen to be upper-bounded by

$$\text{EPT-sum}(aug(T[c_l])) + \text{EPT-sum}(aug(T[c_r])) + \frac{3}{2}n.$$

Therefore

$$\frac{\text{EPT-sum}(aug(T))}{\text{EPT-sum}(T)} \leq \frac{\text{EPT-sum}(aug(T[c_l])) + \text{EPT-sum}(aug(T[c_r])) + \frac{3}{2}n}{\text{EPT-sum}(T[c_l]) + \text{EPT-sum}(T[c_r]) + n}.$$

The right hand side fraction is a mediant of the three fractions $\frac{\text{EPT-sum}(aug(T[c_l]))}{\text{EPT-sum}(T[c_l])}$, $\frac{\text{EPT-sum}(aug(T[c_r]))}{\text{EPT-sum}(T[c_r])}$ and $\frac{3}{2}$. By the induction hypothesis, all these fractions are upper-bounded by $\frac{3}{2}$, from which we conclude that $\frac{\text{EPT-sum}(aug(T))}{\text{EPT-sum}(T)} \leq \frac{3}{2}$.

One should note that this Lemma also holds for EPT's of vertex-weighted trees; this is easily seen by replacing $\frac{n}{2}$ with $\frac{\sum_{v \in V(G)} w_v}{2}$ in the formula.

We introduce one additional notion that will come up in the proof of Theorem 2:

Definition 4 (Splitting). *Given a tree G and an EPT T of G, and an edge $uv \in E(G)$, the splitting of T along uv is a pair of rooted trees T^u, T^v, which are EPT's of the components of $G \setminus uv$, G_u and G_v respectively. T^u and T^v are defined as follows: $L(T^u)$ and $L(T^v)$ are equal to $V(G_u)$ and $V(G_v)$ respectively, and for any node $x \in V(T^u)$ (resp. $V(T^v)$), its parent is equal to the lowest ancestor of x in T whose corresponding edge lies within G_u (resp. G_v). If x has no such ancestor, then it becomes the root of T^u (resp. T^v).*

Remark 1. T^u and T^v are, indeed, EPT's of G_u and G_v.

The proof can be found in the full version.

Lemma 2. *Let G be a tree and T an EPT of G, and let T^u, T^v be a splitting of T along an edge $uv \in E(G)$. Then,* $\mathsf{EPT}\text{-sum}(G_u, T^u) + \mathsf{EPT}\text{-sum}(G_v, T^v) < \mathsf{EPT}\text{-sum}(G, T)$.

Proof. From Definition 1, we see that it is enough to prove that for any internal node x in, say, T^u, $L(T_x^u) \leq L(T_x)$. Due to the way T^u is constructed, every leaf in T_x^u is also a leaf in T_x, and the claim immediately follows. The strict inequality stems from the fact that $L(T_{uv}) > 0$, but uv is not present in either T^u or T^v.

Theorem 2. *The balanced cut-algorithm gives a 1.5-approximation of EPT-sum on trees.*

Proof. Let G be an unrooted tree, with T^* an optimal EPT of G, and T' an EPT of G given by the balanced cut-algorithm. We want to show that $\mathsf{EPT}\text{-sum}(T') \leq \mathsf{EPT}\text{-sum}(aug(T^*))$. We achieve this through a procedure that incrementally modifies $aug(T^*)$ into T'.

During the procedure, we generate a series of trees $T_0, T_1, \ldots, T_t$ where $T_0 = aug(T^*)$, $T_t = T'$ and, for every $0 \leq i < t$, T_{i+1} is made from T_i by way of the following algorithm:

- If $T_i = T'$: set $t := i$ and halt.
- Otherwise: select an internal node $v \in V(T_i)$ that has a subdivided edge to one of its children, but where none of its ancestors have such an edge.
- Modify $T_i[v]$ in a suitable way, according to Cases 1–7 below (depending on where the balanced edge of $G_{T_i}[v]$ is located in $T_i[v]$).
- Set T_{i+1} equal to the modified tree.

We refer to an internal node v in T_i as *correctly placed* when (i) v corresponds to a most balanced edge in $G_{T_i}[v]$, and (ii) every ancestor of v is also correctly placed. Clearly, in an EPT output by the balanced cut algorithm, every node is correctly placed, and vice versa. For every $0 \leq i < t$, the number of correctly placed nodes in T_{i+1} is at least as high as in T_i. When a node has been placed correctly in the EPT, the subdivided edge from it to one of its children is smoothed out; therefore it is gradually modified from an augmented EPT to a non-augmented EPT. The node v that is chosen in each step is a node of minimum depth that is not guaranteed to be correctly placed, since we remove the subdivision node underneath each correctly placed node. Therefore, after replacing v with a node corresponding to a balanced edge, the replacement node is now correctly placed, increasing the number of correctly placed nodes.

The proof hinges on being able to show that the following invariant holds for every $0 \leq i \leq t$:

$$\mathsf{EPT}\text{-sum}(G, T_i) \leq \mathsf{EPT}\text{-sum}(G, aug(T^*))$$

The invariant trivially holds for $i = 0$. The rest of the proof goes by induction; assuming that the invariant holds for some i and showing that it then also must hold for $i + 1$. Given the definition

$$\mathsf{EPT\text{-}sum}(G, T) = \sum_{e \in E(G)} \left(\sum_{v \in L(T[e])} w_v \right)$$

we can see that if the modification from T_i to T_{i+1} is constrained to the rooted subtree $T_i[v]$, then

$$\mathsf{EPT\text{-}sum}(G, T_{i+1}) - \mathsf{EPT\text{-}sum}(G, T_i) = \mathsf{EPT\text{-}sum}(G, T_{i+1}[v]) - \mathsf{EPT\text{-}sum}(G, T_i[v])$$

(this is what Dasgupta [10] refers to as *modularity of cost*).

How to actually modify T_i depends on where the balanced edge is located, and is subject to a lengthy analysis of seven cases, which constitutes the rest of the proof. Each case concentrates on a subtree $T_i[v]$ where v was selected in the manner described above. For visual aid, we include a picture of each case and how that particular tree is modified. We believe that explaining the modification in each case in words does not add any explanatory power over the figures themselves – specifically, the pictures replace sentences like "we smooth out these and these edges, perform such and such rotations on the tree, and subdivide those and those edges". Furthermore, the inferred change in EPT-sum for each case is found by counting up the depth of each leaf in the trees shown in the figures. We will therefore restrict a thorough explanation to Cases 1 and 2.

We will employ the following conventions throughout the case analysis: The balanced edge in question is denoted e, and its corresponding node in $T_i[v]$ is located in the subtree on the left hand side of the root (except in Case 1). The leftmost and rightmost subsets of leaves (corresponding to subsets of $V(G)$) are called A and B, respectively, and the root is called v, its left child l, and l's right child is called b (except in Cases 1–2). Lastly, $A \cup B$ never forms a connected subgraph of G (again, except in Case 1, where $A \cup B$ is the whole of $L(T_i[v])$). Note that since e lies in the left subtree, the right subtree (containing B) must be strictly smaller, and can therefore safely be assumed to be the one having a subdivided edge from the root in $T_i[v]$.

Given these conventions, one should take care to get convinced that in each case, all rooted subtrees (both in the original tree and the modified one) induce connected subgraphs of G; this is important, as it is necessary in order for the arguments to hold water.

In the figures, pink dots signify subdivision nodes in the augmented tree. A pink line between two edges signifies that one of those edges is subdivided (whichever leads to the subtree with the fewest leaves), but we do not know which.

Case 1: $v = e$ (Fig. 2).

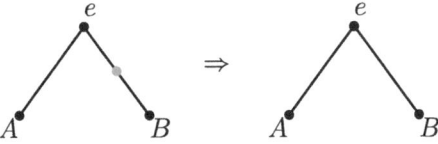

Fig. 2. Case 1.

This is certainly the easiest case. Since e is already correctly placed, the only modification we do is smoothing the subdivided edge from the root. It is trivial to show that $\mathsf{EPT\text{-}sum}(G, T_{i+1}) \leq \mathsf{EPT\text{-}sum}(G, T_i)$ and the invariant holds.

Case 2: $l = e$ (Fig. 3).

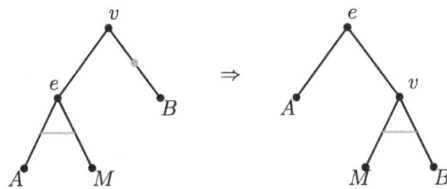

Fig. 3. Case 2.

Both here and in the following cases, we need to remember that A is defined to be separated from B by M, therefore $G[B \cup M]$ is connected; as previously mentioned, this is necessary in order for the modified tree to be an EPT of $G_{T_i}[v]$.

In this case, we perform a binary tree rotation on $T_i[v]$ around v and e, and smooth the subdivided edge from v to B, decreasing the EPT-sum by $|B|$ (but the act of rotating around v and e simultaneously increases EPT-sum by $|B|$, so these cancel each other out). We further smooth the edge from e to the smallest subtree of A and M (decreasing EPT-sum by $\min(|A|, |M|)$, and all leaves in A decrease one in depth, decreasing the EPT-sum by $|A|$ (due to the rotation around v and e), but we subdivide the edge from v to the smallest subtree out of M and B, increasing the EPT-sum with $\min(|B|, |M|)$.

Taking all this into account, the change in EPT-sum therefore simplifies to

$$\min(|B|, |M|) -$$
$$(|A| + \min(|A|, |M|))$$

where all increase is collected on the top line and decrease on the bottom line. We observe that the inequality $|A| \geq |B|$ must hold, as otherwise v would lead to a more balanced cut than e. It is a basic observation that we use throughout the proof, that if $a \leq b$ and $c \leq d$, then $\min(a, c) \leq \min(b, d)$. In this case, we get $\min(|B|, |M|) \leq |B| \leq |A|$. Therefore the change in EPT-sum is non-positive and $\mathsf{EPT\text{-}sum}(G, T_{i+1}) \leq \mathsf{EPT\text{-}sum}(G, T_i)$; thus the invariant holds.

Case 3: e lies inside $G[A]$ (Fig. 4).

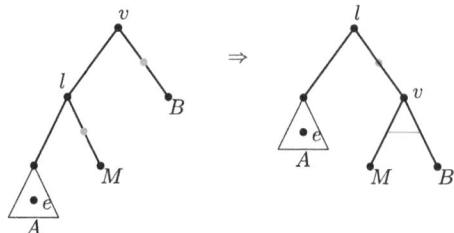

Fig. 4. Case 3.

This is a unique case in the sense that e does not become correctly placed in $T_i[v]$ at this step. Therefore it is important to keep a subdivided edge from the new root l in this case, so that we can balance $T_{i+1}[v]$ at a later step, before moving to its subtrees.

We observe that the inequality $|A| \geq |M| + |B|$ must hold, otherwise l would lead to a more balanced cut than e (this is also why we can be sure that the edge from l to M is subdivided). The change in EPT-sum is

$$(|B| + \min(|M|, |B|)) -$$

$$|A|.$$

From the aforementioned inequality, the change in EPT-sum is non-positive and therefore $\mathsf{EPT\text{-}sum}(G, T_{i+1}) \leq \mathsf{EPT\text{-}sum}(G, T_i)$; thus the invariant holds.

In the next three cases (4–6), we assume that A and B lie on different sides of e, i.e. that there are connected subgraphs $G[L], G[R]$ where $L \cup R = M$ such that the components of $G_{T_i}[v] \setminus e$ are $G[A \cup L]$ and $G[B \cup R]$. In the last case, Case 7, we assume the opposite; that A and B are in the same component of $G \setminus e$.

Case 4: e is the root of the tree containing $L \cup R$ (Fig. 5).

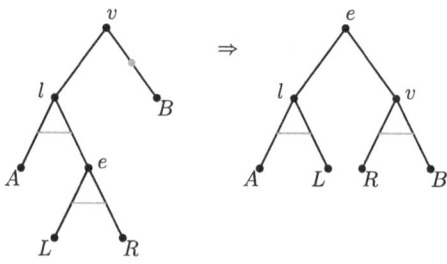

Fig. 5. Case 4.

Here, the change in EPT-sum is

$$(\min(|A|, |L|) + \min(|R|, |B|)) -$$

$$(\min(|A|, |L \cup R|) + \min(|L|, |R|) + |L| + |R|).$$

As the sum of the top row is clearly at most $|L| + |R|$, the change in EPT-sum is non-positive and $\mathsf{EPT\text{-}sum}(G, T_{i+1}) \leq \mathsf{EPT\text{-}sum}(G, T_i)$; thus the invariant holds.

Case 5: One of the children of b (containing e) contains R (Fig. 6).

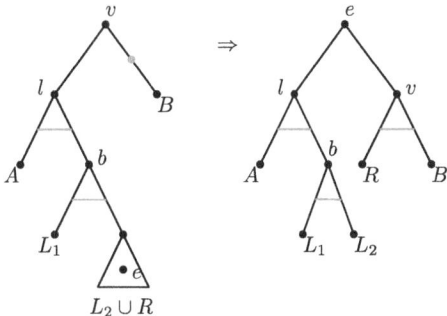

Fig. 6. Case 5.

Both here and in the next case, we look at the splitting of the subtree containing e (in this case $T_i[L_2 \cup R]$). By Lemma 2, it is clear that, also when augmented, the two subtrees $T_{i+1}[L_2]$ and $T_{i+1}[R]$ in sum have lower EPT-sum than $T_i[L_2 \cup R]$ has. Let us argue in more detail: By Lemma 2, the EPT's $T'_{i+1}[L_2]$ and $T'_{i+1}[R]$ (made by smoothing out all the subdivided edges in $T_{i+1}[L_2]$ and $T_{i+1}[R]$) have, in sum, lower EPT-sum than the EPT $T'_i[L_2 \cup R]$ (made by smoothing out the subdivided edges in $T_i[L_2 \cup R]$) has. Subdividing the edges again, we see that every subdivided edge from a node x to its child in (say) $T_{i+1}[L_2]$ corresponds to a subdivided edge from x to its child in $T_i[L_2 \cup R]$. Therefore augmenting $T_i[L_2 \cup R]$ must increase EPT-sum at least as much as augmenting $T_{i+1}[L_2]$ and $T_{i+1}[R]$. So, if we cancel out $\mathsf{EPT\text{-}sum}(T_{i+1}[L_2]) + \mathsf{EPT\text{-}sum}(T_{i+1}[R]) - \mathsf{EPT\text{-}sum}(T_i[L_2 \cup R])$, we can infer that the change in EPT-sum is at most

$$(\min(|A|, |L|) + \min(|L_1|, |L_2|) + \min(|R|, |B|)) -$$

$$(\min(|A|, |L \cup R|) + \min(|L_1|, |L_2 \cup R|) + |R|)$$

where we assume $L = L_1 \cup L_2$. We see that each term in the top row is at most as high as the corresponding term in the bottom row. As such, the change in EPT-sum is non-positive and $\mathsf{EPT\text{-}sum}(G, T_{i+1}) \leq \mathsf{EPT\text{-}sum}(G, T_i)$; thus the invariant holds.

Case 6: One of the children of b (containing e) contains L.

Case 7: e lies within $L \cup R$ and A and B are in the same component of $G \setminus e$.

The proofs of Cases 6 and 7 can be found in the full version.

With the assumptions we put on the orientation of $T_i[v]$, all possible placements of the balanced edge e are covered by one of the Cases 1–7. The procedure must halt after a finite amount of modifications (specifically, $O(n^2)$ modifications), because each internal node (corresponding to an edge in G, of which there are $n-1$) is correctly placed in accordance with T' in $O(n)$ modifications. Specifically, for each subtree, Case 3 may be encountered $O(n)$ times (observe that e has smaller depth in T_{i+1} than in T_i when Case 3 is encountered), while any other case may be encountered maximally one time. From this fact and the fact that the invariant holds, we conclude that the theorem is true.

Corollary 1. *Theorem 2 holds also for vertex-weighted trees.*

Proof. This follows directly from replacing sizes of subsets of vertices ($|A|$, $|B|$, $|M|$ etc.) with their weighted equivalents ($\sum_{a \in A} w_a$, $\sum_{b \in B} w_b$ etc.). All the inequalities used in the proof must still hold, since a balanced edge in a vertex-weighted tree is defined to be balanced with respect to the weights.

4 Conclusion

We have shown that the simple balanced cut algorithm gives a 1.5-approximation of EPT-sum on vertex-weighted trees, and therefore of the minimum average search time in trees.

There are some questions remaining. For the case of weighted trees, Cicalese et al. [7] showed that for the balanced cut algorithm, an approximation ratio of 1.5 must be the best possible, as for any lower ratio, there is an infinite family of graphs failing to achieve that ratio by the balanced cut algorithm. On the other hand, for unweighted trees, we have no such lower bounds, and it should be noted that the construction given by Cicalese et al. does not work in that setting. The highest ratio found is by taking the smallest tree in the family found in [12], a tree with 14 vertices; calculating the EPT-sum of the optimal EPT and the one found by taking balanced cuts, the approximation ratio turns out to be $\frac{65}{58} \approx 1.12$ for this tree (for the family as a whole, the ratio is $(1 + o(1))$). We leave it as an open problem to close the gap for unweighted trees.

It must also be noted that for calculating EPT-sum on unweighted trees, it is still unknown whether the problem is NP-hard or polynomial-time solvable. For the related parameter of edge ranking (equivalent to minimum worst-case search time in trees), there are polynomial- and even linear-time algorithms [2,17,21], but these are non-trivial. The EPT-sum of unweighted trees might be solvable in polynomial time, but an algorithm would probably not be any easier to find.

Finally, in this paper we did not focus on edge-weighted trees. It is not hard to show that Lemma 1 still holds for trees with weights on both vertices and edges, by defining the augmented EPT to have a weight on each of their subdivision nodes equal to that of its parent (corresponding to an edge in the input tree). On the other hand, as the sparsest cut of an edge-weighted tree is more complex than the balanced cut of an unweighted (or vertex-weighted) tree, we did not try to adapt the proof of Theorem 2 to the setting of edge-weighted trees. Of course, the approximation ratio for trees that are both edge- and vertex-weighted cannot be better than that of trees that are only vertex-weighted, but we do not know if there are any strictly larger lower bounds for this case. Cicalese et al. [8] have looked at this problem and provided hardness results for some very restricted classes of trees, as well as an $O(\frac{\log n}{\log \log n})$-approximation algorithm. It does not contradict their results if the balanced cut algorithm finds a constant approximation also in this case: in fact, for trees that are only edge-weighted, the previously mentioned results of Charikar and Chatziafratis [5] imply that this algorithm should give a constant (8) approximation factor. Whether the actual factor for edge-weighted trees (with or without vertex weights) is as low as 1.5 is an interesting question, and a natural next step for anyone interested in the average performance of searching in trees.

References

1. Abboud, A., Cohen-Addad, V., Klein, P.N.: New hardness results for planar graph problems in p and an algorithm for sparsest cut. In: Proceedings of the 52nd Annual ACM SIGACT Symposium on Theory of Computing, STOC 2020, pp. 996–1009 (2020). https://doi.org/10.1145/3357713.3384310
2. Ben-Asher, Y., Farchi, E., Newman, I.: Optimal search in trees. SIAM J. Comput. **28**(6), 2090–2102 (1999). https://doi.org/10.1137/S009753979731858X
3. Berendsohn, B., Golinsky, I., Kaplan, H., Kozma, L.: Fast approximation of search trees on trees with centroid trees (2022). https://arxiv.org/abs/2209.08024, arXiv preprint [cs.DS]
4. Berendsohn, B., Kozma, L.: Splay trees on trees. In: Proceedings of the 2022 Annual ACM-SIAM Symposium on Discrete Algorithms (SODA), pp. 1875–1900 (2022). https://doi.org/10.1137/1.9781611977073.75
5. Charikar, M., Chatziafratis, V.: Approximate hierarchical clustering via sparsest cut and spreading metrics. In: Proceedings of the 28th Annual ACM-SIAM Symposium on Discrete Algorithms (SODA), pp. 841–854 (2017). https://doi.org/10.1137/1.9781611974782.53
6. Cicalese, F., Jacobs, T., Laber, E., Molinaro, M.: On the complexity of searching in trees and partially ordered structures. Theoret. Comput. Sci. **412**(50), 6879–6896 (2011)
7. Cicalese, F., Jacobs, T., Laber, E., Molinaro, M.: Improved approximation algorithms for the average-case tree searching problem. Algorithmica **68**(4), 1045–1074 (2014)
8. Cicalese, F., Keszegh, B., Lidický, B., Pálvölgyi, D., Valla, T.: On the tree search problem with non-uniform costs. Theoret. Comput. Sci. **647**, 22–32 (2016)
9. Cohen-Addad, V., Kanade, V., Mallmann-Trenn, F., Mathieu, C.: Hierarchical clustering: objective functions and algorithms. J. ACM **66**(4), 26:1–26:42 (2019)
10. Dasgupta, S.: A cost function for similarity-based hierarchical clustering. In: Annual ACM Symposium on Theory of Computing (STOC), pp. 118–127 (2016)
11. Hibbard, T.N.: Some combinatorial properties of certain trees with applications to searching and sorting. J. ACM (JACM) **9**(1), 13–28 (1962)

12. Høgemo, S., Bergougnoux, B., Brandes, U., Paul, C., Telle, J.A.: On Dasgupta's hierarchical clustering objective and its relation to other graph parameters. In: Bampis, E., Pagourtzis, A. (eds.) FCT 2021. LNCS, vol. 12867, pp. 287–300. Springer, Cham (2021). https://doi.org/10.1007/978-3-030-86593-1_20
13. Hu, T., Tucker, A.: Optimal computer search trees and variable-length alphabetical codes. SIAM J. Appl. Math. **21**(4), 514–532 (1971)
14. Knuth, D.E.: Optimum binary search trees. Acta Informatica **1**, 14–25 (1971)
15. Laber, E., Molinaro, M.: An approximation algorithm for binary searching in trees. Algorithmica **59**(4), 601–620 (2011)
16. Laber, E.S., Nogueira, L.T.: On the hardness of the minimum height decision tree problem. Discret. Appl. Math. **144**(1–2), 209–212 (2004). https://doi.org/10.1016/j.dam.2004.06.002
17. Lam, T.W., Yue, F.L.: Optimal edge ranking of trees in linear time. Algorithmica **30**(1), 12–33 (2001)
18. Laurent, H., Rivest, R.L.: Constructing optimal binary decision trees is NP-complete. Inf. Process. Lett. **5**(1), 15–17 (1976)
19. Nievergelt, J., Reingold, E.M.: Binary search trees of bounded balance. In: Proceedings of the Fourth Annual ACM Symposium on Theory of Computing, pp. 137–142 (1972)
20. Roy, A., Pokutta, S.: Hierarchical clustering via spreading metrics. In: Proceedings of the 30th International Conference on Neural Information Processing Systems, NIPS 2016, pp. 2324–2332 (2016)
21. de la Torre, P., Greenlaw, R., Schäffer, A.A.: Optimal edge ranking of trees in polynomial time. Algorithmica **13**(6), 592–618 (1995)

Online Deterministic Minimum Cost Bipartite Matching with Delays on a Line

Tung-Wei Kuo

Department of Computer Science, National Chengchi University, Taipei City, Taiwan
twkuo@cs.nccu.edu.tw

Abstract. We study the online minimum cost bipartite perfect matching with delays problem. In this problem, m servers and m requests arrive over time, and an online algorithm can delay the matching between servers and requests by paying the delay cost. The objective is to minimize the total distance and delay cost. When servers and requests lie in a known metric space, there is a randomized $O(\log n)$-competitive algorithm, where n is the size of the metric space. When the metric space is unknown a priori, Azar and Jacob-Fanani proposed a deterministic $O\left(\frac{1}{\epsilon}m^{\log\left(\frac{3+\epsilon}{2}\right)}\right)$-competitive algorithm for any fixed $\epsilon > 0$. This competitive ratio is tight when $n = 1$ and becomes $O(m^{0.59})$ for sufficiently small ϵ.

We improve upon the result of Azar and Jacob-Fanani for the case where servers and requests are on the real line, providing a deterministic $\tilde{O}(m^{0.5})$-competitive algorithm. Our algorithm is based on the Robust Matching (RM) algorithm proposed by Raghvendra for the minimum cost bipartite perfect matching problem. In this problem, delay is not allowed, and all servers arrive in the beginning. When a request arrives, the RM algorithm immediately matches the request to a free server based on the request's minimum t-net-cost augmenting path, where $t > 1$ is a constant. In our algorithm, we delay the matching of a request until its waiting time exceeds its minimum t-net-cost divided by t.

Keywords: Minimum cost bipartite matching with delays · Online algorithm · Deterministic algorithm · One-dimensional metric space

1 Introduction

Consider an online gaming platform where players are paired for gameplay. To improve the gaming experience, players with similar skill ratings should be matched together. However, when a new player joins, a suitable matching may not be available immediately. In this case, it is essential to delay the matching process, in the hope of finding a better matching in the near future. Clearly, the waiting time and the similarity between matched players should be considered jointly, and a natural approach is to minimize the sum of both terms.

The above problem is captured by the Minimum cost Perfect Matching with Delays (MPMD) problem [18]. In the MPMD problem, demands arrive over time, and their similarities are modeled by a metric space. When a demand arrives, an online algorithm has the option to postpone the matching process by incurring a delay cost. The objective is to minimize the sum of the total delay time (i.e., delay cost) and the total distance between matched demands in the metric space (i.e., distance cost).

In numerous matching applications, entities can only be matched if they belong to different types (e.g., teacher-student, donor-donee, buyer-seller, and driver-passenger). These binary classifications motivate the Minimum cost Bipartite Perfect Matching with Delays (MBPMD) problem [3,5]. In the MBPMD problem, there are two types of demands, servers and requests. An online algorithm has to match each request to a server. Like the MPMD problem, the objective is to minimize the total distance and delay cost.

In this paper, we study the MBPMD problem on a line, where requests and servers are positioned on the real line. For example, skiers (requests) should be matched to skis (servers) of approximately their height [2]. Another example is matching buyers (requests) and sellers (servers) based on their stated prices. In these examples, requests and servers are represented as numbers on the real line, corresponding to heights or prices. We analyze our algorithm using the standard notion of competitive ratio. In particular, an online algorithm is said to be c-competitive ($c \geq 1$) if for any input, the cost of the algorithm is at most c times the cost of the optimal offline algorithm.

Background. For the MPMD problem, Emek et al. proposed a randomized $O(\log^2 n + \log \Delta)$-competitive algorithm [18], where n is the number of points in the metric space and Δ is the aspect ratio of the metric space. Azar et al. then proposed a randomized $O(\log n)$-competitive algorithm, and proved that the competitive ratio for any randomized algorithm is $\Omega(\sqrt{\log n})$ [5]. Ashlagi et al. further improved this lower bound to $\Omega\left(\frac{\log n}{\log \log n}\right)$ [3]. All the above randomized algorithms used the celebrated result of Fakcharoenphol et al. [21] to transform the original metric space into a distribution over Hierarchically Separated Trees (HSTs). As a result, these algorithms need to know the metric space in advance.

The algorithms in [18] and [5] are randomized. For offline problems, we can repeatedly execute a randomized algorithm until we find a satisfactory solution. However, for online problems, we can only execute an algorithm once, and the output cannot be changed. Thus, a more robust approach for online problems is to design deterministic algorithms.

For the MPMD problem, Bienkowski et al. first proposed a deterministic $O(m^{2.46})$-competitive algorithm, where m is the number of demands to be matched [14]. Bienkowski et al. then proposed a deterministic $O(m)$-competitive algorithm [13]. Finally, Azar and Jacob-Fanani proposed a deterministic $O(\frac{1}{\epsilon} m^{\log(\frac{3+\epsilon}{2})})$-competitive algorithm for any fixed $\epsilon > 0$ [7]. For small enough ϵ, the competitive ratio becomes $O(m^{0.59})$. Unlike the previous randomized algorithms, the above three deterministic algorithms do not need to know the metric space in advance.

When the metric space is a tree, Azar et al. also proposed a deterministic $O(n)$-competitive algorithm in [5]. Moreover, when $n = 2$, Emek et al. proposed a deterministic 3-competitive algorithm, and proved that 3 is the best possible competitive ratio [19].

For the MBPMD problem, Azar et al. first proposed a randomized $O(\log n)$-competitive algorithm, and proved that any randomized algorithm has a competitive ratio of $\Omega(\log^{1/3} n)$ [5]. This lower bound is further improved to $\Omega\left(\sqrt{\frac{\log n}{\log \log n}}\right)$ [3]. For deterministic algorithms, Bienkowski et al. first proposed an $O(m)$-competitive algorithm [13] for the MBPMD problem. Azar and Jacob-Fanani then proposed an $O(\frac{1}{\epsilon} m^{\log(\frac{3+\epsilon}{2})})$-competitive algorithm for any fixed $\epsilon > 0$ [7]. For small enough ϵ, the competitive ratio becomes $O(m^{0.59})$. All the above algorithms are based on the algorithms for the MPMD problem. Moreover, the competitive ratios in [13] and [7] are tight when the metric space is a line.[1] In summary, prior to our work, the best known competitive ratio for the deterministic MBPMD problem on a line was $O(m^{0.59})$.

Our Contribution. In this paper, we introduce a deterministic $\tilde{O}(m^{0.5})$-competitive algorithm for the MBPMD problem on a line, improving upon the $O(m^{0.59})$-competitive algorithm of [7]. Specifically, we have the following result.

Theorem 1. *There is a deterministic $O(\sqrt{m} \log^2 m)$-competitive algorithm for the MBPMD problem on a line.*

Our algorithm is based on the Robust Matching (RM) algorithm proposed by Raghvendra for the online Minimum cost Bipartite Perfect Matching (MBPM) problem [35]. In the MBPM problem, all servers arrive in the beginning, and an online algorithm must match a request immediately after it arrives. The objective is to minimize the total distance cost of the matching. Nayyar and Raghvendra [33] proved that the competitive ratio of the RM algorithm for any d-dimensional Euclidean metric space is $O(n^{1-1/d} \log^2 n)$. Raghvendra further proved that for one-dimensional Euclidean metric space, RM algorithm is $O(\log n)$-competitive [34]. From a bird's eye view, RM algorithm maintains an offline matching M^{OFF} and an online matching M^{RM}, which is the real output matching. When request r_i arrives, RM algorithm computes an M^{OFF}-augmenting path P_i from r_i to some free server s_j. RM algorithm then adds (r_i, s_j) to M^{RM} and augments M^{OFF} by P_i.

Specifically, P_i is such that minimizes the γ-net-cost[2] among all M^{OFF}-augmenting paths from r_i to a free server. When M^{OFF} is augmented by a path P, edges shared by P and M^{OFF} are removed from M^{OFF}, and other edges in

---
[1] In fact, the competitive ratios in [13] and [7] are tight when $n = 2$ and $n = 1$, respectively.
[2] In [35], this cost is referred to as the t-net-cost. However, because t denotes time in this paper, we change t-net-cost to γ-net-cost.

P are added to M^{OFF}. The γ-net-cost of P is the total distance of the edges added to M^{OFF} (multiplied by γ) minus the total distance of the edges removed from M^{OFF}.

There are two major differences between the MBPM problem and the MBPMD problem. Specifically, in the MBPM problem considered by the RM algorithm:

1. All servers arrive in the beginning.
2. Once a request arrives, the request must be matched immediately. Thus, the objective function does not consider delay cost.

To address the above differences, we first introduce a Moving Virtual (MV) server $\widetilde{s}_i$ for every request r_i. Specifically, we consider the Time-Augmented (TA) plane [7,14] that adds the time axis to the original one-dimensional space. Thus, the TA plane is two-dimensional. When r_i arrives, $\widetilde{s}_i$ and r_i are at the same point in the TA plane, with the time-coordinate being r_i's arrival time. The time-coordinate of r_i is fixed, while the time-coordinate of $\widetilde{s}_i$ is always the current time. Thus, the distance between r_i and $\widetilde{s}_i$ in the TA plane is always r_i's current waiting time. For any real server, its time-coordinate is fixed at its arrival time.

For each request r_i, our algorithm maintains two M^{OFF}-augmenting paths, a *real* augmenting path P_i and a *virtual* augmenting path $\widetilde{P}_i$. P_i is such that minimizes the γ-net-cost among all augmenting paths from r_i to a real free server in the TA plane, and $\widetilde{P}_i$ is such that minimizes the γ-net-cost among all augmenting paths from r_i to an MV server in the TA plane, with the last edge connecting a request r_p (possibly different from r_i) to r_p's MV server $\widetilde{s}_p$.

Initially, the virtual minimum γ-net-cost (i.e., the γ-net-cost of $\widetilde{P}_i$) is 0 (since $\widetilde{P}_i$ contains only r_i and $\widetilde{s}_i$ initially) and is thus less than or equal to the real minimum γ-net-cost (i.e., P_i's γ-net-cost). After a server arrives or after M^{OFF} is augmented by another request's augmenting path, P_i and $\widetilde{P}_i$ may change. Moreover, because the distance between r_p and $\widetilde{s}_p$ in the TA plane increases over time, the virtual minimum γ-net-cost increases over time. When the virtual minimum γ-net-cost is greater than or equal to the real minimum γ-net-cost, we match r_i to the endpoint server of P_i and augments M^{OFF} by P_i.

In Sect. 4, we show that the algorithm can be greatly simplified: we match r_i when its waiting time is greater than or equal to the real minimum γ-net-cost divided by γ. Thus, we no longer need MV servers in our algorithm. To this end, we prove that the virtual minimum γ-net-cost is always γ times r_i's waiting time. Nevertheless, MV servers facilitate the analysis of our algorithm.

In [35], it has been shown that the total distance cost can be upper bounded by the total γ-net-cost. Because we further upper bound the delay cost by the γ-net-cost, our algorithm's total cost is upper bounded by the total γ-net-cost. To prove Theorem 1, we then use the techniques in [33] to relate the total γ-net-cost to the optimal cost in the TA plane (recall that the TA plane is two-dimensional).

1.1 Other Related Work

Without considering waiting times (and thus every request must be matched immediately upon arrival), there has been a considerable amount of research in the literature on how to maximize matching weights or minimize matching costs, considering different arrival patterns for vertices or edges. Relevant literature on these issues can be found in recent years' studies (such as [16,20,23,25,38]), or in the excellent survey by Mehta [31]. On the other hand, some studies have explored settings where recourse is allowed [1,12,22,29,30]. In this subsection, we focus on online problems that allow delays.

Poisson Arrival Processes. In [28], Mari et al. considered the MPMD problem and assumed that the request arrival process follows a Poisson arrival process. Mari et al. considered a simple greedy algorithm: when the total waiting time of two requests exceeds their distance from each other, they are immediately matched. While it has been shown that such an algorithm has a competitive ratio of $\Omega(m^{0.58})$ in instances designed by Reingold and Tarjan [36], Mari et al. proved that when the request arrival process is Poisson, the competitive ratio of this simple greedy algorithm is $O(1)$.

Non-linear Delay Costs. In the MPMD problem, delay cost equals waiting time. Other studies have considered different forms of delay costs. Liu et al. assumed that the delay cost is t^α, where $\alpha > 1$ and t is the waiting time [27]. Liu et al. proposed a deterministic $O(n)$-competitive algorithm for uniform metric space. Azar et al. considered the case where the delay cost is a concave function of waiting time [9]. They proposed an $O(1)$-competitive deterministic algorithm (for $n = 1$) and an HST-based $O(\log n)$-competitive algorithm (for $n > 1$).

In [17], Deryckere and Umboh similarly considered concave functions and designed a deterministic $O(m)$-competitive primal-dual algorithm. Deryckere and Umboh also utilized set delay functions as delay cost. Specifically, at each time t, the algorithm incurs delay cost as a function of the set of unmatched requests. They proposed a deterministic $O(2^m)$-competitive algorithm and a randomized $O(m^4)$-competitive algorithm. Their approach is based on transforming the MPMD problem into a Metrical Task System (MTS) [15]. Deryckere and Umboh also proved that the competitive ratio of any deterministic (respectively, randomized) algorithm is $\Omega(n)$ (respectively, $\Omega(\log n)$).

Other Matching Problems with Delays. In some games, such as poker or mahjong, more than two players are needed. Therefore, we need to match more than two requests at once. [24,32] considered this scenario and designed algorithms based on HST and primal-dual transformation. Another way of allowing delay is setting deadlines. In [4], Ashlagi et al. assumed that each request can wait for δ time units upon arrival and not all requests need to be matched. The algorithm aims to find the maximum weight matching, and Ashlagi et al. proposed an $O(1)$-competitive algorithm.

Other Online Problems with Delays. Many online problems have variants that allow delays. For example, in online network design problems, algorithms need to purchase network links upon request arrival. In the case where delay is allowed, purchased network links may serve multiple requests simultaneously, reducing purchasing costs [10,11]. Another example is online service problems, in which a server can be moved to serve requests. In the case where delay is allowed, we can compute the shortest path based on multiple requests, reducing server movement distance [6,37]. For these problems, randomized $O(\text{poly} \log n)$-competitive algorithms are first proposed based on HST [6,11]. Azar and Touitou then proposed deterministic $O(\text{poly} \log n)$-competitive algorithms [10,37]. Recently, Azar et al. considered the list update with delays problem and designed a deterministic $O(1)$-competitive algorithm [8].

2 Preliminaries

Given two sets A and B, a matching M between A and B is a set of vertex-disjoint edges between A and B. An element v is said to be *saturated* by M (or M saturates v) if v is an endpoint of some edge in M. A matching M between A and B is said to be *perfect* if M saturates all elements in $A \cup B$.

Problem Definition. In the Minimum cost Bipartite Perfect Matching with Delays (MBPMD) problem, there is an underlying metric space $\mathcal{M} = (V, d)$. m servers and m requests from $\mathcal{M}$ arrive over time. Let $R = \{r_1, r_2, \cdots, r_m\}$ and $S = \{s_1, s_2, \cdots, s_m\}$ be the request set and the server set, respectively. For any $u \in R \cup S$, $a(u)$ denotes u's arrival time and $\ell(u) \in V$ denotes u's location in $\mathcal{M}$. When u arrives at time $a(u)$, $\ell(u)$ and u's distances to other requests and servers that arrive by time $a(u)$ are revealed to the algorithm. In this paper, r, r_i, and r_p always refer to some request in R, and s, s_j, and s_q always refer to some server in S.

In the MBPMD problem, an online algorithm computes a perfect matching M between R and S. After a request r arrives, the online algorithm can defer the matching of r by paying the delay cost. If the algorithm matches r and s at time t, then the delay cost is $(t - a(r)) + (t - a(s))$. After r and s are matched, (r, s) is added to matching M and cannot be removed from M afterward. In addition, the algorithm pays for the distance cost $d(\ell(r), \ell(s))$. For simplicity, for any $u_1, u_2 \in R \cup S$, define $d(u_1, u_2) = d(\ell(u_1), \ell(u_2))$.

In this paper, we assume that $\mathcal{M}$ is a line metric. Thus, the location of every server and request is a point on the real line, and the distance between any two elements in $R \cup S$ is their distance on the line. Specifically, for any $u \in R \cup S$, let $\text{pos}(u) \in \mathbb{R}$ be the location of u on the real line. Thus, for any $u_1, u_2 \in R \cup S$, $d(u_1, u_2) = |\text{pos}(u_1) - \text{pos}(u_2)|$.

In summary, an online algorithm for the MBPMD problem has to compute a perfect matching M between R and S. For each edge (r, s) in M, let $\text{mt}(r, s)$ be the time when r and s are matched. Clearly, $\text{mt}(r, s) \geq a(r)$ and $\text{mt}(r, s) \geq a(s)$.

Given a perfect matching M and a matching time function mt, define

$$\text{cost}(M, \text{mt}) = \sum_{(r,s) \in M} \left(|\text{pos}(r) - \text{pos}(s)| + (\text{mt}(r,s) - a(r)) + (\text{mt}(r,s) - a(s)) \right)$$

as the total distance and delay cost of a solution (M, mt). The objective is to minimize $\text{cost}(M, \text{mt})$.

Augmenting Paths. Given a matching M, an M-alternating path is a path that alternates betweens edges in M and edges not in M. An M-alternating path P is said to be an M-augmenting path if both endpoints of P are not saturated by M. For any path P, we use $E(P)$ to denote the set of undirected edges in P. For any two sets A and B, define $A \oplus B = (A \setminus B) \cup (B \setminus A)$ as the symmetric difference between A and B. Observe that for any M-augmenting path P, $M \oplus E(P)$ is a matching of size $|M| + 1$.

While augmenting paths are typically considered as undirected paths, for the sake of convenience, we often view an augmenting path as a directed path from an unsaturated request r to an unsaturated server s. We often refer to an augmenting path by the natural sequence of its vertices. Specifically, an augmenting path P that originates at r and terminates at s can be written in the form of $P = r'_1 s'_1 r'_2 s'_2 \cdots r'_\ell s'_\ell$ for some $\ell \geq 1$, where $r'_1 = r$, $s'_\ell = s$, and $r'_k \in R, s'_k \in S$ for any $1 \leq k \leq \ell$.

Because augmenting paths are directed, edges in augmenting paths can also be viewed as directed edges. We denote by $\overrightarrow{u,v}$ a directed edge from u to v. For any augmenting path P, we use $\overrightarrow{E}(P)$ to denote the set of directed edges in P, (i.e., $\overrightarrow{E}(P) = \{\overrightarrow{r'_1,s'_1}, \overrightarrow{s'_1,r'_2}, \overrightarrow{r'_2,s'_2}, \cdots, \overrightarrow{r'_\ell,s'_\ell}\}$). For any directed edge $\overrightarrow{u,v}$, we say that $\overrightarrow{u,v}$ is in an augmenting path P if $\overrightarrow{u,v} \in \overrightarrow{E}(P)$. We have the following simple fact. Recall that in this paper, r always refers to a request in R and s always refers to a server in S.

Fact 1. *Let M be any matching between R and S. Let P be any M-augmenting path that originates at a request and terminates at a server. Let $M^{aug} = M \oplus E(P)$. Then the following statements hold:*

1. *For any $\overrightarrow{r,s} \in \overrightarrow{E}(P)$, $(r,s) \notin M$ and $(r,s) \in M^{aug}$.*
2. *For any $\overrightarrow{s,r} \in \overrightarrow{E}(P)$, $(r,s) \in M$ and $(r,s) \notin M^{aug}$.*

In other words, for any directed edge from a request to a server in P, it is added to M^{aug}; for any directed edge from a server to a request in P, it is not in M^{aug}.

We can then express the γ-net-cost, which was first introduced by Raghvendra [35], based on the directions of edges. For any $\gamma > 1$ and any augmenting path P, the γ-net-cost of P is defined as

$$\gamma \left(\sum_{\overrightarrow{r,s} \in \overrightarrow{E}(P)} d(r,s) \right) - \sum_{\overrightarrow{s,r} \in \overrightarrow{E}(P)} d(r,s).$$

Time Augmented Plane. Throughout this paper, we use M^{OPT} to denote the optimal matching. Observe that for any $(r,s) \in M^{OPT}$, the optimal solution must match r and s at time $\max(a(r), a(s))$. Thus, the optimal cost can be written as

$$\sum_{(r,s) \in M^{OPT}} (|\operatorname{pos}(r) - \operatorname{pos}(s)| + |a(r) - a(s)|). \qquad (1)$$

The above optimal cost suggests that we can view $S \cup R$ as a set of points in an xy-plane, where each point $v \in S \cup R$ has x-coordinate $\operatorname{pos}(v)$ and y-coordinate $a(v)$. The y-axis can also be viewed as the time axis. Such an xy-plane is called the *Time Augmented (TA) plane* [7,14]. Observe that $|\operatorname{pos}(r) - \operatorname{pos}(s)| + |a(r) - a(s)|$ is the Manhattan distance between r and s in the TA plane.

3 An Online Matching Algorithm with Moving Virtual Servers

In this section, we describe our algorithm, which introduces Moving Virtual (MV) servers into the RM algorithm. We thus call our algorithm the Virtual RM (VRM) algorithm. Due to space limit, all the missing proofs are given in the full version of the paper [26].

3.1 Moving Virtual Servers

In the VRM algorithm, whenever a request r_i arrives, the algorithm creates a Moving Virtual (MV) server $\widetilde{s}_i$, and sets $a(\widetilde{s}_i) = a(r_i)$. Moreover, $\operatorname{pos}(r_i) = \operatorname{pos}(\widetilde{s}_i)$. Therefore, we can also view $\widetilde{s}_i$ as a point in the TA plane. $\widetilde{s}_i$ moves upward in the TA plane. Specifically, at time $t \geq a(r_i)$, the y-coordinate of $\widetilde{s}_i$ in the TA plane is t. A simple property is that the distance between an unmatched request and its MV server in the TA plane is always the request's current waiting time. To differentiate between servers in S and MV servers, we call servers in S the **real** servers. Our algorithm never matches a request to an MV server.

For any $u_1, u_2 \in R \cup S$, the distance between u_1 and u_2 in the TA plane, denoted by $D(u_1, u_2)$, is defined as

$$D(u_1, u_2) = |\operatorname{pos}(u_1) - \operatorname{pos}(u_2)| + |a(u_1) - a(u_2)|.$$

For any $r_i \in R$, the distance between r_i and its MV server $\widetilde{s}_i$ in the TA plane at time t, denoted by $D_t(r_i, \widetilde{s}_i)$, is defined as

$$D_t(r_i, \widetilde{s}_i) = t - a(r_i).$$

An important observation is that if at time t, r_i already arrives but a server s has not arrived yet (i.e., $a(r_i) \leq t < a(s)$), then

$$D(r_i, s) \geq a(s) - a(r_i) > t - a(r_i) = D_t(r_i, \widetilde{s}_i). \qquad (2)$$

3.2 The VRM Algorithm

Like the RM algorithm, the VRM algorithm maintains an offline matching M^{OFF} and an online matching M^{VRM}, which is the real output matching. Unlike the RM algorithm, the VRM algorithm needs to decide the matching time for each edge $(r,s) \in M^{VRM}$, denoted by $\text{mt}^{VRM}(r,s)$. All the servers in M^{OFF} and M^{VRM} are real, and these two matchings saturate the same set of servers and requests. A real server or a request is said to be **free** if it has arrived but not yet matched by M^{OFF}. Initially, both matchings are empty.

We consider two types of augmenting paths, real and virtual, in the TA plane. An augmenting path P is **real** if all servers in P are real. An augmenting path P is **virtual** if the last directed edge of P is from some request r_p to r_p's MV server (i.e., $\overrightarrow{r_p, \widetilde{s}_p}$), and all the other servers in P are real. For any real M^{OFF}-augmenting path P, define the γ-net-cost of P in the TA plane, denoted by $\varphi_\gamma(P)$, as

$$\varphi_\gamma(P) = \gamma \left(\sum_{\overrightarrow{r,s} \in \overrightarrow{E}(P)} D(r,s) \right) - \sum_{\overrightarrow{s,r} \in \overrightarrow{E}(P)} D(r,s).$$

The γ-net-cost of virtual augmenting paths is defined similarly. The only difference is that the distance function of the last directed edge is D_t instead of D. Specifically, for any virtual M^{OFF}-augmenting path $\widetilde{P}$ that terminates at MV server $\widetilde{s}_p$ and any time $t \geq a(r_p)$, define the γ-net-cost of $\widetilde{P}$ at time t in the TA plane, denoted by $\varphi_{\gamma,t}(\widetilde{P})$, as

$$\varphi_{\gamma,t}(\widetilde{P}) = \gamma \left(D_t(r_p, \widetilde{s}_p) + \sum_{\overrightarrow{r,s} \in \overrightarrow{E}(\widetilde{P})} D(r,s) \right) - \sum_{\overrightarrow{s,r} \in \overrightarrow{E}(\widetilde{P})} D(r,s).$$

We stress that in the above definition, s is a real server. In this paper, we assume $\gamma = 3$ and drop the subscript γ in φ_γ and $\varphi_{\gamma,t}$ if it is clear from the context.

Description of the Algorithm. After a request r_i arrives, the VRM algorithm maintains a real M^{OFF}-augmenting path P_i and a virtual M^{OFF}-augmenting path $\widetilde{P}_i$. Specifically, P_i is such that minimizes the γ-net-cost among all real M^{OFF}-augmenting paths that originate at r_i. On the other hand, for any time t, $\widetilde{P}_i$ is such that minimizes the γ-net-cost among all virtual M^{OFF}-augmenting paths that originate at r_i at time t. We call P_i (respectively, $\widetilde{P}_i$) the **real minimum augmenting path** (respectively, **virtual minimum augmenting path**) of r_i. Note that in the absence of free servers, P_i does not exist. If so, we assume $\varphi(P_i) = \infty$.

Fix an offline matching M^{OFF}. Because $\varphi_t(\widetilde{P}_i)$ increases as t increases, $\varphi_t(\widetilde{P}_i)$ exceeds $\varphi(P_i)$ eventually. A request r_i is said to be **ready** at time t if it is free and $\varphi_t(\widetilde{P}_i) \geq \varphi(P_i)$. For any directed path P, denote by $\text{ori}(P)$ and $\text{ter}(P)$ as the first and last vertices in P, respectively. Whenever some request r_i is ready at time t, we first augment M^{OFF} by setting $M^{OFF} \leftarrow M^{OFF} \oplus E(P_i)$. We then add $(r_i, \text{ter}(P_i))$ to M^{VRM}, and set $\text{mt}^{VRM}(r_i, \text{ter}(P_i)) = t$. Finally, we update all free requests' real and virtual minimum augmenting paths (since M^{OFF} is changed).

In the following, we first explain the subroutine that computes P_i and $\widetilde{P}_i$ at any time t. We then discuss the timings for computing P_i and $\widetilde{P}_i$.

Computing Minimum Augmenting Paths. The subroutine for computing P_i and $\widetilde{P}_i$ basically follows that in [35]. Let $\widetilde{S} = \{\widetilde{s}_1, \widetilde{s}_2, \cdots, \widetilde{s}_m\}$. Our algorithm maintains dual variables $z(\cdot)$ for $R \cup S \cup \widetilde{S}$ and the following invariants:

$$z(r) + z(s) \leq \gamma D(r,s), \forall r \in R, s \in S. \tag{I1}$$

$$z(r_p) + z(\widetilde{s}_p) \leq \gamma D_t(r_p, \widetilde{s}_p), \forall r_p \in R, t \geq a(r_p). \tag{I2}$$

$$z(v) = 0, \forall v \in \widetilde{S} \cup \{u | u \in R \cup S, u \text{ is not saturated by } M^{OFF}\}. \tag{I3}$$

$$z(r) + z(s) = D(r,s), \forall (r,s) \in M^{OFF}. \tag{I4}$$

By Invariant (I3), all the dual variables are zero initially. It is easy to see that all invariants hold initially. $z(\cdot)$ is only updated when M^{OFF} is augmented. Fix a time t, a free request r_i, an offline matching M^{OFF}, and $z(\cdot)$, we next explain how to compute P_i and $\widetilde{P}_i$.

Let R_{sat} be the set of requests that are saturated by M^{OFF}, and $\widetilde{S}_{sat}$ be the set of MV servers of the requests in R_{sat}. Let S_t be the set of servers that arrive by time t. We construct an edge-weighted directed bipartite graph $G_{i,t}$ with partite sets $R_{sat} \cup \{r_i\}$ and $S_t \cup \widetilde{S}_{sat} \cup \{\widetilde{s}_i\}$. The edge weight is the edge's slack with respect to dual variables $z(\cdot)$. $G_{i,t}$ is called the **slack graph** of r_i at time t and is constructed as follows:

1. For every server s in S_t and every request r in R_{sat}, if $(s,r) \in M^{OFF}$, we add to $G_{i,t}$ a directed edge $\overrightarrow{s,r}$ with edge weight $sl(\overrightarrow{s,r}) = 0$.
2. For every server s in S_t and every request r in $R_{sat} \cup \{r_i\}$, if $(r,s) \notin M^{OFF}$, we add to $G_{i,t}$ a directed edge $\overrightarrow{r,s}$ with edge weight $sl(\overrightarrow{r,s}) = \gamma D(r,s) - (z(r) + z(s))$.
3. For every request r_p in $R_{sat} \cup \{r_i\}$, we add to $G_{i,t}$ a directed edge $\overrightarrow{r_p, \widetilde{s}_p}$ with edge weight $sl(\overrightarrow{r_p, \widetilde{s}_p}) = \gamma(D_t(r_p, \widetilde{s}_p)) - (z(r_p) + z(\widetilde{s}_p))$.

We then set P_i as the shortest path from r_i to the set of real free server in $G_{i,t}$. This can be done by first computing all the shortest paths from r_i to each free real server in $G_{i,t}$ and then outputting the shortest one among these paths.

Similarly, we set $\widetilde{P}_i$ as the shortest path from r_i to the set of MV servers in $G_{i,t}$. The proof of correctness is given in [26].

Timings for Updating P_i and $\widetilde{P}_i$. For each free request r_i, we compute P_i and $\widetilde{P}_i$ when r_i arrives or whenever one of the following events occurs:

Event SA: A server arrives.
Event AU: M^{OFF} is augmented by another request's augmenting path.

When Event SA occurs, we only need to update P_i, as $\widetilde{P}_i$ cannot change due to the arrival of a new server.

Observe that all virtual augmenting paths increase the γ-net-cost by the same speed. Specifically, for any virtual augmenting path $\widetilde{P}$ at time t and some future time $t' > t$, we have $\varphi_{t'}(\widetilde{P}) = \varphi_t(\widetilde{P}) + \gamma(t'-t)$. Thus, if Event AU does not occur (and thus M^{OFF} does not change), then r_i's virtual minimum augmenting path cannot change. Therefore, we only need to update $\widetilde{P}_i$ when Event AU occurs.

Whenever P_i or $\widetilde{P}_i$ is updated at some time t, we check whether r_i becomes ready (i.e., $\varphi_t(\widetilde{P}_i) \geq \varphi(P_i)$). If not, we compute the **ready timing** t_i^{rdy} for r_i:

$$t_i^{rdy} = t + \frac{\varphi(P_i) - \varphi_t(\widetilde{P}_i)}{\gamma}.$$

At time t_i^{rdy}, r_i becomes ready. Note that P_i and $\widetilde{P}_i$ may change before time t_i^{rdy}. If so, we update t_i^{rdy} again.

Updating the Dual Variables. We update the dual variables $z(\cdot)$ whenever M^{OFF} is augmented by some ready request r_i's augmenting path P_i. Let t be the time when M^{OFF} is augmented by P_i. Let $G^{pre} = G_{i,t}$ be the slack graph of r_i right before M^{OFF} is augmented by P_i. For each vertex v in G^{pre}, define $sl(r_i, v)$ as the shortest distance from r_i to v in G^{pre}. Let $s^* = \text{ter}(P_i)$. We update $z(\cdot)$ in two steps.

Step 1: – For every request r in G^{pre}, if $sl(r_i, r) < sl(r_i, s^*)$, we set $z(r) \leftarrow z(r) + (sl(r_i, s^*) - sl(r_i, r))$.
– For every real server s in G^{pre}, if $sl(r_i, s) < sl(r_i, s^*)$, we set $z(s) \leftarrow z(s) - (sl(r_i, s^*) - sl(r_i, s))$.
Step 2: – For every $\overrightarrow{r,s} \in \overrightarrow{E}(P_i)$, we set $z(r) \leftarrow z(r) - (\gamma - 1)D(r,s)$.

Note that for every directed edge $\overrightarrow{r,s}$ considered in Step 3.2, (r, s) is added to M^{OFF} after r_i is matched. In [26], we prove that all invariants hold throughout the execution of the VRM algorithm.

4 A Simplified Algorithm and Analysis

In this section, we present a simplified algorithm without MV servers. To this end, we prove some properties regarding the change of real and virtual augmenting paths. Throughout the execution of the VRM algorithm, three types of events may occur: a server arrives (Event SA), a request becomes ready (Event AU), and a request arrives (Event RA). The VRM algorithm can be described as an event-driven algorithm:

1. When Event RA occurs, compute the real and virtual minimum augmenting paths for the new request.
2. When Event SA occurs, update the real minimum augmenting paths for all free requests.
3. When Event AU occurs, augment M^{OFF} and update M^{VRM} according to the real minimum augmenting path of the request that becomes ready, and then update the real and virtual minimum augmenting paths for all free requests.

If multiple events occur simultaneously, we can process them in any order.

Fix an arbitrary request r_i. Assume that starting from time $a(r_i)$ to the time when r_i is matched by the VRM algorithm, the algorithm processes events $E_1, E_2, \cdots, E_\nu$ in order, where E_1 is the event that r_i arrives and E_ν is the event that M^{OFF} is augmented by r_i's augmenting path (and thus r_i is matched by the VRM algorithm). For any $1 \leq w \leq \nu$, let t_w be the time when E_w is processed. We introduce the following notations to distinguish the states of the VRM algorithm right before an event occurs and right after the event is processed.

- Let $P_{i,w}^{pre}$ and $P_{i,w}^{post}$ be r_i's real minimum augmenting paths right before E_w occurs and right after E_w is processed, respectively.
- Let $\widetilde{P}_{i,w}^{pre}$ and $\widetilde{P}_{i,w}^{post}$ be r_i's virtual minimum augmenting paths right before E_w occurs and right after E_w is processed, respectively.

Define $\varphi_{i,w}^{post} = \varphi(P_{i,w}^{post})$ and $\varphi_{i,w}^{pre} = \varphi(P_{i,w}^{pre})$. Similarly, define $\widetilde{\varphi}_{i,w}^{post} = \varphi_{t_w}(\widetilde{P}_{i,w}^{post})$ and $\widetilde{\varphi}_{i,w}^{pre} = \varphi_{t_w}(\widetilde{P}_{i,w}^{pre})$.

In [26], we prove that the following two inequalities hold for any $1 \leq w \leq \nu-1$ (Lemma 1). The first one states that after an event is processed, the γ-net-cost of $\widetilde{P}_i$ is at most that of P_i.

$$\widetilde{\varphi}_{i,w}^{post} \leq \varphi_{i,w}^{post}. \tag{3}$$

The next one states that the simplest virtual augmenting path, $r_i \widetilde{s}_i$, is always the the best one. As a result, the virtual minimum γ-net-cost is always γ times r_i's waiting time.

$$\widetilde{\varphi}_{i,w}^{post} = \gamma(t_w - a(r_i)). \tag{4}$$

Lemma 1. *Equation (3) and Eq. (4) hold for any $1 \leq w \leq \nu - 1$.*

4.1 Implications of Lemma 1

Implication 1: A simplified algorithm. Recall that r_i becomes ready at time t if $\varphi_t(\widetilde{P}_i) \geq \varphi(P_i)$, and this is the only reason that we need to compute $\widetilde{P}_i$. By Eq. (4), to determine whether r_i is ready, we only need to compare $\varphi(P_i)$ and $\gamma(t - a(r_i))$. Specifically, whenever P_i is updated at some time t, we check whether r_i becomes ready (i.e., $\gamma(t - a(r_i)) \geq \varphi(P_i)$). If not, we compute the following *ready timing* t_i^{rdy} for r_i:

$$t_i^{rdy} = t + \frac{\varphi(P_i) - \gamma(t - a(r_i))}{\gamma}.$$

As a result, in the simplified algorithm, we do not need to compute $\widetilde{P}_i$ explicitly.

Implication 2: An upper bound for $\mathrm{cost}(M^{VRM}, \mathrm{mt}^{VRM})$. For any matching M, define $D(M) = \sum_{(r,s) \in M} D(r,s)$. Observe that for any request r and server s that are matched together at time $\mathrm{mt}(r,s)$, by the triangle inequality, we have

$$\mathrm{mt}(r,s) - a(s) \leq \mathrm{mt}(r,s) - a(r) + |a(r) - a(s)|.$$

Thus, the distance and delay cost of (r,s) can be upper bounded as follows:

$$|\mathrm{pos}(r) - \mathrm{pos}(s)| + \mathrm{mt}(r,s) - a(r) + \mathrm{mt}(r,s) - a(s)$$
$$\leq |\mathrm{pos}(r) - \mathrm{pos}(s)| + 2(\mathrm{mt}(r,s) - a(r)) + |a(r) - a(s)|$$
$$= D(r,s) + 2(\mathrm{mt}(r,s) - a(r)).$$

Thus,

$$\mathrm{cost}(M^{VRM}, \mathrm{mt}^{VRM}) \leq D(M^{VRM}) + 2 \sum_{(r_i, s_j) \in M^{VRM}} \left(\mathrm{mt}^{VRM}(r_i, s_j) - a(r_i) \right). \tag{5}$$

The following lemma upper bounds $D(M^{VRM})$ by the sum of the minimum γ-net-cost over all requests, and the proof is given in [26]. Let P_i^* be the real minimum augmenting path of r_i when r_i is ready and matched by the VRM algorithm.

Lemma 2. *Let $\gamma > 1$ and M^{VRM} be the final online matching output by the VRM algorithm. Then $D(M^{VRM}) \leq \frac{2}{\gamma - 1} \sum_{i=1}^{m} \varphi(P_i^*)$.*

Next, we prove $\mathrm{mt}^{VRM}(r_i, s_j) - a(r_i) = \frac{1}{\gamma}\varphi(P_i^*)$. By Lemma 1 (Eq. (3)), after $E_{\nu-1}$ is processed, $\varphi_{t_{\nu-1}}(\widetilde{P}_i) \leq \varphi(P_i)$. Because no event is processed between $E_{\nu-1}$ and E_ν, when r_i becomes ready and matched by the VRM algorithm at time t_ν, we must have

$$\varphi(P_i) = \varphi_{t_\nu}(\widetilde{P}_i) = \widetilde{\varphi}_{i,\nu}^{pre}.$$

Moreover, we have $\widetilde{\varphi}_{i,\nu}^{pre} = \widetilde{\varphi}_{i,\nu-1}^{post} + \gamma(t_\nu - t_{\nu-1}) = \gamma(t_\nu - a(r_i))$, where the last equality holds due to Lemma 1 (Eq. (4)). Thus, for any $(r_i, s_j) \in M^{VRM}$, we have $\varphi(P_i^*) = \gamma(t_\nu - a(r_i)) = \gamma(\mathrm{mt}^{VRM}(r_i, s_j) - a(r_i))$, or equivalently, $\mathrm{mt}^{VRM}(r_i, s_j) - a(r_i) = \frac{1}{\gamma}\varphi(P_i^*)$. Combining with Eq. (5), Lemma 2, and $\gamma = 3$, we then have

$$\mathrm{cost}(M^{VRM}, \mathrm{mt}^{VRM}) = O(1) \sum_{i=1}^{m} \varphi(P_i^*). \tag{6}$$

4.2 Proof of Theorem 1

Recall that by Eq. (1), the optimal cost is $D(M^{OPT})$. By Eq. (6), to prove Theorem 1, it suffices to show

$$\sum_{i=1}^{m} \varphi(P_i^*) = O(\sqrt{m}\log^2 m)D(M^{OPT}). \tag{7}$$

The proof of Eq. (7) is similar to that in [33], and we give the proof in [26].

5 Concluding Remarks

A natural open question regarding Theorem 1 is whether the competitive ratio is asymptotically tight. For the MBPM problem, RM algorithm's competitive ratio in [33] is almost tight (up to a polylogarithmic factor). However, because matching can be delayed in the MBPMD problem, the lower bound instance and its analysis in [33] cannot be directly extended to the MBPMD problem.

Acknowledgement. This work was supported in part by the Ministry of Science and Technology of Taiwan under Contract MOST 111-2221-E-004-003-MY2 and by the National Science and Technology Council of Taiwan under Contract NSTC 113-2221-E-004-011-MY2. The author would like to thank the anonymous reviewers for their constructive comments.

Disclosure of Interests. The author has no competing interests to declare that are relevant to the content of this article.

References

1. Angelopoulos, S., Dürr, C., Jin, S.: Online maximum matching with recourse. J. Comb. Optim. **40**(4), 974–1007 (2020). https://doi.org/10.1007/s10878-020-00641-w
2. Antoniadis, A., Barcelo, N., Nugent, M., Pruhs, K., Scquizzato, M.: A $o(n)$-competitive deterministic algorithm for online matching on a line. In: WAOA 2014 (2014)
3. Ashlagi, I., et al.: Min-cost bipartite perfect matching with delays. In: APPROX/RANDOM 2017 (2017)
4. Ashlagi, I., Burq, M., Dutta, C., Jaillet, P., Saberi, A., Sholley, C.: Maximum weight online matching with deadlines. arXiv preprint arXiv:1808.03526 (2018)
5. Azar, Y., Chiplunkar, A., Kaplan, H.: Polylogarithmic bounds on the competitiveness of min-cost perfect matching with delays. In: ACM-SIAM SODA 2017 (2017)
6. Azar, Y., Ganesh, A., Ge, R., Panigrahi, D.: Online service with delay. In: ACM STOC 2017 (2017)
7. Azar, Y., Jacob-Fanani, A.: Deterministic min-cost matching with delays. Theor. Comput. Syst. **64**(4), 572–592 (2020)

8. Azar, Y., Lewkowicz, S., Vainstein, D.: List update with delays or time windows. In: ICALP 2024 (2024)
9. Azar, Y., Ren, R., Vainstein, D.: The min-cost matching with concave delays problem. In: ACM-SIAM SODA 2021 (2021)
10. Azar, Y., Touitou, N.: Beyond tree embeddings–a deterministic framework for network design with deadlines or delay. In: IEEE FOCS 2020 (2020)
11. Azar, Y., Touitou, N.: General framework for metric optimization problems with delay or with deadlines. In: IEEE FOCS 2019 (2019)
12. Bhore, S., Filtser, A., Toth, C.D.: Online duet between metric embeddings and minimum-weight perfect matchings. In: ACM-SIAM SODA 2024 (2024)
13. Bienkowski, M., Kraska, A., Liu, H.H., Schmidt, P.: A primal-dual online deterministic algorithm for matching with delays. In: WAOA 2018 (2018)
14. Bienkowski, M., Kraska, A., Schmidt, P.: A match in time saves nine: deterministic online matching with delays. In: WAOA 2017 (2017)
15. Borodin, A., Linial, N., Saks, M.E.: An optimal on-line algorithm for metrical task system. J. ACM **39**(4), 745–763 (1992)
16. Buchbinder, N., Naor, J., Wajc, D.: Lossless online rounding for online bipartite matching (despite its impossibility). In: ACM-SIAM SODA 2023 (2023)
17. Deryckere, L., Umboh, S.W.: Online matching with set and concave delays. In: APPROX/RANDOM 2023 (2023)
18. Emek, Y., Kutten, S., Wattenhofer, R.: Online matching: haste makes waste! In: ACM STOC 2016 (2016)
19. Emek, Y., Shapiro, Y., Wang, Y.: Minimum cost perfect matching with delays for two sources. Theoret. Comput. Sci. **754**, 122–129 (2019)
20. Fahrbach, M., Huang, Z., Tao, R., Zadimoghaddam, M.: Edge-weighted online bipartite matching. J. ACM **69**(6), 1–35 (2022)
21. Fakcharoenphol, J., Rao, S., Talwar, K.: A tight bound on approximating arbitrary metrics by tree metrics. In: ACM STOC 2003 (2003)
22. Gupta, V., Krishnaswamy, R., Sandeep, S.: Permutation strikes back: the power of recourse in online metric matching. In: APPROX/RANDOM 2020 (2020)
23. Huang, Z., Shu, X., Yan, S.: The power of multiple choices in online stochastic matching. In: ACM STOC 2022 (2022)
24. Kakimura, N., Nakayoshi, T.: Deterministic primal-dual algorithms for online k-way matching with delays. arXiv preprint arXiv:2310.18071 (2023)
25. Kaplan, H., Naori, D., Raz, D.: Online weighted matching with a sample. In: ACM-SIAM SODA 2022 (2022)
26. Kuo, T.W.: Online deterministic minimum cost bipartite matching with delays on a line (2024). arXiv arXiv:2408.02526
27. Liu, X., Pan, Z., Wang, Y., Wattenhofer, R.: Impatient online matching. In: ISAAC 2018 (2018)
28. Mari, M., Pawłowski, M., Ren, R., Sankowski, P.: Online matching with delays and stochastic arrival times. In: AAMAS 2023 (2023)
29. Matuschke, J., Schmidt-Kraepelin, U., Verschae, J.: Maintaining perfect matchings at low cost. In: ICALP 2019 (2019)
30. Megow, N., Nölke, L.: Online minimum cost matching with recourse on the line. In: APPROX/RANDOM 2020 (2020)
31. Mehta, A., et al.: Online matching and ad allocation. Found. Trends Theoret. Comput. Sci. **8**(4), 265–368 (2013)
32. Melnyk, D., Wang, Y., Wattenhofer, R.: Online k-way matching with delays and the h-metric. arXiv preprint arXiv:2109.06640 (2021)

33. Nayyar, K., Raghvendra, S.: An input sensitive online algorithm for the metric bipartite matching problem. In: IEEE FOCS 2017 (2017)
34. Raghvendra, S.: Optimal analysis of an online algorithm for the bipartite matching problem on a line. In: SoCG 2018 (2018)
35. Raghvendra, S.: A robust and optimal online algorithm for minimum metric bipartite matching. In: APPROX/RANDOM 2016 (2016)
36. Reingold, E.M., Tarjan, R.E.: On a greedy heuristic for complete matching. SIAM J. Comput. **10**(4), 676–681 (1981)
37. Touitou, N.: Improved and deterministic online service with deadlines or delay. In: ACM STOC 2023 (2023)
38. Yan, S.: Edge-weighted online stochastic matching: beating $1 - \frac{1}{e}$. In: ACM-SIAM SODA 2024 (2024)

Maximizing Throughput for Parallel Jobs with Speed-Up Curves

Kefu Lu[(✉)] and Mason Marchetti

Washington and Lee University, Lexington, VA 24450, USA
klu@wlu.edu, marchettim24@mail.wlu.edu

Abstract. We consider the problem of scheduling a set of n preemptive jobs with deadlines arriving online on m identical machines with the goal of maximizing weighted throughput. The jobs being scheduled are parallelizable and their parallelism is modelled with the standard speed-up curves model. Each job J_i arrives at time r_i with an associated deadline d_i and profit p_i which is acquired if the job is completed by its deadline. Jobs also have corresponding speed-up functions $\Gamma_i : \Re^+ \to \Re^+$. The speed-up function $\Gamma_i(y)$ describes the rate at which the job is processed when scheduled on y machines and jobs are allowed to have distinct speed-up functions. We give the first result for the throughput scheduling problem for jobs with speed-up curves using resource augmentation, by showing a $O(1+\epsilon)$ speed, $O(\frac{1}{\epsilon^2})$ competitive algorithm.

Keywords: online scheduling · parallel · throughput · speed-up curves

1 Introduction

We study the classic problem of scheduling jobs arriving online to meet their deadlines. Preemption is allowed and the scheduling objective is to maximize the weighted throughput. Each job j has a release time r_j when it arrives to the system, along with a deadline d_i and a profit (or weight) p_i. The scheduler obtains the profit of the job only if it is completed by its deadline. The weighted throughput of the schedule is defined as the total profit obtained by the scheduler. Maximizing the throughput is one popular scheduling objective used for measuring the performance of a scheduler for jobs with deadlines.

In this work, we focus on the throughput maximization problem on *parallel* jobs in the *online* setting. In the online setting, the scheduler is only given knowledge of the job once it arrives to the system. For a maximization problem, an online algorithm is considered α-competitive if, on all instances of the problem, the objective value of the optimal offline solution (adversary) is at most α times the objective value of the solution produced by the online algorithm.

We assume that the system is given m identical machines (or processors) and that the jobs are preemptive. To model the parallelism of the jobs, we use the standard speed-up curves model. Each job j has a corresponding speed-up function (or *curve*) $\Gamma_j : \Re^+ \to \Re^+$. The function $\Gamma_j(y)$ gives the rate at which job

j is processed if given y machines where $0 \leq y \leq m$ but y need not be an integer value. Each job may have a different speed-up curve and the speed-up curve is known to the scheduler once the job arrives. The speed-up curves model for parallel jobs has been utilized for many results in scheduling such as by Turek et al. in [21] and Edmonds in [9]. Typically, it is assumed that the speed-up curve is a non-decreasing, concave down, sublinear function. These assumptions are meant to capture the behavior of jobs in parallel computing - jobs should not slow down when given more machines and a job's parallelism have non-increasing returns when given more and more machines. Speed-up curves capture the variability within the parallelism of jobs, as some jobs may speed up greatly when given additional machines while other jobs may not receive much speed-up. More detailed properties of speed-up curves are given in Sect. 2.

For the objective of maximizing throughput online, there have been many works in the case of sequential jobs on a single processor. For this setting, it is known that the optimal deterministic competitiveness ratio which can be achieved is $\Theta(\Phi)$ [6,7,16] whereas the optimal randomized competitive ratio is $\Theta(\min(\log \Phi, \log \Delta))$ [13,15]. Here, Φ is the ratio of the maximum density of a job to the minimum density of a job and Δ is the ratio of the maximum length of a job to the minimum length of a job. The density of a job is the ratio of its profit to its required processing time. Additionally, the lower bound for the deterministic competitive ratio holds even if all jobs have the same profit, though the randomized competitiveness ratio does not hold [5,12]. These lower bounds can be extended to the speed-up curves setting. We briefly note that the usage of density is a key aspect of works in throughput maximization; our work will also utilize density with a different definition than normal.

To better develop and analyze algorithms when presented with these strong lower bounds for throughput, one common method of analysis is to introduce resource augmentation. Typically, the scheduler is given *speed augmentation* in the form of faster machines, allowing it to do slightly more work per unit time than the adversary. An online scheduler is s-speed c-competitive if it achieves a competitiveness ratio of c while given processors which are s times the speed of the adversary. In the introductory paper for resource augmentation in scheduling by Pruhs and Kalyanasundaram, it is shown that for any fixed constant $\epsilon > 0$, there is a $O(1+\epsilon)$-speed $O(\frac{1}{\epsilon})$-competitive scheduling algorithm on a single processor for the problem of maximizing throughput [14].

In the setting where sequential jobs are to be scheduled on m identical machines, there are also known $O(1+\epsilon)$-speed $O(\frac{1}{\epsilon})$-competitive algorithms for throughput by performing some generalizations of single machine results [4,17]. In this setting, an $O(1)$ competitive algorithm for *unweighted* throughput without resource augmentation has also been recently shown [19]. For parallel jobs modeled by speed-up curves, works in the past have studied different scheduling objectives such as maximum flow time, average flow time, and others, such as in [8,10,18,20]. For these problems, there is also some precedent for considering speed-up curves with restricted forms, for example, by assuming that $\Gamma_j(y) = y^\alpha$

for some constant α, such as the work by Im et al. [11]. However, none of these works consider the problem of maximizing the weighted throughput.

The speed-up curves model is not the only model used to describe parallel jobs. Another popular model for parallel jobs is the directed acyclic graph (DAG) model. In that model, each job j is represented by a DAG and each node in the graph is a task with unit processing time which can be processed. Each node may be run only after all its predecessors have been finished, and the job is only considered finished once all nodes have completed processing. Parallelism is captured by the structure of the DAG. The DAG model differs compared to the speed-up curves model in how it models the parallelism of the jobs - namely, the maximum possible speed-up at each moment is exactly linear up to the number of currently available nodes to run. The DAG model of parallel jobs has also received many works for different scheduling objectives, such as average and max flow time in [1], and [2]. Recently Moseley et al. in [18] considers both the speed-up curves and DAG models and show some fundamental differences in both algorithmic approaches and lower bounds for the two models in the case of flow time objectives. For the objective of maximizing the weighted throughput in the DAG model, there is a known $O(2+\epsilon)$-speed $O(\frac{1}{\epsilon})$-competitive scheduler shown by Agrawal et al. [3]. However, for the speed-up curves model there remained no known scheduler for this problem until our work.

1.1 Our Result

In our work, we show the first algorithm for maximizing the weighted throughput in the speed-up curves model.

Theorem 1. *For jobs with speed-up curves, there is a $O(1 + \epsilon)$-speed $O(\frac{1}{\epsilon^2})$-competitive algorithm for maximizing the weighted throughput.*

We build upon the work of Agrawal et al. [3] in the DAG model and show the first competitiveness result for maximizing the weighted throughput of parallel jobs in the speed-up curves model. Our result does not require the $2+\epsilon$ resource augmentation necessary in the DAG model as our result will only require $1+\epsilon$ speed augmentation. Using $1+\epsilon$ resource augmentation and achieving constant competitiveness is essentially the best possible asymptotic result for problems where resource augmentation is necessary. We believe that the difference in the necessary speed augmentation between the models arises from fundamental differences in how parallelism is captured; our analysis is able to leverage the speed-up curves to reason about the area which jobs occupy in the schedule whereas the structure of DAGs can be difficult to reason about. Our result further differentiates the two common models of parallel jobs.

1.2 Paper Organization

In Sect. 2 we give the notation and describe some features of the scheduling problem along with a few useful properties. The algorithm, the intuition behind the algorithm, and some useful properties are described in Sect. 3. The analysis of the competitiveness is presented in Sect. 4. Some proofs in Subsect. 4.2 are deferred to the full version of the paper due to the page limit.

2 Preliminaries

In the throughput scheduling problem, a set J of n jobs arrive over time to be processed on m identical machines. Each job J_i arrives at time r_i and has an absolute deadline d_i. The relative deadline of each job J_i is calculated as $D_i = d_i - r_i$. Each job has an associated profit or weight p_i which is obtained as long as the job is completed before its deadline expires. Each job requires some W_i total amount of work to be done on it to become completed. The jobs are preemptive, retaining the work done on them if interrupted. The quantities d_i, W_i and the speed-up curve are known to the scheduler once a job arrives and the goal of the scheduler is to maximize the total profit obtained.

The speed-up curve of job J_i is defined as $\Gamma_{i,s} : \Re^+ \to \Re^+$. The function $\Gamma_{i,s}(y)$ gives the work rate per unit time on job J_i when given y machines with speed s. The value of y need not be an integer and $0 \leq y \leq m$. For simplicity of notation, we often omit the speed index s when describing speed-up curves for unit speed machines, i.e. $\Gamma_i(y) := \Gamma_{i,1}(y)$. Like previous works in the speed-up curves model, speed-up functions are assumed to be non-decreasing, concave down, and sublinear in order to capture the features of parallel computations. We also assume that jobs are fully parallelizable up to one machine, that is, $\Gamma_i(y) = y$ for $0 \leq y \leq 1$, like previous works such as [11]. This is to avoid unrealistic parallelization when jobs are assigned to less than a single machine.

We state several properties about the speed-up curves. Property 2 is due to the concavity of the speed-up curves and Property 3 is due to the speed augmentation of machines.

Property 1. For any number of machines y with speed s, if a scheduler assigns job J_i to exactly y machines for $\frac{W_i}{\Gamma_{i,s}(y)}$ time in total, then the job will be finished.

Property 2. For any job J_i and machines $1 \leq m_a \leq m_b$, we have $\frac{m_a}{m_b} \leq \frac{\Gamma_{i,s}(m_a)}{\Gamma_{i,s}(m_b)}$

Property 3. For job J_i and machines with s speed, $\Gamma_{i,s}(y) = s\Gamma_{i,1}(y)$

Throughout the paper, we will use O to denote the optimal scheduler (adversary) and S to denote our scheduler. In the resource augmentation analysis, the optimal scheduler will be given machines with unit speed while our scheduler has machines with $1 + \epsilon$ speed. We can assume that all jobs J_i have relative deadline $D_i \geq \frac{W_i}{\Gamma_i(m)}$. Otherwise, that job is impossible to complete by its deadline even if run on all m machines by O and cannot result in any profit for O.

We define the **total area** of a job J_i in any schedule K on s-speed machines as $A_K(i, s)$. This is the sum of the number of machines J_i is assigned, over all time. Intuitively, this is the amount of scheduling area that the job occupies in schedule K. We also consider the total area used by a job if the job is completed by receiving an unchanging number of machines for its entire runtime. If J_i always receives y' when it is run, it will always be processed at a rate of $\Gamma_{i,s}(y')$. Thus, the time required to complete the job will be $\frac{W_i}{\Gamma_{i,s}(y')}$. We denote the **fixed machines area** as $A_{i,s}(y') = y'\frac{W_i}{\Gamma_{i,s}(y')}$ which is the total area if y' is always the number of machines used to run J_i.

We can compute a **minimum total area** required for a job. For each job J_i, let $m_{i,min}$ be the minimum number of machines such that $D_i\Gamma_{i,1}(m_{i,min}) = W_i$. This is the exact number of machines required to finish J_i right at its deadline if run continuously after it arrives. If the job is complete this way, by using $m_{i,min}$ machines, a total area of $A_{i,1}(m_{i,min})$ will be used in the schedule. If a schedule finishes a job earlier than its deadline, it must use more than $m_{i,min}$ machines at some point. Since the speed-up curves are at most linear, using more machines will not decrease the time required to complete the job by more than a linear amount. Thus, the total area used will not decrease. Hence, there is a minimum total area which must be spent by any unit speed scheduler to complete a job J_i by its deadline. We summarize this minimum total area in the following proposition.

Proposition 1. *In order to finish job J_i by its deadline, a scheduler with unit speed must spend at least $A_{i,1}(m_{i,min}) = m_{i,min}D_i$ area in its schedule on that job.*

3 Algorithm for Maximizing the Throughput

In this section, we will describe the scheduling algorithm S and give important properties useful for proving its competitiveness. We let δ, k, and c be positive constants which we will set to specific values in later sections (in Lemma 1, and Lemma 6). Notably, $\delta < \epsilon/3$, $k < 1$, and $c > 1$.

One key decision of the scheduler is the number of machines to assign to a job J_i when it decides to run J_i. Since there is no additional profit from finishing a job earlier than its deadline and because of the concavity of speed-up curves, the scheduler should try to avoid assigning too many machines to J_i, especially when there are many jobs. As described in Sect. 2, assigning many more machines than necessary can only increase the total area a job uses - and the scheduler has a fixed amount of total area in its schedule for running jobs. However, S should also give a sufficient number of machines so that J_i does not immediately miss its deadline if it is interrupted briefly. This must be balanced with the previous intuition. We will show that with resource augmentation, it is possible to come up with a competitive strategy for the throughput by always assigning a fixed number of machines for a job when it is run.

For each job J_i, we define $x_i = \frac{D_i}{1+2\delta}$, a quantity of time slightly smaller than its relative deadline. Let m_i be the smallest number of machines such that $x_i\Gamma_{i,1+\epsilon}(m_i) = W_i$. That is, if J_i is assigned exactly m_i machines for a total of x_i time by S, the job will be completed by its deadline. When S decides to run a job J_i, it will always allocate exactly m_i machines to it. Intuitively, we choose this number so that the job will be completed by its deadline if run using m_i machines, even if it is interrupted briefly by other jobs. We define the *density* of job J_i as $v_i = \frac{p_i}{x_i m_i}$, this is essentially the amount of profit per unit area of the job. Notice that density is defined based on scheduler S rather than solely based on the total amount of work of the job, in contrast to previous works on throughput.

3.1 The Scheduler

Scheduler S will keep two priority queues of jobs, Q_s and Q_u, in which jobs are ordered from highest to lowest density. First, we give a high level description of the scheduler.

Q_s will contain jobs which are being considered by S for running at the current moment, and Q_u will contain jobs which have arrived but are not in consideration of being run at the moment. We will call the jobs which have been placed in Q_s as *started*; jobs will only ever be moved from Q_u to Q_s (*starting* them), never the other way, though some jobs may be removed from both queues at times when they are completed or have expired deadlines. When jobs are moved into Q_s, the jobs currently being actively run may change and some preemption may occur. Jobs are only considered for placement in Q_s when it arrives or when a job is completed.

Essentially, jobs within Q_s will be processed according to their densities and jobs will only be added to Q_s if there are not already too many other jobs with similar density. Additionally, we will call J_i a *fresh* job at time t if $d_i - t \geq (1+\delta)x_i$. Note that since $D_i = (1+2\delta)x_i$, all jobs are fresh when they arrive as $d_i - r_i = (1+2\delta)x_i > (1+\delta)x_i$.

Now, we will describe the scheduler in detail, focusing on the processing of jobs and then the movement of jobs into the queues.

Processing Jobs: Let $U(t)$ be the amount of unused machines at time t. The scheduler will first remove all jobs in Q_s that are already completed or have expired deadlines. Then the scheduler will consider all jobs in Q_s in order and allocate machines to them. For each job $J_i \in Q_s$, if $U(t) \geq m_i$ then the scheduler will assign it m_i machines and decrease $U(t)$ by m_i, and if there are insufficient machines then no machines are assigned to J_i. The scheduler then moves to consider the next job in Q_s. The scheduler will continue to allocate machines until all jobs in Q_s have been considered.

Job Arrival: When a job J_i arrives, it will be put into Q_s if it fulfills some density conditions (which we describe below), otherwise it will be put into Q_u. Specifically, let $Q'_s = Q_s \cup \{J_i\}$ and consider any job $J_j \in Q'_s$. Let $M(Q'_s, v_j, cv_j)$ be the total amount of machines required by jobs in Q'_s with densities within the range $[v_j, cv_j)$. The scheduler will put J_i in Q_s if, for all jobs J_j, the condition $M(Q'_s, v_j, cv_j) \leq km$ holds. That is J_i is put into Q_s only if after doing so, jobs with similar densities would not use too many of the machines.

Job Completion: When a job is completed, it is removed from Q_s. At this current time t, the scheduler will consider all jobs in Q_u in order and decide whether any should be moved to Q_s. Consider any $J_i \in Q_u$. If the job J_i is not fresh at time t, it is discarded. Otherwise, S checks a similar density condition to when J_i arrives: Let $Q'_s = Q_s \cup \{J_i\}$. The scheduler will only add J_i to Q_s if and only if for all jobs J_j, we have $M(Q'_s, v_j, cv_j) \leq km$.

3.2 Properties of the Scheduler

We will establish some properties of S before proving its competitiveness. We begin by showing that the number of machines assigned to any job J_i is intuitively the right amount. First, we show that for any J_i, the number of machines m_i the algorithm will use assign to the job will be a constant factor less than all the machines, m. This is not immediately obvious as m_i is chosen such that the job would be finished in less time than its relative deadline, which might seem to require many additional machines if the speed-up curve is only a slightly increasing function. However, the resource augmentation of S allows us to show that less than m machines will be assigned to any job.

Lemma 1. *For any job J_i, we have $m_i \leq \frac{(1+2\delta)}{(1+\epsilon)} m = k^2 m$. We set $k = (\frac{1+2\delta}{1+\epsilon})^{1/2}$*

Proof. Consider any job J_i. By definition, m_i is chosen so that $W_i = x_i \Gamma_{i,1+\epsilon}(m_i) = \frac{D_i}{1+2\delta} \Gamma_{i,1+\epsilon}(m_i)$. We can rearrange this equation to $\Gamma_{i,1+\epsilon}(m_i) = \frac{(1+2\delta)W_i}{D_i}$. Now we combine this equation with Property 3 and the fact that $D_i \geq \frac{W_i}{\Gamma_i(m)}$ for all jobs:

$$\Gamma_{i,1+\epsilon}(m_i) = \frac{(1+2\delta)W_i}{D_i} \leq \frac{(1+2\delta)W_i}{\frac{W_i}{\Gamma_i(m)}} = (1+2\delta)\Gamma_i(m)$$

$$\frac{(1+\epsilon)\Gamma_i(m_i)}{\Gamma_i(m)} = \frac{\Gamma_{i,1+\epsilon}(m_i)}{\Gamma_i(m)} \leq (1+2\delta)$$

$$\frac{m_i}{m} \leq \frac{\Gamma_i(m_i)}{\Gamma_i(m)} \leq \frac{(1+2\delta)}{(1+\epsilon)}$$

$$m_i \leq \frac{(1+2\delta)}{(1+\epsilon)} m$$

The final inequalities follow due to the concavity of the speed-up curves in Property 2. Since we set $\delta < \epsilon/3$, we know that $k < 1$ and thus $m_i < m$. □

We will also show that the number of machines m_i assigned by S to any job J_i is no more than the minimum number needed by any scheduler with 1 speed. The resource augmentation of S is the key here.

Lemma 2. *For any job J_i, we have that $m_i \leq m_{i,min}$.*

Proof. By the definition of $m_{i,min}$, it relates to the work of the job by $W_i = D_i \Gamma_{i,1}(m_{i,min})$. In our scheduler S, the value of m_i is chosen such that $W_i = x_i \Gamma_{i,1+\epsilon}(m_i) = \frac{D_i}{1+2\delta} \Gamma_{i,1+\epsilon}(m_i) = \frac{1+\epsilon}{1+2\delta} D_i \Gamma_{i,1}(m_i)$.

Therefore, $D_i \Gamma_i(m_{i,min}) = \frac{1+\epsilon}{1+2\delta} D_i \Gamma_i(m_i)$ with $\frac{1+\epsilon}{1+2\delta} > 1$. Thus, $\Gamma_i(m_{i,min}) \geq \Gamma_i(m_i)$. As speed-up curves are non-decreasing, we have that $m_i \leq m_{i,min}$. □

The previous lemmas show the number of machines assigned by S to a job is not too high which will useful for proving the competitiveness of the scheduler.

Next, we establish some other properties about the behavior of S when it processes jobs.

Lemma 3. *At any time, for any density $v > 0$ and the set of jobs that are being processed, Q_s, it is true that $M(Q_s, v_j, cv_j) \leq km$. That is, the number of machines required to run jobs within any density range is at most km.*

Proof. This is by definition of the algorithm. A job can only be added to Q_s if none of the density ranges for any job requires more than km machines.

Lemma 4. *If any job J_i was added to Q_s but not completed by the scheduler, then that job must have been delayed by other jobs for at least δx_i time while it is in Q_s.*

Proof. By the definition of the algorithm, J_i can only be put in Q_s either when it arrives or when another job is completed. In either case, it must be fresh at that time. Specifically, J_i is started at time t where $d_i - t \geq (1+\delta)x_i$. However, J_i only needs x_i out of this $(1+\delta)x_i$ amount of time to be finished since it will be run using m_i machines when it runs. Therefore, if J_i is not completed, the scheduler must have spent more than δx_i time not running J_i while it is in Q_s. During that amount of time S can only be delaying J_i if S was running other jobs with its machines. □

We now briefly sketch the intuition behind the proof of the competitiveness of S:

1. If a job J_i is put into Q_s at some point by the scheduler S but not finished, then for a nontrivial amount of time while J_i was in Q_s, the scheduler was not running Q_s. At those times, it must be running jobs with density higher than J_i. Hence, the denser jobs can pay for J_i through amortization.
2. Thus, if the scheduler S is far behind the optimal scheduler O in profit, perhaps O obtains a lot of profit from jobs that S never puts into Q_s. However, if there are jobs, say some J_j which S never starts, it must be because S is too busy running jobs within a similar density range, with density of at least v_j/c. Here, those somewhat dense jobs should be able to pay for job J_j.

We will formalize these two statements in the following sections.

4 Analysis of the Algorithm

4.1 Jobs Which S Starts

In this section, we show that the profit of all jobs which scheduler S starts is bounded by the profit of all jobs which S finishes. Recall that a job is considered *started* if S puts it into Q_s at some point. Since we are comparing the jobs done by S against the jobs started by S itself, we avoid utilizing the speed augmentation of S when making comparisons.

We let R be the set of jobs which S starts (puts into Q_s at some point) and let $C \subseteq R$ be the set of jobs completed by S. Also let $U = R \setminus C$ denote the other set of jobs which are started but do not get completed by S. We will call a job J_i to be v-dense if its density is at least v, i.e. $v_i \geq v$. Finally, we will use the notation $||J||$ to denote the total profit of the jobs in set J.

Lemma 5. *For each job J_i in U, it must be the case that S ran jobs with density at least cv_i for a period of δx_i time while J_i was in Q_s, using at least $(1-k)m$ machines during each moment of that time.*

Proof. While J_i was in Q_s, at each moment in time, S either ran J_i or it did not. By Lemma 4 it must have been delayed for more than δx_i time to remain unfinished. Since S runs jobs from high to low density, J_i can only be delayed by jobs with density at least v_i. Jobs with density within the range $[v_i, cv_i)$ use at most km processors by Lemma 3 and the rest of the machines are running cv_i dense jobs. Thus, at least $(1-k)m$ machines ran cv_i dense jobs for δx_i time.

Lemma 6. *For some constant $c > 1 + \frac{2}{\delta k(1-k)}$, the profit of jobs started by S is bounded:*

$$(1 - \frac{k}{(1-k)\delta(c-1)})||R|| \leq ||C||$$

Proof. We will use an amortization scheme to show this result. The jobs which are completed by S will pay for the jobs which it started but did not complete. This is possible because S runs jobs with high density.

Let each $J_i \in R$ possess an account A_i of credits. Each $J_j \in C$ initially starts with a balance of p_j credits; the other jobs $J_k \in U$ initially begin with a balance of 0 credits. Notice that the total amount of starting credit is exactly $||C||$. We will describe a credit transfer scheme based on job processing such that every account A_i for job $J_i \in R$ will end up with at least $(1 - \frac{k}{(1-k)\delta(c-1)})p_i$ credits. A summation over all accounts in R will then prove the lemma.

We first give the transfer scheme, then we show that all jobs receive a sufficient income of credits, and then finally we show that not too many credits are transferred out of each job.

The transfer scheme is as follows: For each moment of time that a job J_i is being run, each machine running J_i will transfer $\frac{v_j m_j}{\delta m(1-k)}$ credits from A_i to all jobs J_j with density $v_j \leq \frac{v_i}{c}$.

Now we show that each job J_i receives at least its profit, p_i, in credits. All $J_i \in C$ receive p_i credits initially, so we focus on jobs in U. Consider some $J_i \in U$. Since J_i is an unfinished job, many dense jobs (with density at least cv_i) must have delayed this job. By Lemma 5, there are at least $(1-k)m$ machines running denser jobs for a time period of at least δx_i. Using this and the transfer scheme, we bound the amount of credits received by J_i from other jobs:

$$\delta x_i (1-k) m \frac{v_i m_i}{\delta m (1-k)} = x_i v_i m_i \frac{1-k}{1-k} = p_i \quad (1)$$

We next show that jobs do not transfer too much credit away. Each job J_i can be run for at most x_i time by the scheduler. During each moment within this time, machines which are running it will transfer credit to all jobs in Q_s with density less than $\frac{v_i}{c}$. We geometrically group these jobs by their density, to be the groups $G(Q_s, \frac{v_i}{c^{l+1}}, \frac{v_i}{c^l})$ for increasing integers $l \geq 1$. By Lemma 3, the total number of machines required by each group is at most km. Therefore, we can bound the amount of credits transferred by J_i to the other jobs in each group:

$$\sum_{J_j \in G(Q_s, \frac{v_i}{c^{l+1}}, \frac{v_i}{c^l})} \frac{v_j m_j}{\delta m(1-k)} \leq \sum_{J_j \in G(Q_s, \frac{v_i}{c^{l+1}}, \frac{v_i}{c^l})} \frac{\frac{v_i}{c^l} m_j}{\delta m(1-k)}$$

$$\leq \frac{v_i}{\delta m(1-k)c^l} \sum_{J_j \in G(Q_s, \frac{v_i}{c^{l+1}}, \frac{v_i}{c^l})} m_j$$

$$\leq \frac{v_i}{\delta m(1-k)c^l} km = \frac{v_i k}{\delta(1-k)c^l}$$

Since $c > 1$, each group has geometrically decreasing densities. We can sum over all geometric groups to obtain the total amount of credits transferred by J_i. Whenever J_i is run, there are exactly m_i machines running it. The credits transferred away from J_i to all groups during each moment it is run is bounded:

$$m_i \sum_{l=1}^{\infty} \frac{v_i k}{\delta(1-k)c^l} = \frac{m_i v_i k}{\delta(1-k)} \sum_{l=1}^{\infty} \frac{1}{c^l} = \frac{m_i v_i k}{\delta(1-k)} \sum_{l=1}^{\infty} (\frac{1}{c})^l = \frac{m_i v_i k}{(1-k)\delta} \frac{1}{c-1}$$

Since J_i is run for no more than x_i amount of time before it completes, the total credits transferred away from is at most:

$$x_i \frac{m_i v_i k}{(1-k)\delta} \frac{1}{c-1} = p_i \frac{k}{(1-k)\delta(c-1)} \quad (2)$$

Recall that the amount of credits each $J_i \in R$ receives has been bounded previously. We combine bounds from (1) and (2) to determine the amount of credits that every job J_i has after the amortization scheme:

$$p_i - p_i \frac{k}{(1-k)\delta(c-1)} = p_i (1 - \frac{k}{(1-k)\delta(c-1)}) \quad (3)$$

By setting $c > (1 + \frac{2}{\delta k(1-k)})$, the expression $\frac{k}{(1-k)\delta(c-1)}$ is always less than $\frac{1}{2}$ and thus every job has some credits. We can sum (3) over every job $J_i \in R$ to obtain the profit $||R||$ multiplied by a constant factor. We began with exactly $||C||$ credits and the total amount of credits is conserved. Hence, the profit of the jobs started by S is bounded by the profits of all jobs completed by S with $(1 - \frac{k}{(1-k)\delta(c-1)})||R|| \leq ||C||$, which is the lemma. □

4.2 Jobs Which O Completes

In this section, we bound the profits of all jobs the optimal scheduler O completes by the profits of all jobs S starts. Here we will be leveraging that S has machines with $1+\epsilon$ speed when compared to the adversary O. We may assume that O completes every job it works on, as it receives no profit from the job if it processes but does not complete it. Furthermore, from the previous Lemma 6, we know that if a job is started by S, then this job's profit can already be accounted for. Hence, our analysis will focus on jobs which O completes but S never starts.

We will first establish some properties about which jobs are being run by the S, and then we will use them to reason about the conditions for jobs which S does not start. The proofs of Lemma 7 and Lemma 8 mainly involve the specific conditions in their statements and are found in the full version of this paper.

Lemma 7. *At any time t for any density v_i, **if** there are less than $k(1-k)m$ machines running jobs with density at least v_i/c, **then** all such v_i/c-dense jobs in Q_s are being run by S.*

Lemma 8. *For any density v_i, **if** there are less than $k(1-k)m$ machines running v_i/c-dense jobs and a new job v_i-dense job J_j arrives or a job completes and there exists a v_i-dense, fresh job J_j, **then** that J_j will be started by the scheduler.*

Let some density be $v > 0$ and some set of jobs be K. We will let $T_O(v, K)$ denote the total area used by the optimal schedule O on jobs in K which are v-dense. We will let $T_S(\frac{v}{c}, K)$ denote the total area used by S on all jobs in K which are v/c-dense. We will bound the total area O uses to run jobs which S never adds to Q_s. Recall that the set J refers to the set of all jobs.

Lemma 9. *Consider all jobs in $J \setminus R$, the jobs which are never started by S (never added to Q_s). For all $v > 0$, we have $T_O(v, J \setminus R) \leq \frac{1+2\delta}{\delta} \frac{1}{k(1-k)} T_S(\frac{v}{c}, J)$.*

Proof. We define a set of time intervals $I_1, I_2, \ldots, I_f$ with each $I_i = [s_i, e_i]$ where, within each time interval, S is using at least $k(1-k)m$ machines to run $\frac{v}{c}$-dense jobs. These intervals place a natural bound on the total area spent on $\frac{v}{c}$-dense jobs and also place some restrictions on the arrival time of the jobs which are never started by S, as we will show.

Recall that the total area of a set of jobs is the sum over time of the number of machines used. Since S uses at least $k(1-k)m$ machines during each interval, summing over the intervals gives a bound on the total area of S's schedule that is spent on $\frac{v}{c}$-dense jobs. Specifically, $\sum_{i=1}^{f}(e_i - s_i)k(1-k)m \leq T_S(\frac{v}{c}, J)$.

Consider some v-dense job J_i which S **never starts**. Let its arrival time be r_i and let r_i lie within the interval $[s_i, s_{i+1}]$. If r_i is within $[e_i, s_{i+1}]$ then, by Lemma 8, J_i would have been started by S at r_i since this is between two intervals and thus less than $k(1-k)m$ machines are used to run $\frac{v}{c}$-dense jobs - a contradiction since J_i is assumed to never start. Therefore, r_i must be within some $[s_i, e_i]$ - the job must have arrived during some interval.

Furthermore, J_i must not be started by S at e_i. At this time, e_i, some job must have just been completed by S because the interval $[s_i, e_i]$ where S uses most of its machines to run $\frac{v}{c}$-dense jobs comes to an end. By Lemma 8, J_i would be added to Q_s here unless J_i is no longer a fresh job at this time.

When J_i arrives at r_i, it is known that it has $D_i = (1+2\delta)x_i$ which means $d_i - r_i = (1+2\delta)x_i$. Since it is no longer fresh at e_i, we have $d_i - e_i < (1+\delta)x_i$, thus a sufficient amount of time must have passed: $e_i - r_i \geq \delta x_i$. Therefore, $d_i - e_i < (1+\delta)x_i = \frac{1+\delta}{\delta}\delta x_i \leq \frac{1+\delta}{\delta}(e_i - r_i) \leq \frac{1+\delta}{\delta}(e_i - s_i)$. This bound is true for every such job.

Now consider the set of all v-dense jobs which arrive during $[s_i, e_i]$ which are not started by S and let this set be denoted U_i. For any job $J_i \in U_i$, its deadline is d_i and thus O can only run them during $[s_i, d_i]$ and O can use at most m machines on them. Therefore, the total area that O can possibly spend on these jobs is be bounded by the following:

$$T_O(v, U_i) \leq (d_i - s_i)m = ((d_i - e_i) + (e_i - s_i))m$$
$$\leq (\frac{1+\delta}{\delta}(e_i - s_i) + (e_i - s_i))m = \frac{1+2\delta}{\delta}(e_i - s_i)m$$

Since v-dense jobs which S never starts must have arrived during the intervals, we sum over all intervals $I_1, \ldots, I_f$ to obtain the lemma:

$$T_O(v, J \setminus R) = \sum_{i=1}^{f} T_O(v, U_i) \leq \frac{1+2\delta}{\delta} \sum_{i=1}^{f}(e_i - s_i)m \leq \frac{1+2\delta}{\delta} \frac{1}{k(1-k)} T_s(\frac{v}{c}, J)$$

□

Now we will use the bound on O's use of total area in order to bound the profit of the optimal solution compared to the profit of R, the jobs S started. Let C^* denote the set of jobs that the adversary O completes.

Lemma 10. *The profit of jobs that the adversary completes is bounded by the profit of jobs S starts:*

$$||C^*|| \leq \left(1 + \frac{c}{\delta k(1-k)}\right)||R||$$

Proof. We assume that O completes every job that it works on, as it receives no profit for incomplete jobs. Let C^* be split into two sets, C_U^* and C_R^* where C_R^* contains jobs which our scheduler also started at some point. Obviously, $||C_R^*|| \leq ||R||$ since $C_R^* \subseteq R$ so we will focus on bounding $||C_U^*||$.

We arrange the jobs in C_U^* by density for analysis; let the densities of jobs in C_U^* be arranged as $\{\phi_1, \phi_2, \ldots, \phi_z\}$ in decreasing order. We set $\phi_0 = \infty$ and $\phi_{z+1} = 0$ for convenience.

The optimal scheduler completed every job that it starts. By Proposition 1, any scheduler with unit speed must spend at least $A_{i,1}(m_{i,min})$ total area on job J_i if it completes the job by its deadline. Therefore, for job J_i with density ϕ_i, the scheduler O must have spent at least $A_{i,1}(m_{i,min})$ area. We will let B_i denote the total area our algorithm spends on jobs with densities within $(\frac{\phi_{i-1}}{c}, \frac{\phi_i}{c}]$, these are jobs with a constant c factor less density than ϕ_i.

From Lemma 9, we know that the total area O uses for jobs with any fixed density is bounded. Specifically, for any fixed density ϕ_y where $1 \leq y < z$ we have

$T_O(\phi_y, J \setminus R) \leq \frac{1+2\delta}{\delta k(1-k)} T_S(\frac{\phi_y}{c}, J)$. Since O must spend at least $A_{i,1}(m_{i,min})$ total area on each job it completes and the densities are in decreasing order, we have that $\sum_{i=1}^{y} A_{i,1}(m_{i,min}) \leq T_O(\phi_y, J \setminus R)$. Also, by definition $T_S(\frac{\phi_y}{c}, J) = \sum_{i=1}^{y} B_i$. Therefore, for each density ϕ_y:

$$\sum_{i=1}^{y} A_{i,1}(m_{i,min}) \leq T_O(\phi_y, J \setminus R) \leq T_S(\frac{\phi_y}{c}, J) = \frac{1+2\delta}{\delta k(1-k)} \sum_{i=1}^{y} B_i \quad (4)$$

We sum the leftmost and rightmost expressions of (4) over all densities in $1 \ldots z$ while ensuring that each density is counted exactly once.

$$\sum_{v=1}^{z} \left[(\phi_y - \phi_{y+1}) \sum_{i=1}^{y} A_{i,1}(m_{i,min}) \right] \leq \sum_{y=1}^{z} \left[(\phi_y - \phi_{y+1}) \frac{1+2\delta}{\delta k(1-k)} \sum_{i=1}^{y} B_i \right]$$

We reorganize each side of the inequality independently. The LHS becomes:

$$\sum_{v=1}^{z} \left[(\phi_y - \phi_{y+1}) \sum_{i=1}^{y} A_{i,1}(m_{i,min}) \right] = \sum_{i=1}^{z} A_{i,1}(m_{i,min}) \sum_{y=i}^{z} (\phi_y - \phi_{y+1})$$

$$= \sum_{i=1}^{z} A_{i,1}(m_{i,min})(\phi_i - \phi_{m+1})$$

$$= \sum_{i=1}^{z} A_{i,1}(m_{i,min})\phi_i$$

We simplify the LHS using the definition of density and minimum total area.

$$\sum_{i=1}^{z} A_{i,1}(m_{i,min})\phi_i = \sum_{i=1}^{z} A_{i,1}(m_{i,min}) \frac{p_i}{x_i m_i} = \sum_{i=1}^{z} (m_{i,min} D_i) \frac{p_i}{x_i m_i}$$

$$= \sum_{i=1}^{z} \frac{m_{i,min}}{m_i} \frac{(1+2\delta)x_i}{x_i} p_i \geq (1+2\delta)||C_U^*|| \quad (5)$$

Here, Lemma 2 is used in the final inequality to conclude that $\frac{m_{i,min}}{m_i} \geq 1$. We next reorganize the RHS similarly to the LHS:

$$\sum_{y=1}^{z} \left[(\phi_y - \phi_{y+1}) \frac{1+2\delta}{\delta k(1-k)} \sum_{i=1}^{y} B_i \right] = \frac{1+2\delta}{\delta k(1-k)} \sum_{i=1}^{z} B_i \phi_i$$

We simplify the RHS by noting that $\sum_{i=1}^{z} B_i \frac{\phi_i}{c} \leq ||R||$.

$$\frac{1+2\delta}{\delta k(1-k)} \sum_{i=1}^{z} B_i \phi_i \leq \frac{1+2\delta}{\delta k(1-k)} c||R|| \quad (6)$$

We put together the LHS (5) and the RHS (6) within the inequality:

$$(1+2\delta)||C_U^*|| \le \frac{1+2\delta}{\delta k(1-k)}c||R|| \implies ||C_U^*|| \le \frac{c}{\delta k(1-k)}||R||$$

Finally, we show Lemma 10 by using substitution for each part of C^*.

$$||C^*|| = ||C_R^*|| + ||C_U^*|| \le \left(1 + \frac{c}{\delta k(1-k)}\right)||R||$$

□

4.3 Completing the Proof

Finally, we prove the main lemma by bounding the profit of O.

Lemma 11

$$||C^*|| \le \frac{1 + \frac{1}{\delta k(1-k)}(1 + \frac{2}{\delta k(1-k)})}{1 - \frac{k}{(1-k)\delta(c-1)}}||C|| \qquad (7)$$

Proof. This is simply by combining Lemma 6 and Lemma 10.

By setting c according to the value given in Lemma 6 and noting that $0 < k < 1$, the denominator of (7) is always larger than $1/2$. Furthermore, since $\delta < \epsilon/3$, the numerator is $O(\frac{1}{\epsilon^2})$. Thus, we have bounded the profit of the optimal scheduler O by our scheduler S which uses $1 + \epsilon$ speed, this proves Theorem 1, the main theorem of the paper.

□

5 Conclusion

We study the classic problem of scheduling jobs with deadlines to maximize the weighted throughput, but in the case that the jobs are parallel with an arbitrary speed-up curve. The speed-up curves model is one of the two main models for parallel jobs along with the DAG model and we give the first result for maximizing throughput in the speed-up curves model. Our result further differentiates the two models in terms of analysis, as we show that only $1 + \epsilon$ speed is required to achieve constant competitiveness in the speed-up curves model.

In future work, we would like to see if this result could be extended to other objectives and whether there exists a scheduler which can handle jobs with multiple parallel phases (with different speed-up functions for each phase) while maximizing the throughput.

References

1. Agrawal, K., Li, J., Lu, K., Moseley, B.: Scheduling parallel DAG jobs online to minimize average flow time. In: Proceedings of the Twenty-Seventh Annual ACM-SIAM Symposium on Discrete Algorithms, pp. 176–189. SIAM (2016)
2. Agrawal, K., Li, J., Lu, K., Moseley, B.: Scheduling parallelizable jobs online to minimize the maximum flow time. In: Proceedings of the 28th ACM Symposium on Parallelism in Algorithms and Architectures, pp. 195–205 (2016)
3. Agrawal, K., Li, J., Lu, K., Moseley, B.: Scheduling parallelizable jobs online to maximize throughput. In: Bender, M.A., Farach-Colton, M., Mosteiro, M.A. (eds.) LATIN 2018. LNCS, vol. 10807, pp. 755–776. Springer, Cham (2018). https://doi.org/10.1007/978-3-319-77404-6_55
4. Bansal, N., Chan, H.-L., Pruhs, K.: Competitive algorithms for due date scheduling. In: Arge, L., Cachin, C., Jurdziński, T., Tarlecki, A. (eds.) ICALP 2007. LNCS, vol. 4596, pp. 28–39. Springer, Heidelberg (2007). https://doi.org/10.1007/978-3-540-73420-8_5
5. Baruah, S.K., Haritsa, J., Sharma, N.: On-line scheduling to maximize task completions. In: 1994 Proceedings Real-Time Systems Symposium, pp. 228–236 (1994)
6. Baruah, S.K., et al.: On the competitiveness of on-line real-time task scheduling. Real-Time Syst. **4**(2), 125–144 (1992)
7. Baruah, S.K., Koren, G., Mishra, B., Raghunathan, A., Rosier, L.E., Shasha, D.E.: On-line scheduling in the presence of overload. In: 32nd Annual Symposium on Foundations of Computer Science, pp. 100–110 (1991)
8. Chan, H., Edmonds, J., Pruhs, K.: Speed scaling of processes with arbitrary speedup curves on a multiprocessor. Theor. Comput. Syst. **49**(4), 817–833 (2011)
9. Edmonds, J.: Scheduling in the dark. Theoret. Comput. Sci. **235**(1), 109–141 (2000)
10. Edmonds, J., Pruhs, K.: Scalably scheduling processes with arbitrary speedup curves. ACM Trans. Algorithms **8**(3), 28:1–28:10 (2012)
11. Im, S., Moseley, B., Pruhs, K., Torng, E.: Competitively scheduling tasks with intermediate parallelizability. ACM Trans. Parallel Comput. (TOPC) **3**(1), 1–19 (2016)
12. Kalyanasundaram, B., Pruhs, K.: Maximizing job completions online. J. Algorithms (1998)
13. Kalyanasundaram, B., Pruhs, K.: Fault-tolerant real-time scheduling. Algorithmica **28**(1), 125–144 (2000)
14. Kalyanasundaram, B., Pruhs, K.: Speed is as powerful as clairvoyance. J. ACM **47**(4), 617–643 (2000)
15. Koren, G., Shasha, D.E.: MOCA: A multiprocessor on-line competitive algorithm for real-time system scheduling. Theoret. Comput. Sci. **128**(1&2), 75–97 (1994)
16. Koren, G., Shasha, D.E.: Dover: an optimal on-line scheduling algorithm for overloaded uniprocessor real-time systems. SIAM J. Comput. **24**(2), 318–339 (1995)
17. Lucier, B., Menache, I., Naor, J.S., Yaniv, J.: Efficient online scheduling for deadline-sensitive jobs: extended abstract. In: Proceedings of the Twenty-Fifth Annual ACM Symposium on Parallelism in Algorithms and Architectures, SPAA 2013, pp. 305–314 (2013)
18. Moselely, B., Zhang, R., Zhao, S.: Online scheduling of parallelizable jobs in the directed acyclic graphs and speed-up curves models. Theoret. Comput. Sci. **938**, 24–38 (2022)
19. Moseley, B., Pruhs, K., Stein, C., Zhou, R.: A competitive algorithm for throughput maximization on identical machines. In: 23rd International Conference on Integer Programming and Combinatorial Optimization, IPCO 2022, pp. 402–414 (2022)

20. Pruhs, K., Robert, J., Schabanel, N.: Minimizing maximum flowtime of jobs with arbitrary parallelizability. In: 8th International Workshop on Approximation and Online Algorithms, WAOA 2010, pp. 237–248 (2010)
21. Turek, J., Wolf, J.L., Yu, P.S.: Approximate algorithms scheduling parallelizable tasks. In: Proceedings of the Fourth Annual ACM Symposium on Parallel Algorithms and Architectures, SPAA 1992, pp. 323–332. Association for Computing Machinery, New York, NY, USA (1992)

Improved Approximation Algorithms for Covering Pliable Set Families and Flexible Graph Connectivity

Zeev Nutov[✉]

The Open University of Israel, Ra'anana, Israel
nutov@openu.ac.il

Abstract. A classic result of Williamson, Goemans, Mihail, and Vazirani [STOC 1993: 708–717] states that the problem of covering an uncrossable set family by a min-cost edge set admits approximation ratio 2, by a primal-dual algorithm with a reverse delete phase. Recently, Bansal, Cheriyan, Grout, and Ibrahimpur [ICALP 2023: 15:1-15:19] showed that this algorithm achieves approximation ratio 16 for a larger class of so called γ-pliable set families, that have much weaker uncrossing properties. In this paper we will improve the approximation ratio to 10. Using this result and other techniques, we also improve approximation ratios for the following two problems related to the CAPACITATED k-EDGE CONNECTED SPANNING SUBGRAPH (CAP-k-ECSS) problem.

- NEAR MIN-CUTS COVER: Given a graph $G_0 = (V, E_0)$ and an edge set E on V with costs, find a min-cost edge set $J \subseteq E$ that covers all cuts with at most $k-1$ edges of the graph G_0. We improve the approximation ratio from 16 to 10. We also obtain approximation ratio $k - \lambda_0 + 1 + \epsilon$, where λ_0 is the edge connectivity of G_0, which is better than ratio 10 when $k - \lambda_0 \leq 8$.
- (k, q)-FLEXIBLE GRAPH CONNECTIVITY $((k, q)$-FGC): Given a graph $G = (V, E)$ with edge costs, a set $U \subseteq E$ of "unsafe" edges, and integers k, q, find a min-cost subgraph H of G such that every cut of H has at least k safe edges or at least $k + q$ edges. We will show that $(k, 1)$-FGC admits approximation ratio $3.5 + \epsilon$ if k is odd (improving previous ratio 4), that $(k, 2)$-FGC admits approximation ratio $7 + \epsilon$ (improving previous ratio 20), and that $(k, 3)$-FGC admits approximation ratio 16 for k even (improving previous ratio 22). We also show that for unit costs, (k, q)-FGC admits approximation ratio $\alpha + \frac{2q}{k}$, where $\alpha \approx 1 + O(1/k)$ is an approximation ratio for the MIN-SIZE k-EDGE-CONNECTED SPANNING SUBGRAPH problem.

Keywords: primal dual method · pliable set family · capacitated k-edge-connected subgraph · flexible graph connectivity · approximation algorithms

1 Introduction

Let $G = (V, E)$ be graph. For $J \subseteq E$ and $S \subseteq V$ let $\delta_J(S)$ denote the set of edges in J with exactly one end in S, and let $d_J(S) = |\delta_J(S)|$ be their number. An edge set J **covers** S if $d_J(S) \geq 1$. The following generic meta-problem captures dozens of specific network design problems, among them STEINER FOREST, (R,k)-CONSTRAINED FOREST, POINT-TO-POINT CONNECTION, STEINER NETWORK AUGMENTATION, and many more.

SET FAMILY EDGE COVER
Input: A graph $G = (V, E)$ with edge costs $\{c_e : e \in E\}$, a set family $\mathcal{F}$ on V.
Output: A min-cost edge set $J \subseteq E$ such that $d_J(S) \geq 1$ for all $S \in \mathcal{F}$.

In this problem the family $\mathcal{F}$ may not be given explicitly, but we will require that some queries related to $\mathcal{F}$ can be answered in time polynomial in $n = |V|$. Specifically, following previous work, we will require that for any edge set I, the inclusion minimal members of the **residual family** $\mathcal{F}^I = \{S \in \mathcal{F} : d_I(S) = 0\}$ of $\mathcal{F}$ (the family of sets in $\mathcal{F}$ that are uncovered by I) can be computed in time polynomial in $n = |V|$.

Agrawal, Klein and Ravi [2] designed and analyzed a primal-dual approximation algorithm for the STEINER FOREST problem, and showed that it achieves approximation ratio 2. A classic result of Goemans and Williamson [11] from the early 90's shows by an elegant proof that the same algorithm applies for proper set families, where $\mathcal{F}$ is **proper** if it is **symmetric** ($A \in \mathcal{F}$ implies $V \setminus A \in \mathcal{F}$) and has the **disjointness property** (if A, B are disjoint and $A \cup B \in \mathcal{F}$ then $A \in \mathcal{F}$ or $B \in \mathcal{F}$). In fact, one of the main achievements of the Goemans and Williamson paper was defining a *generic class* of set families that models a rich collection of combinatorial optimization problems, for which the primal dual algorithm achieves approximation ratio 2. Slightly later, Williamson, Goemans, Mihail, and Vazirani [18] further extended this result to the more general class of uncrossable families, by adding to the algorithm a novel reverse-delete phase; a set family $\mathcal{F}$ is **uncrossable** if $A \cap B, A \cup B \in \mathcal{F}$ or $A \setminus B, B \setminus A \in \mathcal{F}$ whenever $A, B \in \mathcal{F}$. They posed an open question of extending this algorithm to a larger class of set families and combinatorial optimization problems. However, for 30 years, the class of uncrossable set families remained the most general generic class of set families for which the WGMV algorithm achieves a constant approximation ratio.

Recently, Bansal, Cheriyan, Grout, and Ibrahimpur [4] analyzed the performance of the WGMV algorithm [18] for the following generic class of set families that arise in variants of capacitated network design problems.

Definition 1. *Two sets A, B **cross** if all the sets $A \cap B, V \setminus (A \cup B), A \setminus B, B \setminus A$ are non-empty. A set family $\mathcal{F}$ is **pliable** if for any $A, B \in \mathcal{F}$ at least two of the sets $A \cap B, A \cup B, A \setminus B, B \setminus A$ belong to $\mathcal{F}$. We say that $\mathcal{F}$ is γ-**pliable** if it has the following additional property:*

Property (γ): *For any edge set I and sets $S_1 \subset S_2$ in $\mathcal{F}^I$, if an inclusion minimal set C of $\mathcal{F}^I$ crosses each of S_1, S_2, then the set $S_2 \setminus (S_1 \cup C)$ is either empty or belongs to $\mathcal{F}^I$.*

Bansal, Cheriyan, Grout, and Ibrahimpur [4] showed that the WGMV algorithm achieves approximation ratio 16 for γ-pliable families, and that Property (γ) is essential – without it the cost of the solution found by the WGMV algorithm can be $\Omega(\sqrt{n})$ times the cost of an optimal solution. They also considered applications of their result to several variants of capacitated network design problems, as follows.

NEAR MIN-CUTS COVER
Input: A graph $G_0 = (V, E_0)$, an edge set E on V with costs $\{c_e : e \in E\}$, and an integer k.
Output: A min-cost edge set $J \subseteq E$ that covers the set family $\{\emptyset \neq S \subset V : d_{E_0}(S) < k\}$.

It is known that the set family in NEAR MIN-CUTS COVER is pliable, and [4] showed that it satisfies Property (γ), thus obtaining a 16-approximation.

CAPACITATED k-EDGE CONNECTED SPANNING SUBGRAPH (CAP-k-ECSS)
Input: A graph $G = (V, E)$ with edge costs $\{c_e : e \in E\}$ and edge capacities $\{u_e : e \in E\}$, and an integer k.
Output: A min-cost edge set $J \subseteq E$ such that $u(\delta_J(S)) \geq k$ for all $\emptyset \neq S \subset V$.

One can see that NEAR MIN-CUTS COVER is a particular case of CAP-k-ECSS, when all edges in E_0 have cost 0 and capacities 1, and other edges have capacity k. On the other hand, approximation ratio α for NEAR MIN-CUTS COVER implies approximation ratio $\alpha \cdot \lceil k/u_{\min} \rceil$ for CAP-k-ECSS, where $u_{\min}$ is the minimum capacity of an edge; see [4]. This gives approximation ratio $16 \cdot \lceil k/u_{\min} \rceil$ for CAP-k-ECSS.

Adjiashvili, Hommelsheim and Mühlenthaler [1] introduced the following related problem, called (k, q)-FLEXIBLE GRAPH CONNECTIVITY. Suppose that there is a subset $U \subseteq E$ of "unsafe" edges, and we want to find the cheapest spanning subgraph H that will be k-connected even if up to q unsafe edge are removed. Let us say that a spanning subgraph $H = (V, J)$ of G is (k, q)-**flex-connected** if any cut $\delta_H(S)$ of H has at least k safe edges or at least $k + q$ (safe and unsafe) edges. Namely, we require that $d_{H \setminus U}(S) \geq k$ or $d_H(S) \geq k + q$ for all $\emptyset \neq S \subset V$. Summarizing, we get the following problem.

(k, q)-FLEXIBLE GRAPH CONNECTIVITY ((k, q)-FGC)
Input: A graph $G = (V, E)$ with edge costs $\{c_e : e \in E\}$, $U \subseteq E$, and integers $k, q \geq 0$.
Output: A min-cost subgraph H of G such that H is (k, q)-flex-connected.

It is known that $(k,1)$-FGC reduces to Cap-$(k(k+1))$-ECS with $u(e) = k$ if $e \in U$ and $u(e) = k+1$ otherwise, and that $(1,q)$-FGC reduces to Cap-$(q+1)$-ECS with $u(e) = 1$ if $e \in U$ and $u(e) = q+1$ otherwise, c.f. [6]. However, no reduction to Cap-ℓ-ECS is known for other values of k, q. For various approximation algorithms for (k,q)-FGC see recent papers [3–6,15]. Specifically, Bansal [3] showed that if the problem of covering a γ-pliable family achieves approximation ratio α, then $(k,3)$-FGC admits approximation ratio $6 + \alpha$, thus obtaining a 22-approximation for $(k,3)$-FGC.

Another generalization of uncrossable families is considered in [14]. A set family $\mathcal{F}$ is **semi-uncrossable** if for any $A, B \in \mathcal{F}$ we have that $A \cap B \in \mathcal{F}$ and one of $A \cup B, A \setminus B, B \setminus A$ is in $\mathcal{F}$, or $A \setminus B, B \setminus A \in \mathcal{F}$. One can verify that semi-uncrossable families are sandwiched between uncrossable and γ-pliable families. The WGMV algorithm achieves the same approximation ratio 2 for semi-uncrossable families, and [14] shows that many problems can be modeled by semi-uncrossable families that are not uncrossable.

We note that Bansal, Cheriyan, Grout, and Ibrahimpur [4] derived the relevant Property (γ), that enables to obtain a constant approximation ratio for pliable families without excluding known applications. But the proof of the approximation ratio 16 in [4] is somewhat long and complicated, and here by a simpler proof we obtain the following improved result.

Theorem 1. *The* Set Family Edge Cover *problem with a γ-pliable set family $\mathcal{F}$ admits approximation ratio 10.*

Naturally, this also improves approximation ratios for several applications of γ-pliable families discussed in [3,4] – these immediate improvements are summarized in Table 1.

Table 1. Approximation ratios for Near Min-Cuts Cover, Cap-k-ECSS, and $(k,3)$-FGC.

	previous	this paper
Near Min-Cuts Cover	16 [4]	10
Cap-k-ECSS	$16 \cdot \lceil k/u_{\min} \rceil$ [4]	$10 \cdot \lceil k/u_{\min} \rceil$
$(k,3)$-FGC	22 for k even, $11 + \epsilon$ for k odd [3]	16 for k even

In the augmentation version of Cap-k-ECSS we are given a subgraph $G_0 = (V, E_0)$ of G of cost zero and edge-connectivity $\lambda_0 = \min\{d_{E_0}(S) : \emptyset \neq S \subset V\}$ close to k. The goal is to compute a min-cost edge set $J \subseteq E$ such that $G_0 \cup J$ has capacitated edge connectivity k. For $k = \lambda_0 + 1$ this problem is equivalent to the ordinary k-Edge Connectivity Augmentation problem, and when every edge in E has capacity $\geq k - \lambda_0$ it is the Near Min-Cuts Cover problem. Using a 2-approximation algorithm for increasing the edge connectivity by 1, one can obtain approximation ratio $2(k - \lambda_0)$ for this problem. Bansal, Cheriyan, Grout, and Ibrahimpur [4] made a substantial progress by showing that NEAR MIN-CUTS COVER admits a constant approximation ratio 16, and our Theorem 1 improved the approximation ratio to 10. We make an additional small progress by improving over this ratio for the cases when λ_0 is close to k.

Theorem 2. NEAR MIN-CUTS COVER *admits the following approximation ratios:*

- $k - \lambda_0$ *if λ_0, k are both even.*
- $k - \lambda_0 + 1/2 + \epsilon$ *if λ_0, k have distinct parity.*
- $k - \lambda_0 + 1 + \epsilon$ *if λ_0, k are both odd.*

We now describe our additional contribution for the (k,q)-FGC problem. The best approximation ratio for $(k,1)$-FGC was 4 for arbitrary costs and 16/11 for unit costs [5]. The best approximation ratio for $(k,2)$-FGC was $2k+4$ for $k \leq 7$ [6] and 20 for $k \geq 8$ [4]. We improve this as follows, and also give a simple approximation algorithm for (k,q)-FGC with unit costs; see Table 2.

Theorem 3. *(i) $(k,1)$-FGC admits approximation ratio $3.5 + \epsilon$ if k is odd.*
(ii) $(k,2)$-FGC admits approximation ratio $7 + \epsilon$ if k is odd and 6 if k is even.
(iii) For unit costs, (k,q)-FGC admits approximation ratio $\alpha + \frac{2q}{k}$, where α is the best known approximation ratio for the MIN-SIZE k-EDGE-CONNECTED SUBGRAPH *problem.*

Table 2. Approximation ratios for (k,q)-FGC. Here α is the best known approximation ratio for MIN-SIZE k-EDGE-CONNECTED SUBGRAPH; currently $\alpha = 1 + 1/(2k) + O(1/k^2)$ for graphs [8], $\alpha = 1 + \frac{2}{k}$ [9] for multigraphs, and $\alpha = 1.326$ for $k = 2$ [10].

(k,q)	previous	this paper
$(k,1)$	4 [5] 16/11 for unit costs [5]	$3.5 + \epsilon$ for k odd $\alpha + 2/k$ for unit costs
$(k,2)$	20 for k odd, 6 for k even [4]	$7 + \epsilon$ for k odd, 6 for k even
$(k,3)$	$4k+4$ for $k=2,4$ [6] 22 for $k \geq 6$ even, $11 + \epsilon$ for k odd [3]	16 for k even
$(k,4)$	$6k+4$ for k even [6]	
$(1,q)$	$q+1$ [5]	
$(2,q)$	$2q+2$ [6]	
(k,q)	$O(q \log n)$ [5]	$\alpha + \frac{2q}{k}$ for unit costs

We note that the approximation ratio 6 for $(k,2)$-FGC was announced independently both in the preliminary version of this paper [15] and in [4]; both proofs follow from the known observation (c.f. [7,13] and Corollary 1) that in a λ-connected graph with λ even the family of cuts of size $\lambda, \lambda+1$ is uncrossable. On the other hand, our proof of the $(7 + \epsilon)$-approximation for k odd is based on a nontrivial decomposition of the family of relevant cuts into two "convenient" families – an uncrossable family and a symmetric proper crossing family. A recent result of Bansal [3] for $(k,3)$-FGC – approximation ratio 22 for k even and $11+\epsilon$ for k odd, uses our approximation ratios for $(k,2)$-FGC (6 for k even and $7+\epsilon$ for k odd), as well as our idea of decomposing the family of relevant cuts into two "convenient" families.

This paper is organized as follows. In the next section we prove Theorem 1, while Theorems 2 and 3 are proved in Sects. 3 and 4, respectively.

2 A 10-Approximation for Covering γ-Pliable Set Families (Theorem 1)

We start with stating some simple properties of pliable families. One can see that if an edge e covers one of the sets $A \cap B, A \cup B, A \setminus B, B \setminus A$ then it also covers one of A, B. This implies the following.

Lemma 1. *If $\mathcal{F}$ is pliable or is γ-pliable, then so is $\mathcal{F}^I$, for any edge set I.*

An $\mathcal{F}$-**core** is an inclusion minimal member of $\mathcal{F}$; let $\mathcal{C}_\mathcal{F}$ denote the family of $\mathcal{F}$-cores.

Lemma 2. *Let $\mathcal{F}$ be a pliable set family and let $A \in \mathcal{F}$ and $C \in \mathcal{C}_\mathcal{F}$. Then either $C \subseteq A$, or $C \cap A = \emptyset$, or $A \setminus C, A \cup C \in \mathcal{F}$. Consequently, the members of $\mathcal{C}_\mathcal{F}$ are pairwise disjoint.*

Proof. Since $C \in \mathcal{C}_\mathcal{F}$, we can have $C \cap A \in \mathcal{F}$ only if $C \subseteq A$, and we can have $C \setminus A \in \mathcal{F}$ only if $C \cap A = \emptyset$. In any other case, we must have $A \setminus C, A \cup C \in \mathcal{F}$, since $\mathcal{F}$ is pliable. □

We now describe the WGMV algorithm. Consider the following LP-relaxation **(P)** for SET FAMILY EDGE COVER and its dual program **(D)**:

$$\min \sum_{e \in E} c_e x_e \qquad \max \sum_{S \in \mathcal{F}} y_S$$
$$\textbf{(P)} \text{ s.t. } \sum_{e \in \delta(S)} x_e \geq 1 \quad \forall S \in \mathcal{F} \qquad \textbf{(D)} \text{ s.t. } \sum_{\delta(S) \ni e} y_S \leq c_e \quad \forall e \in E$$
$$x_e \geq 0 \quad \forall e \in E \qquad y_S \geq 0 \quad \forall S \in \mathcal{F}$$

Given a solution y to **(D)**, an edge $e \in E$ is **tight** if the inequality of e in **(D)** holds with equality. The algorithm has two phases.

Phase 1 starts with $I = \emptyset$ an applies a sequence of iterations. At the beginning of an iteration, we compute the family $\mathcal{C} = \mathcal{C}_{\mathcal{F}^I}$ of $\mathcal{F}^I$-cores. Then we raise the dual variables of the $\mathcal{F}^I$-cores uniformly (possibly by zero), until some edge $e \in E \setminus I$ becomes tight, and add e to I. Phase I terminates when $\mathcal{C}_{\mathcal{F}^I} = \emptyset$, namely when I covers $\mathcal{F}$.

Phase 2 applies on I "reverse delete", which means the following. Let $I = \{e_1, \ldots, e_j\}$, where e_{i+1} was added after e_i. For $i = j$ downto 1, we delete e_i from I if $I \setminus \{e_i\}$ still covers $\mathcal{F}$. At the end of the algorithm, I is output.

The produced dual solution is feasible, hence $\sum_{S \in \mathcal{F}} y_S \leq \mathsf{opt}$, by the Weak Duality Theorem. We prove that at the end of the algorithm

$$\sum_{e \in I} c(e) \leq 10 \sum_{S \in \mathcal{F}} y_S \ .$$

As any edge in I is tight, the last inequality is equivalent to $\sum_{e \in I} \sum_{\delta_I(S) \ni e} y_S \leq 10 \sum_{S \in \mathcal{F}} y_S$. By changing the order of summation we get:

$$\sum_{S \in \mathcal{F}} d_I(S) y_S \leq 10 \sum_{S \in \mathcal{F}} y_S \ .$$

It is sufficient to prove that at any iteration the increase at the left hand side is at most the increase in the right hand side. Let us fix some iteration, and let $\mathcal{C} = \mathcal{C}_{\mathcal{F}'}$ be the family of cores among the members of $\mathcal{F}$ not yet covered. The increase in the left hand side is $\varepsilon \cdot \sum_{C \in \mathcal{C}} d_I(C)$, where ε is the amount by which the dual variables were raised in the iteration, while the increase in the right hand side is $\varepsilon \cdot 10|\mathcal{C}|$. Consequently, it is sufficient to prove that $\sum_{C \in \mathcal{C}} d_I(C) \leq 10|\mathcal{C}|$. As the edges were deleted in reverse order, the set I' of edges in I that were added after the iteration (and "survived" the reverse delete phase), form an inclusion minimal edge-cover of the family $\mathcal{F}'$ of members in $\mathcal{F}$ that are uncovered at the beginning of the iteration. Note also that $\bigcup_{C \in \mathcal{C}} \delta_I(C) \subseteq I'$. Hence to prove approximation ratio 10, it is sufficient to prove the following purely combinatorial statement, in which due to Lemma 1 we can revise our notation to $\mathcal{F} \leftarrow \mathcal{F}'$ and $I \leftarrow I'$.

Lemma 3. *Let I be an inclusion minimal cover of a γ-pliable set family $\mathcal{F}$ such that every edge in I covers some $C \in \mathcal{C}$. Then:*

$$\sum_{C \in \mathcal{C}} d_I(C) \leq 5 \cdot (2|\mathcal{C}| - 1) \ . \tag{1}$$

In the rest of this section we prove Lemma 3. A set family $\mathcal{L}$ is a **laminar** if any two sets in $\mathcal{L}$ are disjoint or one of them contains the other. Let I be an inclusion minimal edge cover of a set family $\mathcal{F}$. We say that a set $S_e \in \mathcal{F}$ is a **witness set** for an edge $e \in I$ if e is the unique edge in I that covers S_e, namely, if $\delta_I(S_e) = \{e\}$. We say that $\mathcal{L} \subseteq \mathcal{F}$ is a **witness family** for I if $|\mathcal{L}| = |I|$ and for every $e \in I$ there is a witness set $S_e \in \mathcal{L}$. By the minimality of I, there exists a witness family $\mathcal{L} \subseteq \mathcal{F}$. The following lemma was proved in [4].

Lemma 4 ([4]). *Let I be an inclusion minimal cover of a pliable set family $\mathcal{F}$. Then there exists a witness family $\mathcal{L} \subseteq \mathcal{F}$ for I that is laminar.*

Augment $\mathcal{L}$ by the set V. Then $\mathcal{L}$ can be represented by a rooted tree $\mathcal{T}$ with node set $\mathcal{L}$ and root V, where the parent of S in $\mathcal{T}$ is the smallest set in $\mathcal{L}$ that properly contains S. The (unique) edge in I that covers S corresponds to the edge in $\mathcal{T}$ from S to its parent; a node S of $\mathcal{T}$ corresponds to the set of nodes in the set S that do not belong to any child of S.

Definition 2. A set $S \in \mathcal{L}$ **owns** a core $C \in \mathcal{C}$ if S is the inclusion-minimal set in $\mathcal{L}$ that contains C. We say that S is **hollow** if it owns no core. A sequence $(S_0, S_1, \ldots, S_\ell)$ of sets in $\mathcal{L}$ is called a **hollow chain** (of length ℓ) if $S_1, \ldots, S_\ell$ are hollow, $S_\ell \neq V$, and S_{i-1} is the unique child of S_i in $\mathcal{L}$, $i = 1, \ldots, \ell$. For each S_i let $a_i b_i$ be the unique edge in I that covers S_i, where $a_i \in S_i$ (possibly $a_i = b_{i-1}$).

We will use for nodes of $\mathcal{T}$ the same terminology as for sets in $\mathcal{L}$; specifically, a node of $\mathcal{T}$ is hollow if it represents a hollow set, and a hollow chain in $\mathcal{T}$ is a path such that all internal nodes in the path correspond to hollow sets (note that each of these nodes has degree exactly 2 in $\mathcal{T}$). We will need the following well known statement.

Lemma 5. Let $T = (V_T, E_T)$ be a tree rooted at r and let R, W be a partition of V_T such that every $w \in W \setminus \{r\}$ has at least 2 children. Then $|E_T| \leq 2|R| - 1$.

If $\mathcal{T}$ has no hollow chain, then Lemma 5 implies that $|\mathcal{L}| \leq 2|\mathcal{C}| - 1$ – just let W to be the set of hollow nodes in $\mathcal{T}$ in Lemma 5. Since the cores in $\mathcal{C}$ are pairwise disjoint (by Lemma 2), every edge contributes at most 2 to $\sum_{C \in \mathcal{C}} d_I(C)$, hence (since $|I| = |\mathcal{L}|$) $\sum_{C \in \mathcal{C}} d_I(C) \leq 2(2|\mathcal{C}| - 1)$.

Suppose now that every hollow chain has length ℓ. Then the contribution of the edges $a_0 b_0, \ldots, a_\ell b_\ell$ of each chain is at most $2(\ell + 1)$. If we "shortcut" every maximal hollow chain in $\mathcal{T}$, we obtain a tree with at most $2|\mathcal{C}| - 1$ edges. However, every such edge might be a shortcut of a hollow chain, and thus may contribute $2(\ell + 1)$ to $\sum_{C \in \mathcal{C}} d_I(C)$. As the number of edges after the shortcuts is at most $2|\mathcal{C}| - 1$, we get $\sum_{C \in \mathcal{C}} d_I(C) \leq 2(\ell + 1)(2|\mathcal{C}| - 1)$. Bansal et al. [4] showed that the maximum possible length of a hollow chain is 3, which gives the bound $\sum_{C \in \mathcal{C}} d_I(C) \leq 8 \cdot (2|\mathcal{C}| - 1) < 16 \cdot |\mathcal{C}|$.

Let $U = \cup_{C \in \mathcal{C}} C$. To improve this bound of [4], we will show that for $\ell = 2, 3$, among the relevant nodes $a_0, b_0, \ldots, a_\ell, b_\ell$ at most 5 belong to U. This reduces the bound on the contribution of every hollow chain from 8 to 5, and gives the bounds $\sum_{C \in \mathcal{C}} d_I(C) \leq 5 \cdot (2|\mathcal{C}| - 1)$. Specifically, we will prove the following.

Lemma 6. Let $(S_0, S_1, \ldots, S_\ell)$ be a hollow chain and let $a_i b_i$ be as in Definition 2. Then $\ell \leq 3$ and the following holds, see Fig. 1:

(a) If $a_1 \in U$ then $\ell = 1$.
(b) If $b_0 \in U$ then $\ell \leq 2$; if $\ell = 2$ then $a_1 \notin U$ and b_0, b_1, a_2 belong to the same core.
(c) If $b_1 \in U$ then $\ell \leq 3$; if $\ell = 3$ then $a_1, b_0, a_2 \notin U$ and b_1, b_2, a_3 belong to the same core.

Let us show that Lemma 6 implies Lemma 3. Note that at least one of a_1, b_1 is in U, since every edge covers some core $C \in \mathcal{C}$, by the assumption in Lemma 3. Thus $\ell \leq 3$. In the tree representation $\mathcal{T}$ of $\mathcal{L}$ let us "shortcut" all maximal hollow chains, see Fig. 2. This means that we replace the chain – the edges of the chain (that correspond to edges $a_0 b_0, \ldots, a_\ell b_\ell$) and the nodes that correspond to sets $S_1, \ldots, S_\ell$, by a new "shortcut edge" between S_0 and S_ℓ (for illustration, in

Approximation Algorithms for Pliable Set Families   159

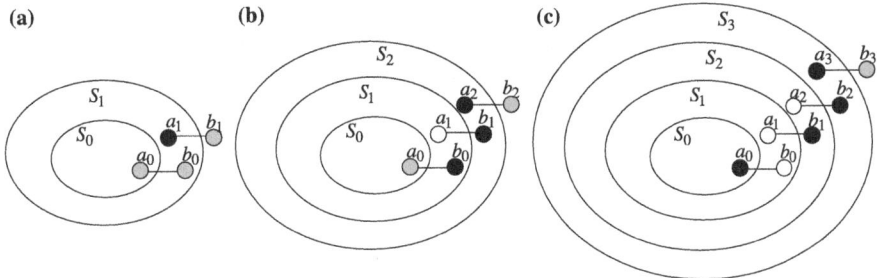

Fig. 1. Illustration to Lemma 6. Black nodes are in U, white nodes are not in U, while gray nodes may or may not be in U.

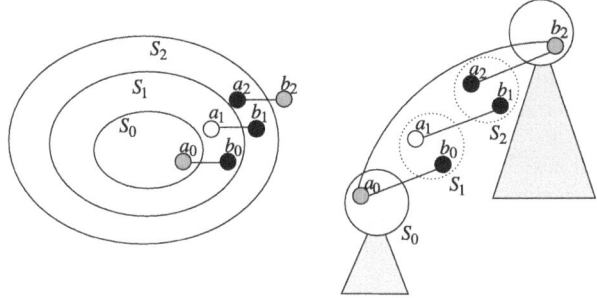

Fig. 2. Illustration of a shortcut of a hollow chain with $\ell = 2$. Black nodes are in U, white nodes are not in U, while gray nodes may or may not be in U.

Fig. 2 we assume that this new edge is $a_0 b_\ell$). This operation also can be viewed as identifying S_0 with S_ℓ, by removing the nodes in $S_\ell \setminus S_0$ and adding to I the edge $a_0 b_\ell$. Every such shortcut edge contributes at most 5 (at most 4 if $\ell = 1$) to $\sum_{C \in \mathcal{C}} d_I(C)$, by Lemma 6. Furthermore, the obtained tree has no hollow chains. By Lemma 5, the total number of edges in the obtained tree is at most $2|\mathcal{C}| - 1$. Each edge of $\mathcal{T}$ contributes to $\sum_{C \in \mathcal{C}} d_I(C)$ at most 2 if it is an ordinary edge and at most 5 if it is a shortcut edge. Consequently, $\sum_{C \in \mathcal{C}} d_I(C) \le 5 \cdot (2|\mathcal{C}| - 1)$. This concludes the proof of Theorem 1, provided that we will prove Lemma 6, which we will do in the rest of this section.

To prove Lemma 6 we will apply Property (γ) on the family $\mathcal{F}$ in Lemma 3 (that is in fact a residual family of the original family that we want to cover), so we restate here Property (γ) only for the current family $\mathcal{F}$ – the one in Lemma 3.

Property (γ): For any sets $S_1 \subset S_2$ in $\mathcal{F}$, if an inclusion minimal set C of $\mathcal{F}$ crosses each of S_1, S_2, then the set $S_2 \setminus (S_1 \cup C)$ is either empty or belongs to $\mathcal{F}$.

In what follows, note that every $C \in \mathcal{C}$ is owned by some set in $\mathcal{L}$, since $V \in \mathcal{L}$. The next two Lemmas is a preparation for using Property (γ).

Lemma 7. *Let S be a hollow set with a unique child A. Then $S \setminus A$ contains no set in $\mathcal{F}$.*

Proof. Suppose to the contrary that $S \setminus A$ contains a set $B \in \mathcal{F}$. Then B contains some core $C \in \mathcal{C}$. Let $S_C \in \mathcal{L}$ be the set that owns C. Then we must have $S_C \subseteq S$ and $S_C \cap A = \emptyset$. Consequently, S has at least 2 children in $\mathcal{L}$, contradicting that S is hollow. □

Lemma 8. *Let $C \in \mathcal{C}$ and let $1 \leq i \leq \ell$. If $(C \cap S_i) \setminus S_0 \neq \emptyset$ then C and S_i cross.*

Proof. We will show that $C \cap S_i, S_i \setminus C, C \setminus S_i$ are nonempty and that $S_i \cup C \neq V$.

- $C \cap S_i \neq \emptyset$ by the assumption.
- $S_i \setminus C \neq \emptyset$ since C does not contain S_0, hence there is $v \in S_0 \setminus C \subset S_i \setminus C$.
- $C \cup S_i \neq V$ since $C \cup S_i \in \mathcal{F}$ by Lemma 2 and since $V \notin \mathcal{F}$.

It remains to show that $C \setminus S_i \neq \emptyset$. Suppose to the contrary that $C \subseteq S_i$. Let S_C be the set that owns C. Since $C \subseteq S_i$, S_C is a descendant of S_i. Since $C \setminus S_0 \neq \emptyset$, S_C is not a descendant of S_0. Thus S_C must be one of the sets $S_1, \ldots, S_i$, which is impossible, since all these sets are hollow. □

Now we will use Property (γ). Note that since the cores are pairwise disjoint (Lemma 2), then by the assumption that every edge covers some $C \in \mathcal{C}$ (Lemma 3), for any edge $uv \in I$ the following holds: $|\{u,v\} \cap U| \geq 1$ and $|\{u,v\} \cap C| \leq 1$ for any $C \in \mathcal{C}$.

Lemma 9. *Let $(S_0, S_1, \ldots, S_\ell)$ be a hollow chain and $a_i b_i$ as in Definition 2.*

(i) *If $a_i \in U$ for some $i \geq 1$ then $\ell = i$.*
(ii) *If $b_{i-1} \in U$ for some $i \geq 0$ then $\ell \leq i+1$ and if $\ell = i+1$ then b_{i-1}, b_i, a_{i+1} belong to the same core.*

Proof. For part (i), suppose to the contrary that S_{i+1} exists. Let $C \in \mathcal{C}$ be such that $a_i \in C$. By Lemma 8, C crosses each of S_i, S_{i+1}. Note that $b_i \notin C$, hence the set $S_{i+1} \setminus (S_i \cup C)$ is non-empty, and thus by Property (γ) is in $\mathcal{F}$. This contradicts Lemma 7.

For part (ii), suppose that C_{i+1} exists and let $C \in \mathcal{C}$ be such that $b_{i-1} \in C$. By Lemma 8, C crosses each of S_i, S_{i+1}. If $b_i \notin C$ or if $a_{i+1} \notin C$ then the set $S_{i+1} \setminus (S_i \cup C)$ is non-empty, and thus by Property (γ) is in $\mathcal{F}$, contradicting Lemma 7. Thus $b_i, a_{i+1} \in C$. Since $a_{i+1} \in U$, we get by part (i) that $\ell = i+1$. □

Lemma 6 easily follows from Lemma 9.

(a) If $a_1 \in U$ then by Lemma 9(i) we have case (a) of Lemma 6.
(b) If $b_0 \in U$ (and $a_1 \notin U$), then by Lemma 9(ii) we have case (b) of Lemma 6.
(c) If $a_1, b_0 \notin U$ then $b_1 \in U$. Then either $\ell = 2$, or $\ell = 3$ and b_1, b_2, a_3 belong to the same core, so we have case (c) of Lemma 6.

This concludes the proof of Lemma 6, and also of Lemma 3 and Theorem 1.

3   Algorithm for NEAR MIN-CUTS COVER (Theorem 2)

We start by giving some (known) "uncrossing" properties of near minimum cuts needed for the proofs of Theorems 2 and 3. Let $H = (V, J)$ be a λ-connected graph and let $A, B \subseteq V$ such that the sets $A \cap B, A \setminus B, V \setminus (A \cup B), B \setminus A$ are all nonempty. Shrinking each of these sets into a single node results in a graph on 4 nodes, and we will further replace parallel edges by a single capacitated edge. We call this graph the **square** of A, B, edges between $A \cap B$ and $V \setminus (A \cap B)$ and between $A \setminus B$ and $B \setminus A$ are **diagonal edges**, while the other edges are **side edges**.

When both A, B are λ-cuts, then it is well known that λ must be even, the square has no diagonal edges, and each side edge has capacity $\lambda/2$. The following two lemmas were implicitly proved in [7,13]; see also explicit proofs in [15].

Lemma 10 ([7,13]). *Suppose that A is a λ-cut and B is a $(\lambda + 1)$-cut. Then the square has no diagonal edges, and the following holds.*

(i) *If λ is odd then the square has one side edge of capacity $(\lambda - 1)/2$ and other three side edges have capacity $(\lambda + 1)/2$.*
(ii) *If λ is even then the square has one side edge of capacity $\lambda/2 + 1$ and other three side edges have capacity $\lambda/2$.*

Lemma 11 ([7,13]). *Suppose that A, B are $(\lambda + 1)$-cuts. If λ is even then one of the following holds.*

(a) *The square has no diagonal edges and has two adjacent side edges of capacity $\lambda/2$ while the other two edges have capacity $\lambda/2 + 1$.*
(b) *The square has one diagonal edge and all side edges have capacity $\lambda/2$;*

If λ is odd then one of the following holds.

(c) *The square has no diagonal edges, two opposite side edges have capacity $(\lambda + 1)/2$, one side edge has capacity $(\lambda + 3)/2$ and its opposite side edge has capacity $(\lambda - 1)/2$.*
(d) *The square has one diagonal edge, two side edge incident to one end of the diagonal edge have capacity $(\lambda - 1)/2$, while the other have capacity $(\lambda + 1)/2$.*
(e) *The square has both diagonal edges of capacity 1 each, and all side edges have capacity $(\lambda - 1)/2$.*
(f) *The square has no diagonal edges and all side edges have capacity $(\lambda + 1)/2$.*

From Lemmas 10 and 11 we immediately get:

Corollary 1. *Let H be a λ-edge-connected graph and let $\mathcal{F} = \{S \subseteq V : d_H(S) \in \{\lambda, \lambda + 1\}\}$. If λ is even then $\mathcal{F}$ is uncrossable.*

It is known that the problem of finding a min-cost cover of cuts of size $< k$ of a λ-connected graph admits the following approximation ratios for $k \leq \lambda + 2$:

- $1.5 + \epsilon$ if $k = \lambda + 1$ [16,17].

Algorithm 1: λ_0, k are both even

1 $F \leftarrow \emptyset$, edges in $E \setminus E_0$ get capacity $k - \lambda_0$
2 **for** $\lambda = \lambda_0$ to $\lambda = k - 2$ **do**
3 $\quad$ Find a 2-approximate cover J of $\{\lambda, \lambda + 1\}$-cuts of $G_0 \cup F$.
4 $\quad$ $F \leftarrow F \cup J$.
5 **return** F

- 2 if $k = \lambda + 2$ and λ is even by Corollary 1, since the problem of covering an uncrossable set family admits approximation ratio 2 [18].

Algorithm 1 computes a solution as required when λ_0, k are both even. The approximation ratio $k - \lambda_0$ (when λ_0, k are both even) follows from the observation that we pay $2 \cdot \mathsf{opt}$ at each iteration, so opt per increasing the connectivity by 1. If λ_0 is odd and k is even, then in the first iteration we apply the $(1.5 + \epsilon)$-approximation algorithm of [17] for covering λ-cuts only and have an extra $1/2 + \epsilon$ term. If λ_0 is even and k is odd, then in the last iteration we apply the $(1.5+\epsilon)$-approximation algorithm of [16] for covering $(k-1)$-cuts only, and also have an extra $1/2 + \epsilon$ term. If both of these occur then the extra term is $1 + 2\epsilon \approx 1 + \epsilon$, concluding the proof of Theorem 2.

4 Algorithms for (K,q)-FGC (Theorem 3)

Let $\langle G = (V, E), U, c, k \rangle$ be an instance of (k, q)-FGC, where $U \subseteq E$ is the set of unsafe edges. Recall that a subgraph H of G is (k, q)-**flex-connected** if $d_{H \setminus U}(S) \geq k$ or $d_H(S) \geq k + q$ for all $\emptyset \neq S \subset V$. We will use the notation $d(S) = d_H(S)$ and $d_U(S) = d_{H \cap U}(S)$. Observing that $d_{H \setminus U}(S) = d_H(S) - d_{H \cap U}(S) = d(S) - d_U(S)$, we get that H is (k, q)-flex-connected if and only if

$$d(S) \geq k + \min\{d_U(S), q\} \qquad \forall\, \emptyset \neq S \subset V.$$

Suppose that H is $(k, q-1)$-flex-connected. Then

$$d(S) \geq k + \min\{d_U(S), q-1\} \quad \forall\, \emptyset \neq S \subset V.$$

If for $\emptyset \neq S \subset V$ the later inequality holds but not the former then $d(S) = k+q-1$ and $d_U(S) \geq q$. Thus to make a $(k, q-1)$-flex-connected H to be (k, q)-flex-connected we need to add to H an edge set that covers the set family

$$\mathcal{F}_q(H) = \{\emptyset \neq S \subset V : d(S) = k + q - 1, d_U(S) \geq q\}\,.$$

Thus the following algorithm computes a feasible solution for (k, q)-FGC.

We can use this observation to prove parts (i) and (iii) of Theorem 3: that $(k, 1)$-FGC admits approximation ratio $3.5 + \epsilon$ if k is odd, and that (k, q)-FGC admits ratio $\alpha + \frac{2q}{k}$ for unit costs, where α is the best known approximation ratio for the MIN-SIZE k-EDGE-CONNECTED SUBGRAPH problem.

Algorithm 2: ITERATIVE-COVER(G, c, U, k, q)

1 Compute a k-edge-connected spanning subgraph $H = (V, J)$ of G.
2 **for** $\ell = 1$ to q **do**
3   $\quad$ Add to H a cover J_ℓ of $\mathcal{F}_\ell(H)$.
4 **return** H.

If k is odd then the family $\{\emptyset \neq S \subset V : d(S) = k\}$ is laminar, and thus $\mathcal{F}_1(H)$ is laminar. Part (i) now follows from the fact that the problem of covering a laminar family admits approximation ratio $1.5 + \epsilon$ [17].

For part (iii) we need the known fact that any inclusion minimal cover J of a set family $\mathcal{F}$ is a forest. To see this, suppose to the contrary that J contains a cycle C. Let $e = uv$ be an edge of C. Since $P = C \setminus \{e\}$ is a uv-path, then for any $A \in \mathcal{F}$ covered by e, there is $e' \in P$ that covers A. This implies that $J \setminus \{e\}$ also covers $\mathcal{F}$, contradicting the minimality of J.

On the other hand, opt $\geq kn/2$, hence $|J_i|/$opt $\leq 2(n-1)/kn < 2/k$. Thus the overall approximation ratio is $\alpha + 2q/k$, as claimed. It remains to show that the algorithm can be implemented in polynomial time. By [5] (see also [6, Lemma 2.8]), if H is $(k, i-1)$-flex-connected then $|\mathcal{F}_i(H)| = O(n^4)$, and the members of $\mathcal{F}_i(H)$ can be listed in polynomial time. Since a $(k, i-1)$-flex-connected H is (k, i)-flex-connected if and only if $\mathcal{F}_i(H) = \emptyset$, we can compute at iteration i an inclusion minimal cover of $\mathcal{F}_i(H)$ in polynomial time

The proof of part (ii) of Theorem 3 relies on several lemmas. For the proof of the next lemma see [15].

Lemma 12. *Let H be $(k, q-1)$-flex-connected, $q \geq 2$, and let $A, B \in \mathcal{F}_q(H)$ cross. Then:*

(i) *If $d_U(A \cap B) \geq 1$ then $d(A \cap B) + d(A \setminus B) \geq 2k + q$.*
(ii) *If $d_U(A \cap B) = 0$ then $A \setminus B, B \setminus A \in \mathcal{F}_q(H)$.*

Lemma 13. *If H is $(k, 1)$-flex-connected for k even then $\mathcal{F}_2(H)$ is uncrossable.*

Proof. Let $A, B \in \mathcal{F}_2(H)$ cross. By Lemma 11(a,b), $d(A \cap B) + d(A \setminus B) = 2k+1$. Thus we must be in case (ii) of Lemma 12, implying $A \setminus B, B \setminus A \in \mathcal{F}_q(H)$. $\square$

We need some definitions to handle the case $q = 2$ and k odd. Let $\mathcal{F}$ be a set family. We say that $\mathcal{F}$ is a **crossing family** if $A \cap B, A \cup B \in \mathcal{F}$ whenever $A \cap B, V \setminus (A \cup B)$ are both non-empty, and if in addition $(A \setminus B) \cup (B \setminus A) \notin \mathcal{F}$ then $\mathcal{F}$ is a **proper crossing family**. By [7,13], the problem of covering a symmetric proper crossing family is equivalent to covering the minimum cuts of a 2-edge connected graph, while [16] shows that the later problem admits a $(1.5 + \epsilon)$-approximation algorithm. As was mentioned, $\mathcal{F}_1(H)$ is laminar if k is odd, and it is also not hard to see that $\mathcal{F}_1(H)$ is uncrossable if k is even, see [5].

Let H be a graph and $\mathcal{F}$ a family of node subsets (or cuts) of H. It is easy to see that the relation $\{(u, v) \in V \times V :$ no member of $\mathcal{F}$ separates $u, v\}$ is an equivalence; the **quotient graph** of $\mathcal{F}$ is obtained by shrinking each equivalence class of this relation into a single node, and replacing every set of parallel edges by a single capacitated edge. We need the following result from [7,13] (specifically, see Lemmas 3.11–3.15 in [13].)

Lemma 14 ([7,13]). *Let H be a λ-edge-connected graph with λ odd, and let $\mathcal{F}$ be the family of $(\lambda+1)$-cuts of H. Then there exists a subfamily $\mathcal{F}'$ of the λ-cuts of H such that $\mathcal{F} \cup \mathcal{F}'$ can be decomposed in polynomial time into parts whose union contains $\mathcal{F}$, such that every cut in $\mathcal{F}$ belongs to at most 2 parts, and such that the cuts in each part correspond to the $(\lambda + 1)$-cuts of its quotient graph, which is either:*

(a) *A cycle of edges of capacity $(\lambda+1)/2$ each.*
(b) *A cycle with one edge of capacity $(\lambda-1)/2$ and other edges of capacity $(\lambda+3)/2$ each.*
(c) *A cube graph, which can occur only if $\lambda = 3$.*

The following lemma gives uncrossing properties of $\mathcal{F}_2(H)$ if $\mathcal{F}_1(H) = \emptyset$.

Lemma 15. *Let k be odd. If H is $(k,1)$-flex-connected then $\mathcal{F}_2(H)$ can be decomposed in polynomial time into an uncrossable family $\mathcal{F}'$ and a symmetric proper crossing family $\mathcal{F}''$ such that $\mathcal{F}' \cup \mathcal{F}'' = \mathcal{F}_2(H)$.*

Proof. Consider a decomposition of $(k+1)$-cuts as in Lemma 14. Note that if two $(k+1)$-cuts belong to the same part then so are their corner cuts. Hence to prove the lemma it is sufficient to provide a proof for each part of $\mathcal{F}$ as in Lemma 14, namely, we may assume that there is just one part which is $\mathcal{F}$.

For an edge e of the quotient graph of $\mathcal{F}$ let u_e be the number of unsafe edges in the edge subset of H represented by e. Let us say that e is **red** if $u_e \geq 2$, **blue** if $u_e = 1$, and **black** otherwise (if $u_e = 0$). Note that since $\mathcal{F}_1(H) = \emptyset$, there cannot be a k-cut in the quotient graph that contain a blue or a red edge. Consequently, only in case (a) the quotient graph may have non black edges, as in cases (b,c) every edge of the quotient graph belongs to some k-cut. On the other hand, every cut in $\mathcal{F}_2(H)$ must contain a blue or a red edge, thus the only relevant case is (a).

Let $\mathcal{F}'$ be the family of cuts in $\mathcal{F}_2$ that contain a red edge and let $\mathcal{F}'' = \mathcal{F} \setminus \mathcal{F}'$. Note that every cut in $\mathcal{F}''$ consists of 2 blue edges. We claim that $\mathcal{F}'$ is uncrossable and that $\mathcal{F}''$ is is a proper crossing family (clearly, $\mathcal{F}''$ is symmetric). Let $A, B \in \mathcal{F}$ cross. If $A, B \in \mathcal{F}'$ then their square has two adjacent red edges, and this implies that $A \cup B, A \cap B \in \mathcal{F}$ or $A \setminus B, B \setminus A \in \mathcal{F}'$. Consequently, $\mathcal{F}'$ is uncrossable. If $A, B \in \mathcal{F}''$ then their square has 4 blue edges, and then all corner cuts are in $\mathcal{F}''$. Thus $\mathcal{F}''$ is a crossing family. Furthermore, the capacity of the cut defined by the set $(A \setminus B) \cup (B \setminus A)$ is $4(k+1)/2 = 2(k+1) > k+1$, and thus $\mathcal{F}''$ is a symmetric proper crossing family. □

We now finish the proof of part (ii) of Theorem 3. We will apply Algorithm 2. At step 1, $c(J) \leq 2\text{opt}$, c.f. [12]. In the loop (steps 2,3) we combine Lemmas 13 and 15 with the best known approximation ratios for solving appropriate set family edge cover problems.

If k is even then each of $\mathcal{F}_1, \mathcal{F}_2$ is uncrossable by Lemma 13 and thus $c(J_1) \le$ 2opt and $c(J_2) \le$ 2opt, by [18]; consequently, $c(J \cup J_1 \cup J_2) \le$ 6opt.

If k is odd then $\mathcal{F}_1$ is laminar and thus $c(J_1) \le (1.5+\epsilon)$opt by [17]. After $\mathcal{F}_1$ is covered, $\mathcal{F}_2$ can be decomposed into an uncrossable family $\mathcal{F}'$ and a symmetric proper crossing family $\mathcal{F}''$, by Lemma 15. We can compute a 2-approximate cover J' of $\mathcal{F}'$ using the algorithm of [18] and a $(1.5 + \epsilon)$-approximate cover J'' of $\mathcal{F}''$ using the algorithm of [16]. Consequently, the approximation ratio of the algorithm is bounded by

$$[c(J) + c(J_1) + c(J') + c(J'')]/\text{opt} \le 2 + (1.5 + \epsilon) + 2 + (1.5 + \epsilon) = 7 + 2\epsilon \approx 7 + \epsilon\,,$$

concluding the proof of part (ii) of Theorem 3.

References

1. Adjiashvili, D., Hommelsheim, F., Mühlenthaler, M.: Flexible graph connectivity: approximating network design problems between 1- and 2-connectivity. Math. Program. **192**(1–2), 409–441 (2022). https://doi.org/10.1007/s10107-021-01664-9
2. Agrawal, A., Klein, P., Ravi, R.: When trees collide: approximation algorithm for the generalized Steiner problem on networks. SIAM J. Comput. **24**(3), 440–456 (1995)
3. Bansal, I.: A constant factor approximation for the $(p, 3)$-flexible graph connectivity problem. CoRR abs/2308.15714, August 2023. https://doi.org/10.48550/arXiv.2308.15714
4. Bansal, I., Cheriyan, J., Grout, L., Ibrahimpur, S.: Improved approximation algorithms by generalizing the primal-dual method beyond uncrossable functions. In: ICALP, pp. 15:1–15:19 (2023)
5. Boyd, S.C., Cheriyan, J., Haddadan, A., Ibrahimpur, S.: Approximation algorithms for flexible graph connectivity. In: FSTTCS, pp. 9:1-9:14 (2021)
6. Chekuri, C., Jain, R.: Approximation algorithms for network design in non-uniform fault models. In: ICALP, vol. 261, pp. 36:1–36:20 (2023)
7. Dinitz, Y., Nutov, Z.: A 2-level cactus model for the system of minimum and minimum+1 edge-cuts in a graph and its incremental maintenance. In: STOC, pp. 509–518 (1995)
8. Gabow, H.N., Gallagher, S.: Iterated rounding algorithms for the smallest k-edge connected spanning subgraph. SIAM J. Comput. **41**(1), 61–103 (2012)
9. Gabow, H.N., Goemans, M.X., Tardos, E., Williamson, D.P.: Approximating the smallest k-edge connected spanning subgraph by LP-rounding. Networks **53**(4), 345–357 (2009)
10. Garg, M., Grandoni, F., Ameli, A.J.: Improved approximation for two-edge-connectivity. CoRR abs/2209.10265v2 (2022). https://arxiv.org/abs/2209.10265v2
11. Goemans, M.X., Williamson, D.P.: A general approximation technique for constrained forest problems. SIAM J. Comput. **24**(2), 296–317 (1995)
12. Khuller, S.: Approximation algorithms for finding highly connected subgraphs, chap. 6. In: Hochbaum, D. (ed.) Approximation Algorithms for NP-hard problems, pp. 236–265. PWS (1995)

13. Nutov, Z.: Structures of cuts and cycles in graphs; algorithms and applications algorithms and applications. Ph.D. thesis, Technion, Israel Institute of Technology (1997)
14. Nutov, Z.: Extending the primal-dual 2-approximation algorithm beyond uncrossable set families. CoRR abs/2307.08270 (2023). https://doi.org/10.48550/arXiv.2307.08270
15. Nutov, Z.: Improved approximation algorithms for some capacitated k edge connectivity problems. CoRR abs/2307.01650, July 2023
16. Traub, V., Zenklusen, R.: A $(1.5+\epsilon)$-approximation algorithm for weighted connectivity augmentation. CoRR abs/2209.07860 (2022). https://doi.org/10.48550/arXiv.2209.07860
17. Traub, V., Zenklusen, R.: Local search for weighted tree augmentation and Steiner tree. In: SODA, pp. 3253–3272 (2022)
18. Williamson, D.P., Goemans, M.X., Mihail, M., Vazirani, V.V.: A primal-dual approximation algorithm for generalized Steiner network problems. Combinatorica **15**(3), 435–454 (1995)

Small Additive Error for Unsplittable Multicommodity Flow in Outerplanar Graphs

Richard Shapley[✉] and David B. Shmoys

Cornell University, Ithaca, USA
{rls499,david.shmoys}@cornell.edu

Abstract. We consider an unsplittable version of the minimum max-load multicommodity flow problem, where each demand must be routed along a single path. The objective is to minimize the maximum load on any edge in the network. In a seminal work, Schrijver, Seymour, and Winkler showed how to efficiently solve this problem on a cycle to within an additive term of $\frac{3}{2}W$ of the optimal value, where W is the largest demand between any two nodes. We extend their result to outerplanar graphs and provide an efficient algorithm for this problem that exceeds the optimal value by an additive term of no more than $O(W \log k)$, where k is the number of faces in the graph. This implies an $O(\log k)$ approximation ratio. We also extend this result to planar graphs with bounded treewidth and demands on the outer face, for which we also achieve an additive $O(W \log k)$ error term.

Keywords: Approximation algorithms · Multicommodity flow · Unsplittable flow · Load balancing

1 Introduction

The minimum max-load multicommodity flow problem is a variant of the multicommodity flow problem, where we are given a graph and a set of flow demands we wish to satisfy. The goal is to minimize the maximum load placed on any edge so that we can route enough flow through the graph to satisfy all the demands. We consider a restricted version of this problem, where we require each demand to be routed along a single path. More precisely, we are given an undirected graph $G = (V, E)$ along with demands $R = \{d_0, \ldots, d_{r-1}\}$, each indicated by a pair of terminals $(s_0, t_0) \ldots (s_{r-1}, t_{r-1})$ and associated demand weights $w_0, \ldots, w_{r-1} > 0$ respectively. We let W be the largest demand weight. Our objective is to route each demand d_i from s_i to t_i along a single path in order to minimize the largest load (in terms of the total weight) placed on any edge in the graph.

When the underlying graph is a ring, this is known as the ring loading problem, first introduced by Cosares and Saniee [2], who showed that it is NP-hard.

© The Author(s), under exclusive license to Springer Nature Switzerland AG 2025
M. Bieńkowski and M. Englert (Eds.): WAOA 2024, LNCS 15269, pp. 167–182, 2025.
https://doi.org/10.1007/978-3-031-81396-2_12

In a seminal work, Schrijver, Seymour, and Winkler [11] provided an efficient algorithm in this setting to come within a $\frac{3}{2}W$ additive error of the optimal solution value. We wish to extend this style of result to a wider class of graphs, including outerplanar graphs and planar graphs.

In subsequent work on the ring loading problem, Khanna provided a polynomial-time approximation scheme [7], and the additive error bound was improved to $\frac{19}{14}W$ [12] and later to $\frac{13}{10}W$ [4].

In this paper, we show how to efficiently approximate the unsplittable minimum max-load multicommodity flow problem on outerplanar graphs, for which we achieve an additive error of no more than $O(W \log k)$, where k is the number of faces in the graph. We also extend this result to planar graphs with demands on the outer face. Given such a planar graph with bounded treewidth, the additive error we achieve is also $O(W \log k)$. Roughly speaking, a graph G has treewidth $\mathsf{tw}(G)$ if we can recursively decompose the graph (into a tree decomposition) using vertex separators of size less than $\mathsf{tw}(G)$. See [3] for a more in-depth introduction to treewidth.

Our additive results imply approximation ratios of $O(\log k)$ in both settings. Notably, our performance guarantees do not depend on the number of demands, or even directly on the size of the graph—they depend only on the shape of the graph through the number of faces and the treewidth.

On more general graphs, with unit-demand weights and assuming the optimal value of the relaxation L^* is large enough, Raghavan and Thompson use randomized rounding to find a solution that exceeds the optimal value by $\sqrt{12L^* \ln n}$ with high probability, where n is the number of vertices [10]. However, applying their techniques to the weighted case gives a multiplicative $O(W \log n)$ approximation ratio. Furthermore, there is no apparent way to improve these bounds by taking advantage of the structure of outerplanar graphs. We should specifically emphasize that even in the worst case where n is $O(k)$, our results supersede the randomized result by removing any dependence on W in the multiplicative approximation guarantee.

More generally, using additional graph structure to understand the feasibility of finding edge-disjoint paths has received significant attention. For example, a celebrated result of Okamura and Seymour states that for a planar graph with terminals on the outer face, we can route edge-disjoint paths between all pairs of terminals if and only if an evenness condition and a cut condition are satisfied [9]. Here, the evenness condition says that the graph is Eulerian if we add edges between all terminal pairs, and the cut condition says the number of demands crossing any cut is at most the number of edges crossing the cut.

Other work on unsplittable multicommodity flow tends to focus on admission control variants, where the goal is to satisfy as many demands as possible within given edge capacities (see for example [1,8]) or single-source variants where all demands originate from a single vertex (as in [5,13]).

In contrast to the problem on a ring where each demand can only be routed in two possible ways, in the outerplanar setting the number of ways a demand can be satisfied may grow exponentially in k. However, despite the added complexity, we still take inspiration from the techniques of Schrijver et al. [11]. They start with a fractional solution to the linear relaxation, make an argument about the structure of a fractional solution, then cleverly round to an integral solution while incurring a small amount of error. We take a similar approach on outerplanar graphs. Beginning with an optimal fractional solution x^* to the linear relaxation of our problem, we first describe how we can perform uncrossings on x^* to simplify our fractional solution. This then allows us to make a similar style of claim about the structure of our uncrossed solution with respect to individual faces in the graph. We consider one face at a time, and decide whether to round each demand so that the load is routed either entirely clockwise or entirely counterclockwise around the face. Due to the structure we imposed through our uncrossing, we are able to write the change in the load on any edge in a closed form that is simple to bound. After we have decided which way to route the demands around a face, we split our remaining problem into multiple smaller subproblems that we can solve independently, and recurse.

We also consider the natural extension to planar graphs with demands on the outer face. Instead of rounding around one face at a time, we round demands around sets of faces. If we have chosen our set of faces well, we will again be left with smaller independent versions of the problem. We use the treewidth of the graph to determine how large our sets of faces need to be.

The rest of the paper is organized as follows. In §2, we set up some terminology and describe how to simplify a fractional solution. The main results for how to achieve an $O(W \log k)$ additive error bound on outerplanar graphs are in §3, and in §4, we provide some final remarks. Details on the extension to planar graphs are deferred to the extended online version.

2 Preliminaries

We begin by letting $\mathcal{P}_i$ be the set of all paths from s_i to t_i, then formulate our problem as the integer program:

$$
\begin{aligned}
\text{minimize} \quad & L \\
\text{subject to} \quad & \sum_{p:p \in \mathcal{P}_i} x_p = w_i, && \forall i \in [r]; \\
& \sum_{p:e \in p} x_p \leq L, && \forall e \in E; \\
& x_p \geq 0, && \forall p \in \mathcal{P}_i, \forall i \in [r].
\end{aligned}
$$

Consider the fractional version of this problem where we replace the last constraint with $x_p \geq 0$; call this program (P). This matches the more typical multicommodity flow formulation, where we do not require that our flows are unsplittable. We can solve this relaxation exactly in polynomial time. This is because although we have exponentially many paths to consider, we can solve the dual problem via the ellipsoid method, where the separation oracle is solving a shortest path problem. (We can also solve an edge-based formulation, and then decompose that fractional solution into distinct paths.) Our strategy takes this fractional solution, x^*, performs some "uncrossing" to improve the structure, and then rounds each demand to be fully satisfied by a single path.

First, we define some terminology and notation. We let k be the number of faces in our graph G (not counting the outer face). And we will use $|D|$ to denote the number of demands in D.

Each planar graph has a dual; we need to briefly discuss the dual graph to better understand our strategy. In particular, we will use the so-called "weak dual".

Definition 1 (Weak dual). *We can construct the dual G^d of G by creating a vertex for every face of G, and connecting two vertices in G^d if the corresponding faces share an edge in G. The weak dual G^* of G is a subgraph of G^d with the vertex corresponding to the outer face of G removed.*

For any 2-vertex connected outerplanar graph G (i.e., the graph remains connected even if we delete any vertex), G^* is a tree. A classic result of Jordan [6] states that every tree has a single vertex whose removal divides the graph into components with no more than $n/2$ vertices. This idea allows us to operate recursively on trees (and therefore on outerplanar graphs). In particular, by removing vertices from our tree (or equivalently removing faces from our outerplanar graph), we are able to separate our graph into components that we can handle independently. When we consider general planar graphs, we can extend this idea via graph separators, enabling us to recursively handle those instances as well. The performance of our algorithm will depend on the size of these graph separators, which we can relate to the treewidth of the graph.

2.1 Uncrossing

Before we start rounding, we would like to impose some additional structure on our fractional solution x^*. Here, we argue that we can find an optimal fractional solution to (P) in which there is relatively little crossing, and we will exploit this property in our rounding procedures.

Definition 2 (Crossing). *Given two paths p_1 and p_2, we say that they cross if there is some Jordan curve C enclosing a region A such that $A \cap p_1$ and $A \cap p_2$ are both non-empty and continuous, and the endpoints strictly alternate around C.*

The number of crossings is the number of disjoint regions A we can find that satisfy the above criteria (see Fig. 1a).

We say that two paths are parallel if they do not cross. Similarly, we say that two demands d_a and d_b are parallel if there are paths p_a from s_a to t_a and p_b from s_b to t_b such that p_a and p_b do not cross.

However, we need to be careful with how crossing works at the endpoints of our paths to ensure that our future arguments hold. For the purposes of determining crossings, we consider a slightly modified graph where all terminals are moved to unique degree 1 vertices, extending into the outer face from their original vertex. When we have terminals from multiple demands at a single vertex, we arrange their degree 1 representations in the reverse order of the corresponding terminals. For example, suppose terminals s_1, s_2, and s_3 are all at a vertex v and their opposing terminals are in the order t_1, t_2, t_3 traveling clockwise around the outer face of the graph from v. Then in clockwise order, the added vertices extending from v should represent terminals s_3, s_2, then s_1. With this representation, any two demands that share an endpoint are parallel, but paths that share a terminal may not be.

We can reduce the number of total crossings in our fractional solution in the following way. Take any pair of paths p_a and p_b (satisfying demands d_a and d_b respectively) that cross twice or more, and say they have been assigned fractional loads x_a and x_b, where $x_a \geq x_b$ without loss of generality. Fix two points as follows: starting from s_a, let v_s be the last vertex before p_a crosses p_b (i.e., the subpath from s_a to v_s does not cross p_b, but including one more vertex from p_a would). And let v_t be the last vertex before p_a crosses p_b a second time. Now, we can exchange the paths between v_s and v_t to eliminate both crossings. If $x_a > x_b$, then we can first decompose p_a into two paths along the same route with loads x_b and $x_a - x_b$ and exchange between the pair of paths with equal load. See Fig. 1 for an example of uncrossing two paths.

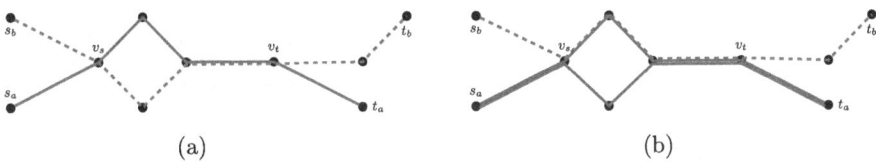

(a) (b)

Fig. 1. In (a), the solid blue path (p_a) and the red dashed path (p_b) cross twice, with the regions A defining the crossing highlighted in yellow. In (b), the paths have been uncrossed. In this example, the blue path carried more load than the red path, so it has been decomposed into two parallel paths. (Color figure online)

Doing this repeatedly will eventually ensure that no two paths cross more than once, as each time the above operation is performed, the total load over all crossings decreases. However, it is not clear that this will terminate efficiently, since each uncrossing operation may reduce this total load by a very small amount, and can actually increase the total number of crossings. The proof of the following uncrossing lemma is left to extended online version.

Lemma 1. *Given a solution x to (P), there is a fractional solution x' with objective no worse than x in which no pair of paths $p_1, p_2 \in \{p \in \cup_{i \in [r]} \mathcal{P}_i, x_p > 0\}$ cross more than once, and we can find such an x' in polynomial time.*

2.2 Demand Splitting

Given an uncrossed solution x to (P), let $G[D]$ be the subgraph of G induced by all paths satisfying demands in D with nonzero load (that is, paths in the set $\{p \in \cup_{i \in [r]} \mathcal{P}_i, x_p > 0\}$). When considering a single demand d_i, we will write $G[d_i]$ to mean $G[\{d_i\}]$. Similarly, for a set of paths $\mathcal{P}$, let $G[\mathcal{P}]$ be the subgraph of G induced by the paths in $\mathcal{P}$.

Definition 3 (Splitting). *We say a demand d_i splits around a face if the face is in the interior of $G[d_i]$.*

Equivalently, if we orient all paths so that they travel from s_i to t_i, a demand d_i splits around a face F if some path in $G[d_i]$ travels clockwise around F, and some path travels counterclockwise around F. It is not hard to see that for any face of G, no two demands d and d' that split around it can be parallel. Otherwise, there must be paths $p \in G[d]$ and $p' \in G[d']$ that cross twice.

Therefore, all demands split around a face F must be pairwise mutually non-parallel. This imposes rather heavy restrictions on the set D of demands that split around the face. First, no vertex can be a terminal for more than one demand. Furthermore, there are exactly $|D| - 1$ terminals from D along each outside path between the two ends of any demand. This implies that there is a consistent ordering of these terminals around the outside of the graph. That is, if s_j is the first terminal in D clockwise of s_i, then t_j is the first terminal in D clockwise of t_i. (Note that due to the undirected nature of our problem, we can freely interchange the labels s_i and t_i. We will, however, see later that we do need to remain consistent within each rounding step.) See Fig. 2 for an example of how pairs of terminals are arranged around the outer face of a graph.

For a set of demands D that split around some face F, we will choose an arbitrary terminal of one of the demands to be $s_0(D)$ and its paired terminal to be $t_0(D)$. We then label the remaining terminals increasing clockwise so that when we read all the terminals of the demands in D clockwise around the outer face of the graph, we have the sequence

$$s_0(D), s_1(D), \ldots, s_{|D|-1}(D), t_0(D), t_1(D), \ldots, t_{|D|-1}(D)$$

before returning back to $s_0(D)$. We also define the demands $d_i(D)$ and the demand weights $w_i(D)$ so that they correspond. Furthermore, we extend this notation so that $d_i(D) = d_j(D)$ if $i \equiv j \pmod{|D|}$.

3 The Rounding Algorithm

In this section, we describe how to round an uncrossed fractional solution to an integral solution. For every demand that is satisfied fractionally, we will choose to round up exactly one of the fractional paths to fulfill the demand. At a high level, we take a face of the graph and divide the graph into two parts via a vertex cut through two vertices on the face. Then any demand that is separated by the cut must pass through one of the two vertices. We then argue that we can find an equivalent fractional solution by uncrossing paths to be laminar on each side of the cut. Finally, we can use this laminarity property to bound the rounding error on each edge in the graph.

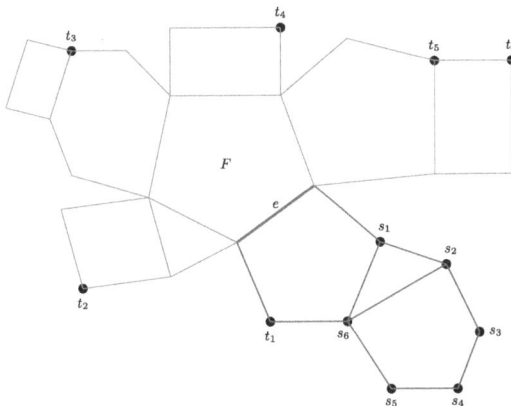

Fig. 2. All terminals shown belong to a set of demands D that split around the face F. The subgraph highlighted in red is $P_e(D, F)$. We can note that the set D is not well-spaced around F, because both s_1 and t_1 are in $P_e(D, F)$. (Color figure online)

We begin by noting that if all paths for a demand pass through some vertex v, we can round the partial paths on both sides of v independently of each other. In a more general sense, this means that we can assume our graph is 2-vertex connected. Additionally, we can assume that for any demand d_i, $G[d_i]$ is 2-vertex connected, because if not, we could replace the demand with demands between sequential pairs of cut vertices (these are the vertices we could remove to disconnect the graph).

Given a face F and a set of demands D that split around F, we will divide $G[D]$ into subgraphs $P_e(D, F)$ for each e on F so that $\cup_{e \in F} P_e(D, F) = G[D]$ and $E(P_{e_1}(D, F)) \cap E(P_{e_2}(D, F)) = \emptyset$ for all $e_1 \neq e_2$. This is done as follows.

For any edge e on F, notice that removing the endpoints of e but not the edges incident to it (so in effect, replacing a vertex of degree d with d distinct endpoints) may separate $G[D]$ into either two or three connected components, due to outerplanarity and 2-vertex connectivity. These components are: one including the rest of F, one including just the edge e, and one including everything else which we will call R_e (which may be empty). Then we define $P_e(D, F)$ to be the set of edges $\{e\} \cup R_e$. We will also abuse notation and let $P_e(D, F)$ also refer to the corresponding edge-induced subgraphs of $G[D]$. Importantly, the intersection of two subgraphs $P_{e_1}(D, F)$ and $P_{e_2}(D, F)$ is a single vertex v when e_1 and e_2 are both incident to v; otherwise the intersection is empty.

Definition 4 (Well-spaced). *A set of demands D that split around a face F is well-spaced if for each demand $d_i \in D$, there are distinct parts $P_{e_1}(D, F)$ and $P_{e_2}(D, F)$ such that $s_i \in P_{e_1}(D, F)$ and $t_i \in P_{e_2}(D, F)$.*

Intuitively, demands are well-spaced roughly if we can disconnect each pair of terminals by removing two vertices on the face. For an example of a set of demands that are not well-spaced around some face, see Fig. 2. For now, we will assume that the set of demands that split around F is well-spaced and later revisit what to do if this is not the case.

For a path p on a face F, let $Q(D, F, p) = \cup_{e \in p} P_e(D, F)$ (note that path here does not refer to a path between terminals as it did in all previous instances). We also let $u(p)$ and $v(p)$ represent the endpoints of p, where the path travels clockwise around F from $u(p)$ to $v(p)$. Observe that for any non-trivial choice of p, if p' is the path going counterclockwise from $u(p)$ to $v(p)$, then $Q(D, F, p) \cap Q(D, F, p') = \{u(p), v(p)\}$.

Lemma 2. *If the set of demands D that split around F are well-spaced, then we can partition them into two sets A and B and associated paths on F, p^A and p^B so that every demand in A has exactly one terminal in $Q(D, F, p^A)$, and similarly that every demand in B has exactly one terminal in $Q(D, F, p^B)$.*

Proof. We first find the path p^A. Let p^A be a minimal length path on F such that $Q(D, F, p^A)$ contains both terminals of some demand. Let A be all demands with exactly one terminal in $Q(D, F, p^A)$.

Note that all demands should have at least one terminal in the set. What remains as B is all demands that had both terminals in the set.

Since the demands are well-spaced, we can divide p^A into shorter paths p and p'. For any choice of p and p', all demands in B will have exactly one terminal in $Q(D, F, p)$, because otherwise, we contradict the minimality of our choice of p^A. So we can let $p^B = p$. □

Note that when we say exactly one terminal of d_i is in $Q(D, F, p)$, we also assume exactly one terminal of d_i is in $Q(D, F, p')$, where p' is the path going counterclockwise from $u(p)$ to $v(p)$. While this is not strictly true if a terminal is at $u(p)$ or $v(p)$, we can correct this by replacing the terminal with two vertices and a dummy edge (as in vertex splitting).

Lemma 3. *Suppose we have a set of demands D that split around F and a path p on F. If every demand in D has exactly one terminal in $Q(D, F, p)$, then every path satisfying a demand in D enters (or exits) $Q(D, F, p)$ through exactly one of $u(p)$ and $v(p)$.*

Proof. If any path used both, that would require either both endpoints or neither endpoint to be in $Q(D, F, p)$. □

Let $\mathcal{P}_u(D, i, p)$ be the set of paths satisfying $d_i(D)$ entering (or exiting) $Q(D, F, p)$ through $u(p)$. And similarly, let $\mathcal{P}_v(D, i, p)$ be the set of paths satisfying $d_i(D)$ entering (or exiting) $Q(D, F, p)$ through $v(p)$.

Using this notation, we will now impose some structure on our demands. For any edge, naively we must consider the effect of rounding any subset of demands. But if we restrict what subsets of demands can affect each edge, our problem becomes much easier. This inspires the following lemma, which roughly states that the demands we need to worry about are consecutive.

Lemma 4. *Suppose we have a set of demands D that split around F and an associated path p^D on F such that every demand in D has exactly one terminal in $Q(D, F, p^D)$. We can perform uncrossings on the set of demands so that for any edge e in G, the following conditions hold:*

- *The set of demands $\{d_i \in D : \{p \in \mathcal{P}_u(D, i, p^D) : p \cap e \neq \emptyset\} \neq \emptyset\}$ is given by $\cup_{i=\ell}^{h}\{d_i(D)\}$ for some choice of ℓ and h (where $h - \ell < |D|$). Moreover, if $\ell < i < h$, all paths $p \in \mathcal{P}_u(D, i, p^D)$ use the edge e.*
- *The set of demands $\{d_i \in D : \{p \in \mathcal{P}_v(D, i, p^D) : p \cap e \neq \emptyset\} \neq \emptyset\}$ is given by $\cup_{i=\ell}^{h}\{d_i(D)\}$ for some (different) choice of ℓ and h (where $h - \ell < |D|$). Moreover, if $\ell < i < h$, all paths $p \in \mathcal{P}_v(D, i, p^D)$ use the edge e.*

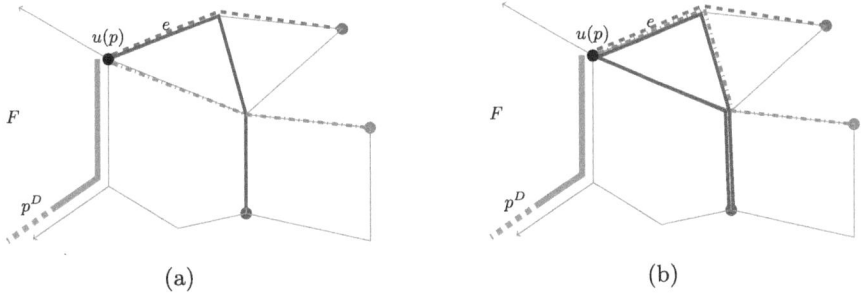

Fig. 3. Both figures show three paths satisfying demands in D that enter $Q(D, F, p^D)$ through $u(p)$. (We only show the paths on one side of $u(p)$.) In (a), notice that the red and blue paths use e, but the green edge does not, so the demands that affect it are not consecutive. In (b), we have done some uncrossing as per Lemma 4 so that a consecutive set of demands all use e. (Color figure online)

While notationally cumbersome, the above nested set notation describes the set of demands fractionally satisfied by paths that use both $u(p^D)$ and the edge e. See Fig. 3 for an application of this lemma.

Proof. To show this result, we first describe our method to uncross the demands so that the paths in $\mathcal{P}_u(D, i, p^D)$ do not cross each other at all in $Q(D, F, p^D)$. If they cross, we would like to reallocate the demand along paths using the same edges so that they no longer cross. We later argue that this structure requires all paths using any given edge to come from a consecutive set of demands.

Once again, we borrow from the techniques of Wagner and Weihe [14] to perform this uncrossing. We restrict our focus to the part of the paths $\mathcal{P}_u(D, i, p^D)$

in $Q(D, F, p^D)$ and orient them so that they travel from $u(p^D)$ to their terminal in $Q(D, F, p^D)$. We perform right-first searches as normal, beginning at the demand with the clockwise-most terminal first, and proceeding counterclockwise through the rest of the demands. As in the proof of Lemma 1, we subdivide edges and demands adaptively to ensure the process terminates efficiently. As a result, no paths should cross in $Q(D, F, p^D)$.

Consider paths p and p', where p was drawn first. The path p' can never be on the right of p, because that would violate the right-first choices of p.

Let p'^D be the path from $u(p^D)$ to $v(p^D)$ going counterclockwise around F. Since the uncrossing operation in one part $G[D]$ has no effect on any other parts, we can achieve the above property simultaneously for $Q(D, F, p'^D)$.

Now we must argue that with this structure, paths that make use of an edge e must satisfy consecutive demands. We start by fixing some $e \in Q(D, F, p^D)$, and make the argument for paths entering $Q(D, F, p^D)$ through $u(p^D)$. The arguments for $e \in Q(D, F, p'^D)$ and for $v(p^D)$ instead of $u(p^D)$ are symmetric.

If the set of demands that make use of e and $u(p^D)$ (that is, $\{d_i \in D : \{p \in \mathcal{P}_u(D, i, p^D) : p \cap e \neq \emptyset\} \neq \emptyset\}$) has cardinality either zero or one, then the statement is trivially true.

Otherwise, take two distinct demands $d_\ell(D)$ and $d_h(D)$. (Without loss of generality, we may say that $\ell < h$ and $h - \ell < |D|$.) If $h = \ell + 1$, again the statement is trivially true, so we can assume that $\ell + 1$ is strictly between ℓ and h. Suppose that some path in $\mathcal{P}_u(D, \ell + 1, p^D)$ does not use the edge e. We will argue that this would require two paths to cross somewhere in $Q(D, F, p^D)$.

Take paths $p_\ell \in \mathcal{P}_u(D, \ell, p^D)$ and $p_h \in \mathcal{P}_u(D, h, p^D)$ that both use e, and take path $p_{\ell+1} \in \mathcal{P}_u(D, \ell + 1, p^D)$ that does not. As e is not on $p_{\ell+1}$, we know that at least one endpoint of e (say w) is not on $p_{\ell+1}$. Suppose w is on the right of $p_{\ell+1}$—that is, on $p_{\ell+1}$ from $u(p^D)$ to $t_{\ell+1}$, we would arrive at w by exiting some vertex using an edge clockwise of the one that $p_{\ell+1}$ uses. Then $p_{\ell+1}$ crosses p_ℓ in $Q(D, F, p^D)$, where our region A that defines the crossing is $p_{\ell+1}[u(p^D), t_{\ell+1}] \cup p_\ell[w, t_\ell]$ (here, we use $p[a, b]$ to mean the subpath of p between vertices a and b). If instead, w is on the left of $p_{\ell+1}$, then $p_{\ell+1}$ crosses p_h, where the region A is $p_{\ell+1}[u(p^D), t_{\ell+1}] \cup p_h[w, t_h]$.

Therefore, if there are paths in $\mathcal{P}_u(D, \ell, p^D)$ and in $\mathcal{P}_u(D, h, p^D)$ that include e, all paths entering $Q(D, F, p^D)$ through $u(p^D)$ satisfying demands d_i for $\ell < i < h$ must also include e. □

For any demand $d_i \in D$ that splits around a face F with one endpoint in $Q(D, F, p)$, we wish to round all of the load on a single path by determining whether we should route it through $u(p)$ or $v(p)$. When performing this rounding around a single face, our task is straight forward if there are only two paths that fractionally satisfy the demand: we can just choose one of the paths to send all the demand through, while sending none along the other. But if there are more than two paths, it is not so simple. By focusing on one face at a time, our decision can only be whether we want to route the demand clockwise or counterclockwise around F, (that is, through $u(p)$ or $v(p)$) but not which specific path to take. To handle this, for each demand split around F, we decide to either route all the load clockwise, or all counterclockwise. After rounding a demand (say we route it clockwise) all paths traveling counterclockwise around F will be set to zero. And the extra clockwise demand will be proportionally divided among all clockwise paths satisfying the demand, relative to their fractional load before the rounding.

For a set of demands D that split around a face F, let $\alpha_i(D)$ be the amount of fractional load routed clockwise (when going from $s_i(D)$ to $t_i(D)$) around F to satisfy $d_i(D)$, and let $\beta_i(D) = w_i(D) - \alpha_i(D)$ be the amount routed counterclockwise around F. (Clearly, any demand that splits around F will have these both be positive values.) We also let $\gamma_i(D)$ be the change in the amount routed clockwise around F as a result of our rounding. Depending on which way we decide to round, we have either $\gamma_i(D) = \beta_i(D)$ or $\gamma_i(D) = -\alpha_i(D)$.

We proceed with a technical lemma, which will allow us to bound the cost of our rounding scheme.

Lemma 5. *Given some set of demands D that are pairwise mutually non-parallel, we can assign either $\gamma_i(D) = \beta_i(D)$ or $\gamma_i(D) = -\alpha_i(D)$ for each i so that for $0 \leq k \leq |D| - 1$, $\left|\sum_{i=0}^{k} \gamma_i(D) - \lambda \gamma_k(D)\right| \leq W/2$ for $0 \leq \lambda < 1$.*

Proof. First, we deal with $\lambda = 0$. The argument for this case is taken directly from Schrijver et al. [11]. We can greedily choose $\gamma_i(D) = -\alpha_i(D)$ or $\gamma_i(D) = \beta_i(D)$ in order, so that $\left|\sum_{i=0}^{k} \gamma_i(D)\right| \leq W/2$. Since the two choices for $\gamma_i(D)$ differ by $\beta_i(D) + \alpha_i(D) \leq W$, at least one of them must keep the running sum in the desired range.

Furthermore, since we have both $\left|\sum_{i=0}^{k-1} \gamma_i(D)\right| \leq W/2$ and $\left|\sum_{i=0}^{k} \gamma_i(D)\right| \leq W/2$, and since $\sum_{i=0}^{k} \gamma_i(D) - \lambda \gamma_k(D)$ is a convex combination of $\sum_{i=0}^{k-1} \gamma_i(D)$ and $\sum_{i=0}^{k} \gamma_i(D)$, the same bound holds when λ is positive. □

We are now ready to consolidate our results into a rounding algorithm.

Lemma 6. *Suppose we have a set of demands D that split around F, and an associated path p^D on F such that every demand in D has exactly one terminal in $Q(D, F, p^D)$. We can route all demands in D either entirely clockwise or entirely counterclockwise around F so that the load on any edge increases by at most $3W$.*

The proof relies on noting that the change in load on any edge can be written as a nice summation of consecutive demands due to Lemma 4, which we can then bound via Lemma 5. The full proof can be found in the extended version of this paper.

Lemma 7. *Let D be a set of demands that split around F and suppose D is well-spaced around F. We can route all demands in D either entirely clockwise or entirely counterclockwise around F while increasing the load on any edge by at most $6W$.*

Proof. Since D is well-spaced around F, we can use Lemma 2 to partition D into A and B, with associated paths p^A and p^B. Then, by Lemma 6, we can route the demands in A and B independently, each increasing load by no more than $3W$ on any edge. So we can bound the total increase on any edge by $6W$. □

With this lemma, we can see how our entire rounding algorithm should work. If we could guarantee that for every face of G we choose, the set of demands that split around the face is well-spaced, then our algorithm is clear. Each time we choose a face F, we can apply Lemma 7 to completely route all demands around F so that no more are split around F. We can then handle the remaining parts of the graph independently. By recursively choosing faces and rounding around them, we round each edge $O(\log k)$ times, so the total load on any edge increases by at most $O(W \log k)$. The only loose end is our assumption that the demands will be well-spaced.

3.1 Handling Demands That Are Not Well-Spaced

We now address the scenario where our set of demands are not well-spaced. To achieve an $O(W)$ bound in this case, we proceed by rounding the demands around a second face, which we will show suffices to round most of the demands around the original face.

Lemma 8. *Suppose a set of demands D that split around F are not well-spaced. Let $\tilde{D} \subseteq D$ be the set of demands that violate well-spacing. Then we can find another face F' in $G[D]$ such that $\tilde{D}$ is well-spaced around F'.*

Proof. We find this face F' by "walking" in the direction that violates well-spacing until all demands are well-spaced.

First, we note that all demands that violate well-spacing have terminals in one part, say $P_{e_0}(D, F)$: otherwise, since the demands split around F, some paths would cross too many times. Consider the face $F_1 \subseteq G[D]$ that shares the edge e_0 with F, and compute the parts $P_1(D, F_1)$. If $\tilde{D}$ is not well-spaced around F_1, we can find the part $P_{e_1}(D, F_1)$ that contains the violations, and consider the adjacent face F_2 in $G[D]$. We repeat until all demands in $\tilde{D}$ are well-spaced around the chosen face. This process cannot continue indefinitely, because we must eventually reach a leaf in the dual graph $G[D]^*$.

So we would like to show two things: that the number of demands in $\tilde{D}$ that are well-spaced around F_j monotonically increases (as long as we have not finished), and that every demand must become well-spaced by the time we reach a face corresponding to a leaf in $G[D]^*$.

Consider some demand $d_i \in \tilde{D}$ that at iteration j has terminals in different parts. Since the process is not yet finished at iteration j, there must be some demand in $\tilde{D}$ with both terminals in some part $P_{e_j}(D, F_j)$ and therefore, one terminal of d_i (say s_i) must also be in $P_{e_j}(D, F_j)$. Then, from the perspective of face F_{j+1}, t_i must be in $P_{e_j}(D, F_{j+1})$. So once a demand in $\tilde{D}$ becomes well-spaced around some face F_j, it must remain well-spaced for all subsequent faces until termination.

Suppose that while running this process, we consider the sequence of faces $F_1, \ldots, F_\ell$ where F_ℓ corresponds to a leaf node in $G[D]^*$. Then if some demand $d_i \in \tilde{D}$ was not well-spaced around any of $F_1, \ldots, F_{\ell-1}$, then both terminals of d_i must be on F_ℓ, so d_i will be well-spaced around F_ℓ.

Therefore, this process terminates, and we indeed find some face that all demands in $\tilde{D}$ are well-spaced around. See Fig. 4 for an illustration of this process. □

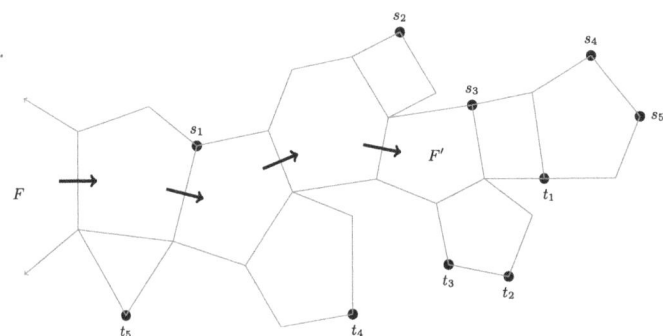

Fig. 4. An illustration of the sequence of faces we consider in Lemma 8. While some terminal pairs become well-spaced around earlier faces, d_3 is not well-spaced around any face until F'.

Now, we must argue that we can round the demands $\tilde{D}$ that violate well-spacing without incurring much additional cost.

Lemma 9. *Suppose the set of demands D that split around F are not well-spaced. Let $\tilde{D}$ be the set of demands that violate well-spacing. We can uncross or route the demands in $\tilde{D}$ so that all but one will also be routed with respect to F, and the load on any edge increases by no more than $6W$.*

Proof. Let F' be the closest face to F that $\tilde{D}$ is well-spaced around, as found in Lemma 8. Let $P_{e'}(D, F')$ be the part around F' that includes F. First, notice that every demand in $\tilde{D}$ needs to have a terminal not in $P_{e'}(D, F')$. Since $G[d_i]$ is 2-vertex connected for all $d_i \in \tilde{D}$, we know all demands in $\tilde{D}$ must split around both F and F', as well as all faces in the unique sequence from F to F' (this corresponds to the unique path between the vertices in $G[D]^*$). Let H be the subgraph induced by this sequence of faces.

Let $d_\ell \in \tilde{D}$ be a demand that is well-spaced around F', but not around any faces closer to F. Then d_ℓ must have no terminals in $P_{e'}(D, F')$, so it must contain a path going the long way around H (say p_ℓ). Thus, any path satisfying another demand that includes a chord of H can be uncrossed with p_ℓ. As a result, either the load on the chord will now belong to a path satisfying d_ℓ, or d_ℓ will no longer split around F. In the second case, d_ℓ no longer belongs to $\tilde{D}$, so we recompute F' for the remaining demands in $\tilde{D}$. We continue uncrossing in this manner, until d_ℓ is the only demand in $\tilde{D}$ that has paths using chords of H.

For all other demands, we can round as in Lemma 7, increasing the load on any edge by no more than $6W$. And since these demands use no chords of H, rounding around F' is equivalent to rounding around F. □

Lemma 10. *Let D be a set of demands that split around F. We can route all demands in D either entirely clockwise or entirely counterclockwise around F so that the load on any edge increases by no more than $\frac{19}{2}W$.*

Proof. We suppose D is not well-spaced around F (otherwise, Lemma 7 suffices). Let $\tilde{D}$ be the set of demands that violate well-spacing. We route these demands according to Lemma 9 for an increase of no more than $6W$.

Let $P_e(D, F)$ be the part that contains the terminals of the demands in $\tilde{D}$. For the demands in $D \setminus \tilde{D}$, all have exactly one terminal in $P_e(D, F)$, so we can apply Lemma 6 where the associated path is the edge e. This increases load on every edge by no more than $3W$.

Finally, for the one demand remaining in $\tilde{D}$, we can round it greedily, which will increase the load again by at most $W/2$. So in total, the load on any edge increases by no more than $\frac{19}{2}W$. □

Since we have addressed how to handle demands that are not well-spaced, we are finally ready to prove our main result.

Theorem 1. *Given a fractional solution to the unsplittable minimum max-load multicommodity flow problem on an outerplanar graph, we can route each demand along a single path so that the load on any edge increases by no more than $\frac{19}{2}W \log k$.*

Proof. We start with an uncrossed optimal fractional solution to the relaxation. First, consider what happens after we round the demands that split around some face F, as in Lemma 10. Since no more demands split around F, the remaining parts of the graph are independent: for a particular demand d_i that is not yet rounded, any paths satisfying d_i must enter (or exit) each part through the same vertex. So we could equivalently consider dividing d_i into independent demands between these vertices, which we handle separately. Thus, each remaining part is a fully self-contained instance of our problem along with a current uncrossed fractional solution.

The independent parts of the graph correspond to the connected components in G^* after we have removed the vertex corresponding to F. So by Jordan's tree separator theorem [6], we can find a face such at most half of the faces are in any part. Therefore, by recursively choosing faces in this manner and routing the demands that split around it, every edge will be rounded no more than $\log k$ times. So in total, the load on any edge increases by no more than $\frac{19}{2}W \log k$. □

3.2 Planar Graphs With Terminals on the Outer Face

Our techniques on outerplanar graphs can be generalized to planar graphs with the additional restriction that all terminals must be on the outer face. Our arguments, which are formalized in the extended version of this paper, culminate in the following theorem.

Theorem 2. *Given a fractional solution to the unsplittable minimum max-load multicommodity flow problem on a planar graph with demands on the outer face, we can route all demands along a single path so that the load on any edge increases by no more than $96W \log k(\mathsf{tw}(G) + 3)2^{\mathsf{tw}(G)}$.*

4 Concluding Remarks

We have provided an algorithm to find a feasible solution to the unsplittable minimum max-load multicommodity flow problem on an outerplanar graph with k faces that exceeds the optimal value by an additive term of no more than $O(W \log k)$. We further extend this result to planar graphs with demands on the outer face with an $O(W \log k)$ bound on the additive term. An immediate open question is whether it is possible to do better. For instance, is there an algorithm for outerplanar graphs that can match the result of Schrijver et al. [11] and remove the dependence on k in the additive bound? Or can we find a lower bound suggesting that no bound depending on only W is possible? Efforts in both directions have not yielded results, suggesting that perhaps different techniques are required to resolve this gap.

Another natural question is whether we can find a polynomial-time approximation scheme for this problem, as Khanna did on the ring [7]. For the ring, it is not difficult to extend an additive approximation to an approximation scheme, but our setting and algorithms are structurally more sophisticated, and we see no clear way to leverage our results to do the same.

Disclosure of Interests. The authors have no competing interests to declare that are relevant to the content of this article.

References

1. Chakrabarti, A., Chekuri, C., Gupta, A., Kumar, A.: Approximation algorithms for the unsplittable flow problem. Algorithmica **47**(1), 53–78 (2007)
2. Cosares, S., Saniee, I.: An optimization problem related to balancing loads on SONET rings. Telecommun. Syst. **3**, 165–181 (1994)
3. Cygan, M., et al.: Parameterized Algorithms. Springer, Cham (2015). https://doi.org/10.1007/978-3-319-21275-3
4. Däubel, K.: An improved upper bound for the ring loading problem. SIAM J. Discret. Math. **36**(2), 867–887 (2022)
5. Dinitz, Y., Garg, N., Goemans, M.X.: On the single-source unsplittable flow problem. Combinatorica **19**(1), 17–41 (1999)

6. Jordan, C.: Sur les assemblages de lignes. J. für die reine und angewandte Mathematik **70**, 185–190 (1869)
7. Khanna, S.: A polynomial time approximation scheme for the SONET ring loading problem. Bell Labs Tech. J. **2**(2), 36–41 (1997)
8. Kolman, P., Scheideler, C.: Improved bounds for the unsplittable flow problem. J. Algorithms **61**(1), 20–44 (2006)
9. Okamura, H., Seymour, P.D.: Multicommodity flows in planar graphs. J. Comb. Theor. Ser. B **31**(1), 75–81 (1981)
10. Raghavan, P., Thompson, C.D.: Randomized rounding: a technique for provably good algorithms and algorithmic proofs. Combinatorica **7**(4), 365–374 (1987)
11. Schrijver, A., Seymour, P., Winkler, P.: The ring loading problem. SIAM J. Discret. Math. **11**(1), 1–14 (1998)
12. Skutella, M.: A note on the ring loading problem. SIAM J. Discret. Math. **30**(1), 327–342 (2016)
13. Traub, V., Koch, L.V., Zenklusen, R.: Single-source unsplittable flows in planar graphs. In: Proceedings of the 2024 Annual ACM-SIAM Symposium on Discrete Algorithms (SODA), pp. 639–668. SIAM (2024)
14. Wagner, D., Weihe, K.: A linear-time algorithm for edge-disjoint paths in planar graphs. Combinatorica **15**(1), 135–150 (1995)

Complexity of Fixed Order Routing

Steven Miltenburg[✉], Tim Oosterwijk, and René Sitters

Vrije Universiteit Amsterdam, De Boelelaan 1105, 1081 Amsterdam, HV,
The Netherlands
{s.j.g.miltenburg,t.oosterwijk,r.a.sitters}@vu.nl

Abstract. We consider the classic Vehicle Routing Problem with the additional property that all requests are ordered and the subtour of each server (or vehicle) must obey the fixed order. A scheduling version of this problem was introduced by Bosman et al. (2019). We study several metric spaces and objective functions and our results show that in some settings such a fixed order simplifies the problem, while in others it makes an easy problem become NP-hard. For general metrics, we show that c-capacitated VRP remains APX-hard in the fixed order setting for $c = 3$ and show that the well-known iterated tour partitioning algorithm yields a $(2 - 1/c)$-approximation. When all points are on the line, we show that the fixed order restriction makes VRP NP-hard to solve for minimizing total completion time or maximum completion time, in contrast to standard VRP. We also sketch how to obtain a PTAS in these settings for general metrics.

Keywords: Routing · Complexity · Approximation Algorithms

1 Introduction

Routing problems are present in modern society in a wide range of applications. Finding satisfactory solutions to them, or preferably optimal ones, can therefore have a big impact. A classic variant is the vehicle routing problem (VRP) in which a fleet of servers (or vehicles), initially situated at a depot, needs to transport goods from the depot to a set of requests (or customers). Usually the goal is to minimize the total distance traveled while adhering to several constraints, such as capacities of the servers or time windows of the requests [7,23]. For practical reasons, the order in which these requests need to be served is sometimes not very flexible and reordering may be difficult or expensive. This certainly happens in an online setting where requests need to be served one by one in the order of arrival as in the k-server problem [15,17]. But reordering may even be difficult in an offline setting, where goods are already ordered in a certain way due to the production process or the loading dock and where a lack of time or space makes reordering expensive. Fixed order problems with some room for reordering in the form of a buffer have been studied extensively recently both in an online and offline setting but mainly for a single server, where the problem is easy to solve if reordering is not allowed [1]. In this article, we consider this extreme case, where reordering is not allowed, in a parallel setting.

We consider an (offline) *fixed order routing* (FOR) perspective in which the requests are given a priori in a specified order. After assigning requests to servers, the order in which the assigned requests are served needs to match the fixed order. We refer to the route of one server along its associated requests as a subtour. We distinguish between the variant where the number of subtours is bounded by some given number k (the k-FOR problem) and the variant where the number of requests in each subtour is bounded by some given number c (the c-capacitated FOR problem). The scheduling variant with total completion time objective was introduced by Bosman et al. [6] and inspired the generalization to routing discussed here.

A fixed order framework has also been studied recently in a game theoretic version of a scheduling problem [21]. In their model, there are parallel machines and agents (jobs) aim to minimize their completion time. The machines are endowed with machine-dependent priority lists, dictating the order in which the jobs assigned to it are processed. The authors study the equilibrium inefficiency in terms of makespan and sum of completion times.

The fixed order perspective has interesting connections to many different optimization problems. We already mentioned the well-studied k-server problem [15,17]. In this problem, there are k servers in some initial configuration, usually in a metric space, which need to serve requests that appear sequentially. When a request appears, one of the servers moves to the location of the new request, and then the next request is revealed. The goal is to move the servers around, fulfilling all requests, while minimizing the total distance travelled. Note that the offline version is exactly the k-FOR problem with the objective of minimizing total traveled distance. Moreover, in a uniform metric space the k-FOR problem equals the offline paging problem [17]. It is well-known that the offline k-server problem can be solved by a minimum-cost flow computation [9]. However, as we show in this paper, most other variants are surprisingly hard to solve.

For many interesting optimization problems the fixed order condition leads to a generalization instead of a restriction of the problem. For example, when minimizing total weighted completion time on parallel machines ($P||\sum_j w_j C_j$), the optimal order per machine is dictated by Smith's ratio rule [22]. That means that the fixed order scheduling problem as studied in [6] equals the scheduling problem $P||\sum_j w_j C_j$ if the jobs arrive in Smith's order. While the classic scheduling problem admits a simple PTAS, the more general fixed order version turned out to be much harder [6]. Here we see a similar behavior for routing: The classic VRP on the line is trivial, while in Sect. 3 we prove this to be NP-hard for fixed order routing. The line is an example of a metric that has the so called *master tour* property [10]. Hence, for parallel scheduling and routing problems with this property, one only needs to decide on the partition and not on the ordering. Allowing for any fixed order yields an interesting generalization of these problems.

For general metrics, a fixed order seems to make approximation easier. Though we show that c-capacitated VRP remains APX-hard in the fixed order setting for $c = 3$. Table 1 presents the complexity results for fixed order routing (FOR) compared with VRP. New results are indicated by a †. Here, k is assumed a constant. The APX-hardness for k-VRP follows directly from APX-hardness of TSP and TRP. The k-VRP on the line with total weighted completion time objective can be solved by a simple dynamic program for constant k [2]. Note that the other three variants of VRP on the line are trivial (and no refence is given). The complexity of c-capacitated FOR on the line remains an open problem. Our algorithmic results follow easily from known techniques while the complexity theorems are the more involved results. Due to space constraints, we defer some proofs to the full version.

Table 1. Complexity results of fixed order routing (FOR) compared with VRP.

	Line metric		General metric space	
	k-FOR	k-VRP	k-FOR	k-VRP
Total dist.	∈ P [8]	∈ P	∈ P [8]	APX-hard [20]
$\sum_j w_j C_j$	NP-hard†	∈ P [2]	PTAS†	APX-hard [16]
$C_{\max}$	NP-hard†	∈ P	FPTAS†	APX-hard [20]
	c-cap. FOR	c-cap VRP	c-cap. FOR	c-cap VRP
Total dist.	?	∈ P	APX-hard $c = 3$†	APX-hard $c = 3$ [11]
			$(2 - 1/c)$-approx.†	2.4997-approx. [5]

1.1 Model

An instance of the Fixed Order Routing (FOR) problem is given by a sequence of n requests. Each request i occurs at location r_i, where locations are defined as points in a metric space endowed with distance function $d(\cdot, \cdot)$. Generally, points in the metric space are denoted by v_i, with a special point v_0 which is called the origin. We will usually use graphs, where the distance between any two vertices is the length of the shortest path between them. A feasible solution consists of a partition P of the request sequence into subsequences. Each subsequence corresponds to a server traversing a tour that starts and ends in v_0 and visits the locations of its associated requests in ascending order of their index. To make this precise, let ρ be such a subsequence consisting of the indices $\rho_1 < \ldots < \rho_c \in \{1, \ldots, n\}$. Then ρ defines a subtour starting in v_0, visiting the locations $r_{\rho_1}, \ldots, r_{\rho_c}$ in that order, and ending in v_0, with length $\ell(\rho) = d(v_0, r_{\rho_1}) + \sum_{j=1}^{c-1} d(r_{\rho_j}, r_{\rho_{j+1}}) + d(r_{\rho_c}, v_0)$. We distinguish two server models: In the k-FOR model, an integer k is given and a solution must consist of a partition of the requests into at most k subsequences of arbitrary length. In the c-capacitated FOR model there is no bound on the number of subsequences, but each should contain at most c requests. Unless stated otherwise, we consider unit demands.

We consider several objective functions. When minimizing *total length* we minimize the sum of the lengths of all subtours of the partition P: $\sum_{\rho \in P} \ell(\rho)$. Another objective is minimizing *total weighted completion time*. The completion time of request i, denoted by C_i, is defined as the length of the path from v_0 to r_i in the subtour that contains request i. (We assume that any server moves at unit speed and starts in the origin at time 0.) The sum of weighted completion times then refers to $\sum_{i=1}^{n} w_i C_i$, where w_i is the weight of request i. A third objective is minimizing the *maximum completion time* $C_{\max} = \max_{i=1}^{n} C_i$. Observe that for the latter two objectives, the distance the server drives back from its last request to the origin is irrelevant. If there is no bound on the number of servers, then minimizing the sum of completion times or the maximum completion time is trivial by creating a separate subsequence for each request. Therefore, we only consider the total distance objective in the capacitated model.

Relation to Fixed Order Scheduling. In [6] the authors defined the fixed order scheduling problem (FOS) where jobs need to be scheduled on identical parallel machines in a fixed order π. They gave an $O(1)$-approximation algorithm for the total weighted completion time objective and a quasi-PTAS for the unweighted version. Neither of these results is easily obtained. The authors observe that the weighted version is NP-hard but leave the complexity of the unweighted version as an open problem. The routing version is in general more complicated than the scheduling version. In fact, the fixed order scheduling problem from [6] is

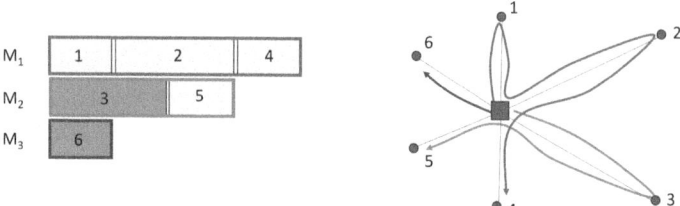

Fig. 1. The fixed order scheduling problem $P||\sum w_j C_j$ reduces to k-FOR on star graphs. In this example $p_1 = p_4 = p_5 = p_6 = 1$, $p_2 = p_3 = 2$, $w_1 = w_2 = w_4 = 1$, $w_5 = 2$, $w_3 = w_6 = 10$. For each job there is an edge with a request at the leaf. Weights are the same and the lengths of the edges are proportional to the processing times. The number of servers equals the number of machines.

the special case of FOR on star graphs with the additional property that there is exactly one request in every leaf. To see this, consider an instance of the FOS problem with processing times p_j and weights w_j for $j = 1, 2, \ldots, n$ (see Fig. 1). For each job j we define an edge (v_0, v_j) in the star graph of length $p_j/2$ and put a request r_j in point v_j of weight w_j. For any partition P of the request sequence $(1, 2, \ldots, n)$ into subsequences, let C_j be the completion time of job j

for the corresponding schedule and let C'_j be the completion time of request j in the corresponding routing solution. Then, $C'_j = C_j - p_j/2$ and for the total completion time we have

$$\sum_j w_j C'_j = \sum_j w_j C_j - \frac{1}{2} \sum_j w_j p_j \, .$$

Hence, these objectives only differ by a constant and the complexity of computing optimal solution is equivalent. Also, since $\sum_j w_j p_j/2$ is an obvious lower bound on the total completion time of either of the two, any α-approximation algorithm for one implies an $(\alpha + 1)$-approximation algorithm for the other. In particular, the $O(1)$-approximation for FOS from [6] implies an $O(1)$-approximation for FOR on star graphs with the additional restriction that each leaf receives exactly one request. Allowing more than one request per leaf (as we do in the FOR model) makes the problem much harder (see Sect. 4).

For the scheduling problem [6] only the total completion time objective is of interest among the three objectives that we consider in this paper. The scheduling equivalence of the total tour length is the total processing time which is constant and therefore irrelevant. Furthermore, minimizing maximum completion time in scheduling is the classic load balancing problem $P||C_{\max}$.

We conclude that the FOR problem is a rich generalization of the fixed order scheduling problem from [6] but also of many routing problems without a fixed order or where an order is fixed in some way. For example, requests may arrive one-by-one online, the problem may have the master tour property or in a game theoretic setting where players choose a resource endowed with a priority list [21].

2 FOR for General Metrics

Intuitively, vehicle routing problems where the order is fixed are easier than problems where the order is part of the decision. Indeed, without capacity constraints, metric k-VRP reduces to a single TSP tour (cf. Fig. 2) but is still a hard problem. With a fixed order constraint (k-FOR) it reduces to a minimum cost flow problem and can be solved efficiently [9]. For the total weighted completion time and maximum completion time objectives, k-FOR admits a (quasi) polynomial time approximation scheme (cf. Sect. 2.3). Fixed order routing in general metrics appears to be easier than its non-fixed order counterparts, while the opposite is true for special metrics such as the line or tree.

We start here with the capacitated variant and our main result in this section is that unit demand c-capacitated VRP remains APX-hard for any $c \geq 3$ even when a fixed order is given. This is in sharp contrast to the uncapacitated version.

2.1 A $(2 - \frac{1}{c})$-Approximation for c-Capacitated FOR

The well-known tour partitioning algorithm by Haimovich and Rinnooy Kan [12] and by Altinkemer and Gavish [3] has been the best approximation algorithm for

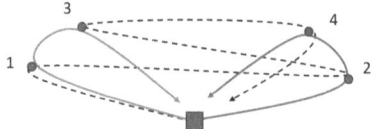

Fig. 2. Without capacity constraints a single TSP tour is obviously optimal for minimizing total distance in the metric Vehicle Routing Problem. However, in the FOR model, a single tour (dotted line) is usually far from optimal. Here, the total distance is minimized using two tours.

capacitated VRP for more than 30 years until Blauth et al. [5] reduced the ratio recently with an $\varepsilon > 1/3000$. The classic algorithm first computes a TSP tour solution on all requests and then splits it into sections of c requests each. It has an approximation ratio of $1+\alpha-1/c$ for the unit demand c-capacitated VRP and $2+\alpha$ for arbitrary non-splittable demand, where α is the approximation ratio of the TSP solution. As mentioned before, this uncapacitated solution can be computed optimally in the fixed order setting. Intuitively this should then give a $(2-1/c)$-approximation for c-capacitated FOR. Note, however, that this first step solution is composed of multiple subtours which can have any number of request, also less than c. However, by adding extra 'dummy' requests in the origin that need to be done before the given requests we may assume that any subtour in the uncapacitated solution contains at least c requests. For completeness we do give the algorithm and proof here, which is only a slight modification of the classic result. Let n be the number of requests. For simplicity we denote the request and their location by $1, 2, \ldots, n$.

Tour Partitioning Algorithm:

Step 1: Solve the uncapacitated FOR instance. Let $T_1, T_2, \ldots, T_q \subset \{1, \ldots, n\}$ be the partition into subsequences (subtours) and OPT' be its total length.

Step 2: Take $h \in \{1, 2, \ldots, c\}$ uniformly at random. Partition each subtour T_j in sets of c consecutive requests each, except for each first set which has exactly h requests and the last set which contains the remaining requests.

Theorem 1. *The tour partitioning algorithm gives a $2-1/c$-approximation for the c-capacitated FOR problem with total length objective.*

Proof. Let $\ell(T_j)$ be the length of subtour T_j in the solution of Step 1. Let d_i be the distance of a point i to the origin. Then, two lower bounds on the optimal value are

$$\text{OPT} \geq \sum_{j=1}^{q} \ell(T_j) \text{ and } \text{OPT} \geq \frac{2}{c}\sum_{i=1}^{n} d_i.$$

The first holds since the optimal value for the capacitated problem is at least that of the uncapacitated problem. The second bound holds since the length of any subtour is at least twice the average distance to the origin. Now consider one of the subtours T_j and its requests $j_1, j_2, \ldots, j_p$ for some $p \leq c$. For two consecutive requests j_i and j_{i+1} on the tour let $d(j_i, j_{i+1})$ be their distance. The edge from the origin to the first request on the tour j_1 appears with probability 1 in the algorithm's solution and the edge from request j_1 back to the origin appears with probability exactly $1/c$.

Similarly, the solution of the algorithm contains the edge from request j_p to the origin with probability 1 and the edge from the origin to request j_p with probability $1/c$. Furthermore, for any two consecutive requests j_i and j_{i+1} the edge between them appears in the solution with probability $(c-1)/c$. For any $i \in \{2, 3, \ldots, p-1\}$, the path between the origin and j_i appears with probability $1/c$ in each direction. Hence, the expected length of the part of the algorithm's solution for this subtour is *exactly*

$$\frac{c-1}{c}\ell(T_j) + \frac{2}{c}\sum_{i \in T_j} d_i.$$

Using the lower bounds above, the total expected length of the solution of the algorithm is

$$\frac{c-1}{c}\sum_{j=1}^{k}\ell(T_j) + \frac{2}{c}\sum_{i=1}^{n} d_i \leq \left(2 - \frac{1}{c}\right)\text{OPT}.$$

Clearly the analysis for the expected value is tight. Consider an instance with c requests in a node at distance 1 from the origin. Then the optimal solution uses 1 vehicle for all nodes, with a value of 2 while the algorithm will use 2 tours in $c-1$ out of the c solutions. With a slight modification it follows that the ratio is also tight w.r.t. the best choice for h.

2.2 APX-Hardness for c-Capacitated FOR

The c-capacitated VRP is well-known to be APX-hard even for $c = 3$ [4]. The proof follows easily from the fact that partitioning the vertices into a maximum number of subsets such that each subset induces a path of length 2 or a triangle is an APX-hard problem [14]. This result, however, does not carry over to the fixed order setting. For example, a path of 3 vertices indexed in the order $\{3, 1, 2\}$ cannot be traversed efficiently if the vertices with index 2 and 3 are not connected. To overcome this issue we reduce from the bounded 3-Dimensional Matching problem, which was proven to be APX-complete by Kann [13]. In the analysis we use that the optimal value is at least $n/7$. This follows directly from the fact that each triple in T intersects with at most 7 triples (including itself), and each element is in at least one triple.

MAXIMUM BOUNDED 3-DIMENSIONAL MATCHING (MAX-3DM-B):

Instance: Sets $A = \{a_1, a_2, \ldots, a_n\}$, $B = \{b_1, b_2, \ldots, b_n\}$, and $C = \{c_1, c_2, \ldots, c_n\}$. A subset T of $A \times B \times C$ such that any element of A, B, and C occurs in exactly one, two, or three triples in T.

Goal: Find a subset $T' \subseteq T$ of maximum cardinality such that no two triples of T' agree in any element.

Theorem 2. *Minimizing total length for c-capacitated fixed order routing is APX-hard for any $c \geq 3$. This holds even for unweighted graph metrics.*

Proof. We first prove NP-hardness and then extend the reduction to prove APX-hardness. Given an instance I_{3DM} of MAX-3DM-B we define an instance I_{FOR} of the fixed order routing problem as follows. For each element a_i, we define vertices a_i^1, a_i^2, a_i^3. We do the same for all elements b_i and c_i. We call these the a-vertices, b-vertices, and c-vertices and together these are the *element* vertices. For each element we define 4 *dummy* vertices as shown in Fig. 3. The total number of vertices is $7 \cdot 3n = 21n$. Next to these we define the root r.

For each triple $t = (a_i, b_j, c_k) \in T$ we define edges (a_i^x, b_j^y) and (b_j^y, c_k^z) for some $x, y, z \in \{1, 2, 3\}$. We say that vertices a_i^x, b_j^y and c_k^z *represent* triple t. We do this such that each vertex represents at most one triple from T. All vertices except for the heart-shaped vertices are connected to the root. All edges have length one and each vertex has exactly one request, except for the root which has no request. We order the requests (vertices) arbitrarily from left to right in the figure. This completes the description of the reduction.

We say that a tour is *efficient* if it contains 3 points and has length 4. Since hearts are not connected to the root, any efficient tour either has a heart in the middle or has no heart. That implies that it either consists of a triple or it contains two dummies followed by an element vertex as shown in the figure. Let's call these two options, respectively, *triple tours* and *dummy tours*.

Assume there is a matching of size n. Then we can make a solution of n triple tours and $6n$ dummy tours. This gives $7n$ efficient tours of total length $28n$. On the other hand, if we have a solution of length $28n$ then we must have $7n$ efficient tours and this can only be obtained by $6n$ dummy tours and n triple tours. Hence, the optimal tour length is $28n$ if there is a perfect matching and is strictly less otherwise. This completes the proof of NP-hardness.

To prove APX-hardness, we use the same FOR-instance and extend it with vertices v_i and w_i and edges (v_i, w_i) for $i \in \{1, 2, \ldots, Q\}$ where $Q = 2n$ (any larger integer $Q = \Omega(n)$ would work too). Additionally, for each element in $A \cup B \cup C$ there are 6 more vertices and 6 more edges as shown in Fig. 4. Let's call the 13 vertices corresponding to an element a *block*. To keep the figure simple, we only show the blocks for a_1, b_1 and c_1. Each star-shaped vertex in a block is connected to all vertices v_i and each pentagon-vertex in a block is connected to all vertices w_i. All vertices in this FOR-instance are connected to the root, except for the hearts. The ordering is again from left to right in the figure. Due to space constraints, the proof of this reduction we defer to the full version.

Note that the proof of NP-hardness also holds without a fixed order. In Fig. 3 we can see that the VRP without fixed order would give the same solution, because the heart-nodes still always have to be in the middle, in order to be efficient. As a consequence, VRP is even NP-hard when the order of the optimal solution is given. It is not clear if this is also APX-hard, since our APX-hardness proof does not have this property.

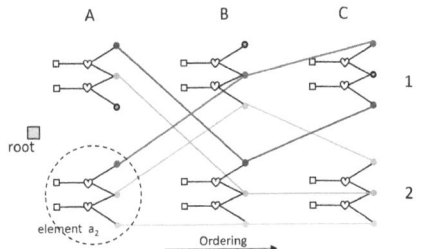
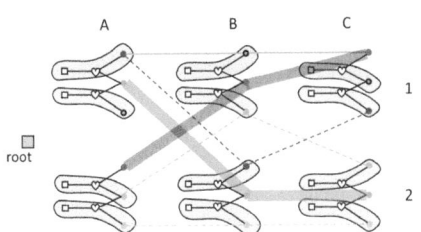
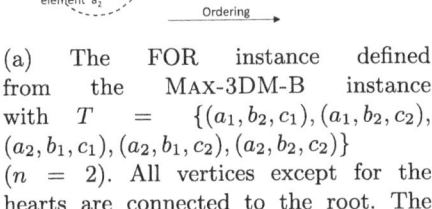

(a) The FOR instance defined from the MAX-3DM-B instance with $T = \{(a_1, b_2, c_1), (a_1, b_2, c_2), (a_2, b_1, c_1), (a_2, b_1, c_2), (a_2, b_2, c_2)\}$ ($n = 2$). All vertices except for the hearts are connected to the root. The order is from left to right.

(b) A perfect 3-dimensional matching partitions the graph into directed paths of 3 vertices, whose endpoints are connected to the root.

Fig. 3. Reducing c-capacitated FOR to MAX-3DM-B.

Corollary 1. *c-capacitated vehicle routing with unit demands is NP-hard for $c = 3$ even when given an optimal ordering of the points.*

2.3 Approximation Schemes for k-FOR in General Metrics

We now turn our attention from the c-capacitated FOR problem to k-FOR. Recall that k-FOR is well-known to be solvable in polynomial time for minimizing total length [8]. Conversely, we will show in the next section that k-FOR is already strongly NP-hard on the line for $k \geq 3$ when minimizing the total completion time or the maximum completion time (and weakly NP-hard for $k = 2$). These results are quite strong for two reasons. Firstly, they are in sharp contrast with k-VRP on the line, which is trivial. Secondly, the polynomial time approximation schemes hold for any general metric space (the proofs we defer to the full version).

Theorem 3. *For k-FOR on general metric spaces there exists an FPTAS for minimizing maximum completion time. For minimizing total weighted completion time there is a PTAS when weights are polynomially bounded.*

For both schemes k is assumed a constant. For non-constant k, the approximability is still open. We shortly discuss this in Sect. 4.

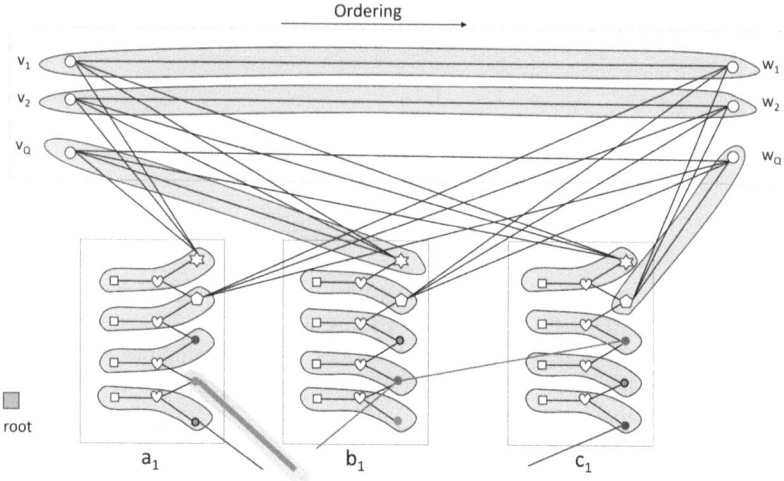

Fig. 4. An extended reduction to prove APX-hardness. The FOR instance now has additional vertices v_i and w_i for $i \in \{1, 2, \ldots, Q\}$. Each element from $A \cup B \cup C$ is now represented by a block of 13 vertices. (The figure only shows these for elements a_1, b_1, c_1.) All vertices except for the hearts are connected to the root. The ordering is from left to right. An optimal FOR solution consists of triples and pairs of vertices.

3 FOR on Line Graphs and Trees

Trees, and in particular line graphs, are examples of metrics where an optimal TSP solution is easy to find, since any depth first search order is optimal on trees. Hence, the fixed order restriction really complicates things for such metrics. We shall see here that the FOR problem is non-trivial even on the line. We prove strong NP-hardness for k-FOR on the line for the total completion time and maximum completion time objective and where k is part of the input. For $k = 2$ both versions are weakly NP-hard. (Note that when there is no order restriction then all completion times are minimized simultaneously by letting 2 servers move in opposite direction in a straight line.)

To prove strong NP-hardness for k-FOR on the line we reduce from the strongly NP-hard numerical 3-dimensional matching problem [11].

NUMERICAL 3-DIMENSIONAL MATCHING (NUM-3DM):

Instance: Sets of positive integers $A = \{a_1, \ldots, a_m\}$, $B = \{b_1, \ldots, b_m\}$, and $C = \{c_1, \ldots, c_m\}$. Denote $q = (1/m) \cdot \sum_i (a_i + b_i + c_i)$.

Question 1.
:
Does there exist a *partition* of $A \cup B \cup C$ into m triples (x_j, y_j, z_j) such that

$x_j \in A$, $y_j \in B$, $z_j \in C$ for $j \in \{1, \ldots, m\}$ and such that $x_j + y_j + z_j = q$ for all $j \in \{1, \ldots, m\}$?

Theorem 4. *k-FOR is strongly NP-hard on the line for both of the following objectives:*

(i) Minimizing the sum of weighted completion times.
(ii) Minimizing the maximum completion time.

Proof. We prove (i); the proof of (ii) we defer to the full version. We reduce from the numerical 3-dimensional matching problem as defined above. We define an instance of k-FOR on the line with $k = m$ servers and $6m$ requests: two requests for each number in $A \cup B \cup C$. For each $a_i \in A$ we define an odd request r_{2i-1} in point v_{2i-1} and an even request r_{2i} in point $v_{2i} < v_{2i-1}$, as shown in Fig. 5. The weight of any odd request is 1 and the weight of any even request r_{2i} is a_i. We refer to two requests r_{2i-1} and r_{2i} as *request couple i*. The length of the edge between points v_{2i} and v_{2i-1} of request couple i is defined as a_i. All other edges are of large length L. For B we define a similar sequence of requests $r_{2m+1}, \ldots, r_{4m}$ and for C we define a similar sequence of requests $r_{4m+1}, \ldots, r_{6m}$. The length of any edge that is not between a request couple is L, so in particular $v_{2m} = L$. Denote this instance by I.

To simplify the computation, we define another instance I' which is the same as I except that each request r_{2i-1} is given in the point v_{2i} instead of the point v_{2i-1} and there is no request in v_{2i-1}. That means the two requests of each request couple i can be done simultaneously while visiting point v_{2i}.

For each partition of the 3DM instance into triples (x_j, y_j, z_j) we define a corresponding routing solution where each server j visits those 3 request couples corresponding to the numbers $x_j \in A$, $y_j \in B$, and $z_j \in C$. Let us call such a solution a *partition* solution. Note that any partition solution is optimal for I' since the completion time of each request is exactly its distance to the origin. Let C' be the optimal sum of weighted completion times for instance I'. Note that for any solution for I' that is not a partition solution, there must be a server that moves towards the origin over one of the long edges of length L, while it still needs to serve at least one request. Hence, the sum of weighted completion times for non-partition solutions is at least $C' + L$.

Now consider the original instance I again. As before, for each partition into triples (x_j, y_j, z_j) we define the partition solution where each server j visits the request couples corresponding to (x_j, y_j, z_j). To be precise, assume that i is the request couple corresponding to x_j. Compared to instance I', this implies that when server j visits request couple i, it needs to move to location v_{2i-1}, move back a distance of a_i to arrive at location v_{2i}, and cover the distance a_i anew when moving towards the request couple corresponding to y_j. A similar behavior happens when visiting the request couple corresponding to y_j. Therefore, for any partition, the corresponding routing solution has sum of weighted completion times exactly

$$C_{\text{tot}} = C' + \sum_{j=1}^{n}(2x_j(x_j + y_j + 1 + z_j + 1) + 2y_j(y_j + z_j + 1) + 2z_j z_j)$$
$$= C' + C_{\text{tot}}^{(1)} + C_{\text{tot}}^{(2)}, \text{ with}$$
$$C_{\text{tot}}^{(1)} = 4\sum_{j=1}^{n} x_j + 2\sum_{j=1}^{n} y_j + \sum_{j=1}^{n}(x_j^2 + y_j^2 + z_j^2), \text{ and } C_{\text{tot}}^{(2)} = \sum_{j=1}^{n}(x_j + y_j + z_j)^2.$$

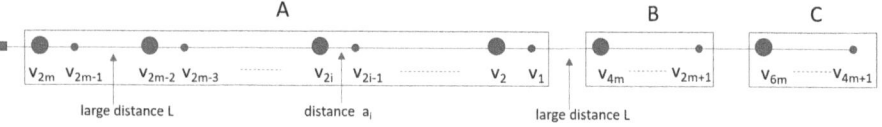

Fig. 5. The instance for the reduction for the sum of weighted completion times on the line.

Note that C' and $C_{\text{tot}}^{(1)}$ only depend on the instance and not on the solution. Further, for any solution corresponding to a perfect partition we have $C_{\text{tot}}^{(2)} = nq^2$ and $C_{\text{tot}}^{(2)}$ is strictly larger for any non-perfect partition solution. Finally, for any solution that does not correspond to a partition solution the sum of weighted completion times is at least $C' + L$. Hence, by taking L large enough, for example $L = C_{\text{tot}}^{(1)} + (\sum_{i=1}^{n}(a_i + b_i + c_i))^2$, there is a solution for I with sum of weighted completion times equal to $C' + C_{\text{tot}}^{(1)} + nq^2$ if and only if there exists a perfect partition in the corresponding 3-dimensional matching instance.

Using a similar reduction, one can prove weak NP-hardness for the special case of $k = 2$. The reduction is from the EVEN-ODD PARTITION problem, which is weakly NP-hard [11]. The proof follows from similar arguments as the proof above.

EVEN-ODD PARTITION:

Instance: Positive integers $\{a_1, \ldots, a_{2n}\}$ with $\sum_{i=1}^{2n} a_i = 2A$.

Question 2. Does there exist a partition of $[2n] = \{1, \ldots, 2n\}$ into two sets I and $[2n] \setminus I$ such that $|I \cap \{2i-1, 2i\}| = 1$ for all $i = 1, \ldots, n$, and such that $\sum_{i \in I} a_i = \sum_{i \notin I} a_i = A$?

Theorem 5. *2-FOR is weakly NP-hard on the line for each of the following variants:*

(i) Minimizing the sum of weighted completion times.
(ii) Minimizing the maximum completion time.

The complexity results of FOR on the line immediately carry over to trees. One might expect that the $(2 - 1/c)$-approximation for capacitated FOR on general metrics can easily be improved for trees and that a PTAS may even be possible. However, such a result would be far from trivial, since on trees, c-capacitated FOR is a generalization of c-capacitated VRP, for which a PTAS was only recently found by Mathieu et al. [18]. Stated otherwise, VRP is the special case of FOR where the order is any depth first search order. Hence, the best we can say at this point is that c-capacitated FOR on trees admits a PTAS for any DFS order.

4 Open Problems

We showed that the FOR problem has interesting relations to well-known optimization problems and in several cases it is a generalization rather than a restriction of these problems. On one hand, we showed that c-capacitated FOR is APX-hard even for unweighted graphs and $c = 3$. But given the recent improvements by Blauth et al. [5] and Mömke and Zhou [19] on the classic tour partitioning algorithm for VRP, one might suspect that an improvement is also possible in the FOR setting. However, the improvement there is made on the TSP tour and this is not possible in our setting, since this first step of the algorithm is already solved optimally. Reducing the gap between APX-hardness and the ratio of $(2 - 1/c)$ is a challenging open problem that requires a new approach.

Another gap that remains is for k-FOR with total completion time objective. Although our PTAS works in any metric space, we do not have any constant factor approximation without the assumption that k is constant, not even for the uniform metric space. Note that in a uniform metric space, k-FOR is the offline version of the well known paging problem but with the total completion time objective. Requests for a page arrive in a fixed order. Reading a page takes no time but can only be done from the cache which can store at most k pages. Bringing a page in the cache takes one time unit and we want to minimize the total latency.

A related problem is that of sorting a buffer [1]. Here, the input is given by a sequence of requests in a metric space and a buffer size k. Requests need to be served in the given order but the buffer can be used to delay (store) k requests simultaneously. This problem has been well-studied, but mainly in the online setting or with a single server and mainly with total length objective. The combination, i,e., the k-FOR problem with an additional buffer (or buffers) makes an interesting but complex problem which has, to the best of our knowledge, not been studied before. A problem setting with partially ordered requests forms another natural and interesting generalization. Below, we list some of the most interesting open problems.

- Obtain a $(2-1/c-\varepsilon)$-approximation for c-capacitated FOR in general metrics for some $\epsilon > 0$ or prove that this is not possible (assuming $P \neq NP$).
- Obtain a polynomial time constant factor approximation for k-FOR with total (weighted) completion time objective and where k is part of the input. Even for the uniform metric space this is open.

- Obtain a PTAS for c-capacitated FOR on trees (c part of input) or show that this is not possible (assuming $P \neq NP$). As a first step a PTAS on the line seems plausible, though still not straightforward.

References

1. Adamaszek, A., Czumaj, A., Englert, M., Räcke, H.: Almost tight bounds for reordering buffer management. SIAM J. Comput. **51**(3), 701–722 (2022)
2. Afrati, F.N., Cosmadakis, S.S., Papadimitriou, C.H., Papageorgiou, G., Papakostantinou, N.: The complexity of the travelling repairman problem. RAIRO Theor. Inform. Appl. **20**, 79–87 (1986)
3. Altinkemer, K., Gavish, B.: Heuristics for unequal weight delivery problems with a fixed error guarantee. Oper. Res. Lett. **6**, 149–158 (1987)
4. Asano, T., Katoh, N., Tamaki, H., Tokuyama, T.: Covering points in the plane by k-tours: a polynomial approximation scheme for fixed k. Technical report RT0162, IBM Tokyo Research Laboratory (1996)
5. Blauth, J., Traub, V., Vygen, J.: Improving the approximation ratio for capacitated vehicle routing. Math. Program. **197**(2), 451–497 (2023)
6. Bosman, T., Frascaria, D., Olver, N., Sitters, R., Stougie, L.: Fixed-order scheduling on parallel machines. In: Lodi, A., Nagarajan, V. (eds.) IPCO 2019. LNCS, vol. 11480, pp. 88–100. Springer, Cham (2019). https://doi.org/10.1007/978-3-030-17953-3_7
7. Braekers, K., Ramaekers, K., Van Nieuwenhuyse, I.: The vehicle routing problem: state of the art classification and review. Comput. Ind. Eng. **99**, 300–313 (2016)
8. Chrobak, M., Karloff, H., Payne, T., Vishwanathan, S.: New results on server problems. In: Proceedings of the First Annual ACM-SIAM Symposium on Discrete Algorithms, SODA '90, pp. 291–300, USA. Society for Industrial and Applied Mathematics (1990)
9. Chrobak, M., Karloof, H., Payne, T., Vishwnathan, S.: New results on server problems. SIAM J. Discret. Math. **4**(2), 172–181 (1991)
10. Deineko, V., Rudolf, R., Woeginger, G.: Sometimes travelling is easy: the master tour problem. SIAM J. Discrete Math. **11**, 81–93 (1998)
11. Garey, M.R., Johnson, D.S.: Computers and Intractability: A Guide to the Theory of NP-Completeness. Freeman and Company, San Francisco (1979)
12. Mordecai Haimovich and Alexander Rinnooy Kan: Bounds and heuristics for capacitated routing problems. Math. Oper. Res. **10**(4), 527–542 (1985)
13. Kann, V.: Maximum bounded 3-dimensional matching is MAX SNP-complete. Inf. Process. Lett. **37**(1), 27–35 (1991)
14. Kann, V.: Maximum bounded h-matching is MAX SNP-complete. Inf. Process. Lett. **49**(6), 309–318 (1994)
15. Koutsoupias, E.: The k-server problem. Comput. Sci. Rev. **3**(2), 105–118 (2009)
16. Koutsoupias, E., Papadimitriou, C., Yannakakis, M.: Searching a fixed graph. In: Meyer, F., Monien, B. (eds.) ICALP 1996. LNCS, vol. 1099, pp. 280–289. Springer, Heidelberg (1996). https://doi.org/10.1007/3-540-61440-0_135
17. Manasse, M.S., McGeoch, L.A., Sleator, D.D.: Competitive algorithms for on-line problems. In: Symposium on the Theory of Computing (1988)
18. Mathieu, C., Zhou, H.: A PTAS for capacitated vehicle routing on trees. ACM Trans. Algorithms **19**, 12 (2022)

19. Mömke, T., Zhou, H.: Capacitated vehicle routing in graphic metrics. In: Symposium on Simplicity in Algorithms (SOSA), pp. 114–123 (2023)
20. Papadimitriou, C.H., Yannakakis, M.: The Traveling Salesman Problem with distances one and two. Math. Oper. Res. **18**, 1–11 (1993)
21. Vijayalakshmi, V.R., Schröder, M., Tamir, T.: Scheduling games with machine-dependent priority lists. Theoret. Comput. Sci. **855**, 90–103 (2021)
22. Smith, W.E.: Various optimizers for single-stage production. Nav. Res. Logist. Q. **3**, 59–66 (1956)
23. Toth, P., Vigo, D., (eds.): Vehicle Routing, volume 18 of MOS-SIAM Series on Optimization. SIAM (2014)

Approximate Min-Sum Subset Convolution

Mihail Stoian[✉]

University of Technology Nuremberg, Nuremberg, Germany
mihail.stoian@utn.de

Abstract. Exponential-time approximation has recently gained attention as a practical way to deal with the bitter NP-hardness of well-known optimization problems. We study for the first time the $(1+\varepsilon)$-approximate min-sum subset convolution. This enables exponential-time $(1+\varepsilon)$-approximation schemes for problems such as minimum-cost k-coloring, the prize-collecting Steiner tree, and many others in computational biology. Technically, we present both a weakly- and strongly-polynomial approximation algorithm for this convolution, running in time $\widetilde{O}(2^n \log M/\varepsilon)$ and $\widetilde{O}(2^{\frac{3n}{2}}/\sqrt{\varepsilon})$, respectively. Our work revives research on tropical subset convolutions after nearly two decades.

Keywords: subset convolution · min-plus convolution · min-max convolution · dynamic programming · approximation algorithm

1 Introduction

Fast subset convolution is one of the tools in parameterized algorithms which made their way in many of the dynamic programming solutions to well-known NP-hard problems [11]. Given functions f and g defined on the subset lattice of order n, their *sum-product* subset convolution is defined for all $S \subseteq [n] := \{1,\ldots,n\}$ as

$$h(S) = (f * g)(S) = \sum_{T \subseteq S} f(T) g(S \setminus T).$$

Its prominence does not yet come to a surprise since many computationally hard problems accept a convolution-like shape. The remarkable reduction in time-complexity from $O(3^n)$ to $O(2^n n^2)$ by Björklund et al. [3] represented indeed a breakthrough, allowing these problems to be solved faster. These problems, however, do indeed reduce to a slightly different subset convolution, namely the *min-sum* subset convolution, defined for all $S \subseteq [n]$ as

$$h(S) = (f \star g)(S) = \min_{T \subseteq S} \bigl(f(T) + g(S \setminus T)\bigr). \tag{1}$$

The naïve algorithm inherently takes $O(3^n)$-time, as in the case of any subset convolution: For each set S, iterate over all its subsets T.[1] In contrast to the

---
[1] The running time can be derived from $\sum_{k=0}^{n} \binom{n}{k} 2^k = (1+2)^n = 3^n$.

© The Author(s), under exclusive license to Springer Nature Switzerland AG 2025
M. Bieńkowski and M. Englert (Eds.): WAOA 2024, LNCS 15269, pp. 198–212, 2025.
https://doi.org/10.1007/978-3-031-81396-2_14

speedup achieved by Björklund et al. [3] for the sum-product subset convolution, the naïve algorithm is still the fastest known for the min-sum subset convolution. This is also the reason why min-sum subset convolution is not used "as is", but in a two-step approach: (i) embed the min-sum semi-ring into the sum-product ring, (ii) perform the fast subset convolution in this ring instead. This results in an $\widetilde{O}(2^n M)$-time algorithm [3], where M is the largest input value.

This workaround has two limitations: First, the input functions must have a bounded integer range $\{-M, \ldots, M\}$, and second, the running time of the final algorithm – $\widetilde{O}(2^n M)$ – depends on M, making the algorithm a pseudo-polynomial one.[2] This is also where the story behind the min-sum subset convolution ends. Is that all? In what follows, we argue that there is more to it.

1.1 Approximating Min-Sum Subset Convolution

Exponential-time approximation algorithms for NP-hard problems have recently attracted much attention [2,5,16–19,30]. In the light of this development, we propose the $(1+\varepsilon)$-*approximate* variant of the min-sum subset convolution as a "Swiss Army knife" that enables exponential-time approximation for many NP-hard optimization problems: Compute the set function $\widetilde{h}$ such that for all $S \subseteq [n]$ the following holds:

$$(f \star g)(S) \leq \widetilde{h}(S) \leq (1+\varepsilon)(f \star g)(S).$$

If we could devise a faster-than-naïve algorithm for the approximate counterpart, this would enable $(1+\varepsilon)$-approximation schemes for a plethora of computationally hard problems that use min-sum subset convolution as a primitive. Typically, these are problems on graphs such as the minimum Steiner tree problem [13], coloring [34], and spanning problems on hypergraphs [33,39]. On the application side, min-sum subset convolution and its counterpart, *max-sum* subset convolution,[3] are also present in computational biology [35,36]. This leads us to our driving research question:

Are there faster-than-naïve $(1+\varepsilon)$-approximation algorithms for min-sum subset convolution?

On another note, Björklund et al. [3] showed that, assuming a bounded input, one can improve on decades-old stagnant results. The most notable of these was the $\widetilde{O}(2^k n^2 M + nm \log M)$-time algorithm for the minimum Steiner tree in graphs with n vertices, k terminals, and m edges with integer weights bounded by M. This is indeed the case for all other hard problems that reduce to the min-sum subset convolution: They are all "scapegoats" of the bounded-input assumption, having to accept an additional factor of M in their time-complexities. This leads us to our next question:

Are there faster-than-naïve $(1+\varepsilon)$-approximation schemes for convolution-like NP-hard optimization problems with running time independent of M?

---

[2] Note that the input size itself is on the order of $O(2^n)$.
[3] The operations are performed in the $(\max, +)$-semi-ring.

1.2 Our Results

Overview. We answer our driving questions in the affirmative. The way we answer them is also intriguing in itself. Let us review it briefly. First, we bring together two lines of research that have been so far considered separately: *sequence* and *subset* convolutions in semi-rings. This gives us both the weakly- and the strongly-polynomial algorithms for the $(1+\varepsilon)$-approximate min-sum subset convolution. Then, using these algorithms and a standard approximation technique, we obtain $(1+\varepsilon)$-approximation algorithms for several convolution-like NP-hard optimization problems. This enriches the current toolbox in exponential-time approximation [2,5,16–19,30].

Min-Sum Subset Convolution. Moreover, we revive the line of research on min-sum subset convolution after nearly two decades, by presenting both a weakly- and a strongly-polynomial $(1+\varepsilon)$-approximation algorithm for this problem:

Theorem 1. $(1+\varepsilon)$-*Approximate min-sum subset convolution can be solved in* $\widetilde{O}(2^n \log M/\varepsilon)$-*time.*

Theorem 2. $(1+\varepsilon)$-*Approximate min-sum subset convolution can be solved in* $\widetilde{O}(2^{\frac{3n}{2}}/\sqrt{\varepsilon})$-*time.*

Note that the strongly-polynomial algorithm (Theorem 2) outperforms its weakly-polynomial counterpart (Theorem 1) for huge co-domains.[4] While our weakly-polynomial algorithm uses the scaling technique, as does the fastest algorithm for the $(1+\varepsilon)$-approximate min-plus *sequence* convolution [31], obtaining the strongly-polynomial algorithm requires a detour in another semi-ring. Namely, in the course of proving it, we had to design an algorithm for a more curious convolution, namely the *min-max* subset convolution, defined for all $S \subseteq [n]$ as
$$h(S) = (f \varovee g)(S) = \min_{T \subseteq S} \max\{f(T), g(S \setminus T)\}.$$

To the best of our knowledge, the *min-max* subset convolution has not yet been present in the literature. As a by-product, we show that its exact evaluation does indeed break the natural $O(3^n)$-time barrier:

Theorem 3. *Exact min-max subset convolution can be solved in* $\widetilde{O}(2^{\frac{3n}{2}})$-*time.*

This generalizes Kosaraju's algorithm [27] for the min-max sequence convolution to the subset lattice. We need this intermediate result so that we can instantiate the recent framework of Bringmann, Künnemann, and Węgrzycki [7], used to obtain the first strongly-polynomial algorithm for the $(1+\varepsilon)$-approximate min-plus sequence convolution. In particular, they also showed the equivalence between exact min-max sequence convolution and the latter.[5] We show that this

---
[4] Indeed, a sufficient condition is $M = \omega(2^{\sqrt{\varepsilon 2^{n+\delta}}})$, for any $\delta > 0$.
[5] The current version of [7, Thorem 8.2] contains a typo, relating the min-plus sequence convolution to the min-max *product*. The theorem actually refers to convolution, as can be seen from its proof.

Table 1. Reviving research on tropical subset convolutions.

Reference	Type	(Semi-)ring	Running Time
ad-hoc	exact	any	$O(3^n)$
Björklund et al. [3]	exact	$(+, \times)$	$O(2^n n^2)$
Björklund et al. [3]	exact	$(\min, +)$	$\widetilde{O}(2^n M)$
Theorem 3, *this work*	exact	$(\min, \max)$	$\widetilde{O}(2^{\frac{3n}{2}})$
Theorem 1, *this work*	$(1+\varepsilon)$-approx.	$(\min, +)$	$\widetilde{O}(2^n \log M/\varepsilon)$
Theorem 2, *this work*	$(1+\varepsilon)$-approx.	$(\min, +)$	$\widetilde{O}(2^{\frac{3n}{2}}/\sqrt{\varepsilon})$

result naturally holds in the subset setting as well, a missing piece in their work that has immediate application to well-known NP-hard optimization problems.

Theorem 4 (Extension of [7, Thorem 1.5]). *For any $c \geq 1$, if one of the following statements is true, then both are:*

- $(1 + \varepsilon)$-*Approximate min-sum subset convolution can be solved in time $\widetilde{O}(2^{cn}/\mathsf{poly}(\varepsilon))$,*
- *Exact min-max subset convolution can be solved in time $\widetilde{O}(2^{cn})$.*

Simply using the main backbone of their framework, the Sum-to-Max-Covering lemma [7, Thorem 1.7] (we define it later in Sect. 4), one could obtain an $\widetilde{O}(2^{\frac{3n}{2}}/\varepsilon)$-time algorithm for the approximate min-sum subset convolution; note that the running time is independent of M. However, the dependence on the parameter ε can indeed be improved using a refined analysis. In Table 1, we summarize our results, which revive the research on tropical subset convolutions after nearly two decades and enable out-of-the-box exponential-time $(1 + \varepsilon)$-approximation schemes for several hard optimization problems, as outlined in the following.

Applications. Min-sum subset convolution is present in the dynamic programming formulation of many NP-hard optimization problems. Thus, we obtain $(1+\varepsilon)$-approximation schemes for these problems as well, such as the minimum-cost k-coloring [10], the prize-collecting Steiner tree, and two applications in computational biology: the maximum colorful subtree problem [35], which instead requires the *max-sum* subset convolution – for which we devise an $(1 - \varepsilon)$-approximation scheme – and a problem on protein interaction networks [36] (already proposed as an application by Björklund et al. [3]). We outline two of these results here:

Theorem 5. $(1 + \varepsilon)$-*Approximate minimum-cost k-coloring can be solved in $\widetilde{O}(2^{\frac{3n}{2}}/\sqrt{\varepsilon})$-time.*

Theorem 6. $(1 + \varepsilon)$-*Approximate prize-collecting Steiner tree can be solved in $\widetilde{O}(2^{\frac{3s^+}{2}}/\sqrt{\varepsilon})$-time.*[6]

---
[6] The parameter s^+ is the number of proper potential terminals.

1.3 Related Work

Our work builds a new bridge from sequence convolutions to subset convolutions. The sequence convolution of two sequences a and b in the $(\oplus, \otimes)$-semi-ring is computed as $c_k = \oplus_{i+j=k}(a_i \otimes b_j)$, while the subset convolution of two set functions f and g in the same semi-ring is defined as $h(S) = \oplus_{T \subseteq S}(f(T) \otimes g(S \setminus T))$. Independently on the structure of the convolution, i.e., either on sequences or on subsets, there are three types of sequence convolution which predominate in the literature, corresponding to three different (semi-)rings: (a) $(+, \times)$-ring, (b) $(\min, \max)$-semi-ring, and (c) $(\min, +)$-semi-ring, respectively. In the following, we outline previous work for both convolution types.

Sequence Convolution

(a) The sequence convolution in the $(+, \times)$-ring can be solved in time $O(n \log n)$ via FFT.
(b) In the $(\min, \max)$-semi-ring, Kosaraju presented an $\widetilde{O}(n\sqrt{n})$-time algorithm [27], and even conjectured that his algorithm can be improved to $\widetilde{O}(n)$. However, no improvement has been reported so far [7].
(c) Whether Min-Plus Sequence Convolution can be computed in time $O(n^{2-\delta})$ for any $\delta > 0$ is still an open problem [12,28]. The fastest algorithm to date runs in time $n^2/2^{\Omega(\sqrt{\log n})}$, by combining the reduction to Min-Plus Matrix Product by Bremner et al. [6] and an algorithm for the latter due to Williams [41], subsequently derandomized by Chan and Williams [9]. In case the values are bounded by a constant W, the convolution can be performed in $\widetilde{O}(nW)$-time [40, Lemma 5.7.2]. The first $(1+\varepsilon)$-approximation algorithm has been presented by Backurs et al. [1], as an application for the Tree Sparsity problem. Their algorithm runs in time $O(\frac{n}{\varepsilon^2} \log n \log^2 W)$ and has been improved to $O(\frac{n}{\varepsilon} \log(n/\varepsilon) \log W)$ [31]. Bringmann et al. [7] finally provided a strongly-polynomial algorithm, i.e., independent of W, in time $\widetilde{O}(n^{3/2}/\sqrt{\varepsilon})$. Their result is indeed more general, obtaining a framework that can also be applied to All-Pairs Shortest Pairs (APSP). In this paper, we extend their framework to the context of subset convolutions, with immediate application in $(1+\varepsilon)$-approximation schemes of several convolution-like NP-hard problems.

Subset Convolution

(a) Prior to the work of Björklund et al. [3], the best algorithm for any subset convolution was the straightforward $O(3^n)$-time evaluation. They provided an $O(2^n n^2)$-time algorithm for the subset convolution in the $(+, \times)$-ring, by relating the FFT to the Möbius transform on the subset lattice. This result found applications in many problems, among them the Domatic Number and the Chromatic Number (and many others) in $2^n n^{O(1)}$-time [4].
(b) As far as the $(\min, \max)$-semi-ring is concerned, there is, to our best knowledge, no algorithm that runs in time better than $O(3^n)$. In this paper, we provide such an algorithm, as a by-product.

(c) On the other hand, the convolution in the (min, +)-semi-ring received more attention, as it is implicitly present in many hard optimization problems, a prominent example being the minimum Steiner tree. While so far no $o(3^n)$-time algorithm is known to exist for this convolution, a simple embedding technique can leverage the speedup obtained for the sum-product subset convolution. The caveat is that the input values are required to be bounded by a constant M. In this case, the convolution runs in time $\widetilde{O}(2^n M)$ [3] (compare to the $\widetilde{O}(nW)$-time algorithm for the corresponding sequence convolution). In particular, no $(1+\varepsilon)$-approximation algorithm has been known until our work.

2 Preliminaries

Notation. We use the $\widetilde{O}$-notation to suppress poly-logarithmic factors in the input size and in ε, but never in M (or in W in the sequence setting). We denote by $[n]$ the set $\{1, \ldots, n\}$ for n a natural number. To be consistent with the notation used in previous work in *both* research fields, we refer to the (min, +)-semi-ring as *min-sum* for subset convolutions and as *min-plus* for sequence convolutions. The same distinction is done between the terms M and W, corresponding to the largest (finite) input value in the subset and sequence setting, respectively. We refer to the $(+, \times)$-ring as the sum-product ring, and to the (min, max)-semi-ring as the min-max semi-ring. We will later need Iverson's bracket notation: Given a property P, $[P]$ takes the value 1 whenever P is true and 0 otherwise.

Machine Model and Input Format. We follow the same setup as assumed by Bringmann et al. [7], namely: The input numbers in *approximate* problems are represented in floating-point, whereas those in *exact* problems are integers in the usual bit representation. This particular setup is needed when extending the equivalence between exact min-max and approximate min-plus to the subset setting. To economically use the space for the actual applications, we postpone the details on the setup to the appendix of the full paper.[7]

3 Exact Min-Max Subset Convolution

Both our strongly-polynomial algorithms for evaluating the $(1+\varepsilon)$-approximate *min-sum* subset convolution heavily rely on the *min-max* subset convolution, defined for all $S \subseteq [n]$ as

$$h(S) = (f \varotimes g)(S) = \min_{T \subseteq S} \max\{f(T), g(S \setminus T)\}. \tag{2}$$

In this section, we provide an $\widetilde{O}(2^{\frac{3n}{2}})$-time algorithm for Eq. (2), inspired by Kosaraju's $\widetilde{O}(n\sqrt{n})$-time algorithm for min-max *sequence* convolution [27]. We note that similar techniques to Kosaraju's have been considered for the min-max matrix product by Shapira et al. [37], Williams et al. [38], and Duan and Pettie [14], as applications to the all-pairs bottleneck pairs problem.

---
[7] The full version of the paper is accessible at: https://arxiv.org/abs/2404.11364.

Theorem 3. *Exact min-max subset convolution can be solved in $\widetilde{O}(2^{\frac{3n}{2}})$-time.*

Proof. We closely follow Kosaraju's algorithm for the min-max *sequence* convolution [27]: First, we collect all values of f and g in a list $\mathcal{L} := \{(f(S), S), (g(S), S) \mid S \subseteq [n]\}$ and sort it by its first argument (ties are broken arbitrarily). We then divide the sorted list into $O(\sqrt{2^n})$ chunks. Let $\mathcal{C}_i$ be the current chunk, and $\mathcal{C}_i^1$ and $\mathcal{C}_i^2$ be its projections onto the first and second arguments, respectively. We apply fast (boolean) subset convolution on $[f \leq \max\mathcal{C}_i^1]$ and $[g \leq \max\mathcal{C}_i^1]$ and obtain $\hat{h} = [f \leq \max\mathcal{C}_i^1] * [g \leq \max\mathcal{C}_i^1]$. In particular, $\hat{h} \equiv [h \leq \max\mathcal{C}_i^1]$, and thus a non-zero $\hat{h}(S)$ tells us whether the actual $h(S)$ is bounded above by $\max\mathcal{C}_i^1$. At this point, we only need to figure out what the *actual* value of $h(S)$ is.

To this end, we consider only those sets S for which $\mathcal{C}_i$ is the *first* chunk that made $\hat{h}(S)$ positive. Let us call this set $\mathcal{S}_i$. We thus consider each $S \in \mathcal{S}_i$ separately and compute its exact $h(S)$ value by iterating over the $O(\sqrt{2^n})$ sets $U \in \mathcal{C}_i^2$. Formally, we set

$$h(S) = \min_{\substack{U \in \mathcal{C}_i^2 \\ U \subseteq S}} \max \{f(U), g(S \setminus U)\}. \tag{3}$$

The correctness follows from construction. Let us analyze the running time. Sorting the list takes $\widetilde{O}(2^n)$-time. Then, it takes $O(|\mathcal{S}_i|\sqrt{2^n})$-time to evaluate Eq. (3) for each chunk $\mathcal{C}_i$, since $|\mathcal{C}_i| = O(\sqrt{2^n})$ by construction. Since we run $O(\sqrt{2^n})$ convolutions in the sum-product ring and Eq. (3) is run only once per $S \subseteq [n]$, the total running time reads $O(2^n n^2 \sqrt{2^n} + 2^n \sqrt{2^n}) = \widetilde{O}(2^{\frac{3n}{2}})$.

Notably, the running time given in Thorem 3 is faster than $O(3^n)$. This result will serve as the basis for our *strongly-polynomial* approximation algorithms for *min-sum* subset convolution.

4 Approximate Min-Sum Subset Convolution

We now turn to our original goal: *approximating the min-sum subset convolution*. We present two ways to do this, either via a weakly-polynomial time algorithm or a strongly-polynomial time algorithm. Let us start with the former.

4.1 Weakly-Polynomial Algorithm

Our weakly-polynomial algorithm is based on the scaling technique, which has been successfully used in several graph problems [15,23–25,32]. In the context of sequence convolution, it has been used by Backurs et al. [1] to provide the first $(1+\varepsilon)$-approximation scheme for the min-plus sequence convolution, running in time $O(\frac{n}{\varepsilon^2} \log n \log^2 W)$. This has been later improved by Mucha et al. [31] to $O(\frac{n}{\varepsilon} \log(n/\varepsilon) \log W)$-time.

We can use a method similar to that in Ref. [31] and solve the $(1+\varepsilon)$-approximate min-sum subset convolution in weakly-polynomial time. The key insight is that we can replace the "heart" of their algorithm, namely using the

Algorithm 1 APXMINSUMSUBSETCONV(f, g, ε) [Theorem 1] – Running time: $\widetilde{O}(2^n \log M/\varepsilon)$

1: **procedure** SCALE(f, q, ε)
2:     **return** $S \mapsto \begin{cases} \lceil \frac{2f(S)}{\varepsilon q} \rceil, & \text{if } \lceil \frac{2f(S)}{\varepsilon q} \rceil \leq \lceil \frac{4}{\varepsilon} \rceil, \\ \infty, & \text{otherwise.} \end{cases}$
3: **end procedure**
4:
5: Set $\widetilde{h}(S) \leftarrow \infty$ for all $S \subseteq [n]$
6: **for** $q = 2^{\lceil \log 2M \rceil}, \ldots, 4, 2, 1$ **do**
7:     $f_q, g_q \leftarrow$ SCALE(f, q, ε), SCALE(g, q, ε)
8:     $h_q \leftarrow$ EXACTMINSUMSUBSETCONV(f_q, g_q)
9:     $\widetilde{h}(S) \leftarrow \min\{\widetilde{h}(S), h_q(S) \cdot \frac{\varepsilon q}{2}\}$ for all $S \subseteq [n]$
10: **end for**
11: **return** $\widetilde{h}$

exact $\widetilde{O}(nW)$-time min-plus sequence convolution for bounded input, with the subset counterpart running in time $\widetilde{O}(2^n M)$, due to Björklund et al. [3]. To our best knowledge, this observation had not appeared in the literature before. In particular, the applications themselves had no $(1 + \varepsilon)$-approximation schemes prior to our work. We outline our algorithm in Algorithm 1 and prove Theorem 1 in the appendix of the full paper.

Theorem 1. *$(1+\varepsilon)$-Approximate min-sum subset convolution can be solved in $\widetilde{O}(2^n \log M/\varepsilon)$-time.*

Striving for Strongly-Polynomial Time. While the weakly-polynomial algorithm allows us to reduce the linear dependence on M in the running time of the algorithm of Björklund et al. [3] to a logarithmic one, the dependence on M still remains. The caveat is that this dependence will carry over in the actual applications. *Is this the best we can hope for?* In the following, we will show that we can *completely* discard this dependence.

In a recent breakthrough, Bringmann et al. [7] introduced the first strongly-polynomial algorithm for evaluating the approximate min-plus *sequence* convolution: Given two sequences a and b, their min-plus convolution[8] is a sequence c, where $c_k = \min_{i \leq k}(a_i + b_{k-i})$. As in our case, its approximate variant asks for a sequence $\widetilde{c}$ such that $c_k \leq \widetilde{c}_k \leq (1+\varepsilon)c_k$. Notoriously, previous work on this problem had employed the scaling trick [42], which relies on the largest input value W and introduces an additional $\log W$ into the running time.[9] The approximate convolution finds a natural application in the Tree Sparsity problem [8], where its exact algorithm is prohibitively expensive.

---
[8] In this work, since we are dealing with two types of convolution, on sequences and subsets, respectively, we will consistently add the specification "sequence" to avoid any confusion.
[9] As noted above, the term W in the literature on sequence convolutions corresponds to M in subset convolutions.

The aforementioned authors asked a fundamental question: *Is it possible to completely avoid the scaling trick?* They indeed answered this question affirmatively, by designing the first *strongly-polynomial* $(1+\varepsilon)$-approximation schemes for the all-pairs shortest pairs (APSP) problem and many others, including the min-plus sequence convolution. Their cornerstone result is the Sum-to-Max-Covering lemma:

Lemma 1 (Sum-to-Max Covering [7]). *Given vectors $A, B \in \mathbb{R}_+^d$ and a parameter $\varepsilon > 0$, there are vectors $A^{(1)}, \ldots, A^{(s)}, B^{(1)}, \ldots, B^{(s)} \in \mathbb{R}_+^d$ with $s = O(\frac{1}{\varepsilon} \log \frac{1}{\varepsilon} + \log d \log \frac{1}{\varepsilon})$ such that for all $i, j \in [d]$:*

$$A[i] + B[j] \leq \min_{\ell \in [s]} \max\{A^{(\ell)}[i], B^{(\ell)}[j]\} \leq (1+\varepsilon)(A[i] + B[j]).$$

The vectors $A^{(1)}, \ldots, A^{(s)}, B^{(1)}, \ldots, B^{(s)}$ can be computed in time $O(\frac{d}{\varepsilon} \log \frac{1}{\varepsilon} + d \log d \log \frac{1}{\varepsilon})$.

The motivation behind their lemma is that min-max sequence convolution can be solved in $\widetilde{O}(n\sqrt{n})$-time (using Kosaraju's algorithm [27]), in contrast to the best-known algorithm for min-plus convolution which runs in $O(n^2)$-time.[10] In this light, one can first compute the auxiliary vectors $A^{(1)}, \ldots, A^{(s)}, B^{(1)}, \ldots, B^{(s)}$ via Lemma 1 and then solve the $(1+\varepsilon)$-approximate min-plus sequence convolution by performing s min-max sequence convolutions.

4.2 Simple Strongly-Polynomial Approximation Algorithm

The reader can already see the applicability to our setting: We can apply the Sum-to-Max Covering lemma in min-sum *subset* convolution. Indeed, this is the cornerstone neglected by Bringmann et al. [7]. This key insight is the building block behind our new results in Sect. 5.

Algorithm 2 APXMINSUMSUBSETCONV(f, g, ε) [Lemma 2] – Running time: $\widetilde{O}(2^{\frac{3n}{2}}/\varepsilon)$

1: $\{f^{(1)}, \ldots, f^{(s)}, g^{(1)}, \ldots, g^{(s)}\} \leftarrow$ SUMTOMAXCOVERING(f, g, ε) (Lemma 1)
2: $h^{(\ell)} \leftarrow$ MINMAXSUBSETCONV$(f^{(\ell)}, g^{(\ell)})$ for each $\ell \in [s]$
3: $\widetilde{h}(S) \leftarrow \min_{\ell \in [s]} \{h^{(\ell)}[\texttt{idx}(S)]\}$ for each $S \subseteq [n]$
4: **return** $\widetilde{h}$

Lemma 2. $(1+\varepsilon)$-*Approximate min-sum subset convolution can be solved in time $\widetilde{O}(2^{\frac{3n}{2}}/\varepsilon)$.*

---

[10] Indeed, it is an open problem whether this is the best that can be achieved [12,28].

Note that this *is still an intermediate result*. In the following, we will use it to prove the equivalence between exact min-max subset convolution and approximate min-sum subset convolution. This represents the extension of the same result at the level of sequence convolutions.

Theorem 4 (Extension of [7, Thorem 1.5]). *For any $c \geq 1$, if one of the following statements is true, then both are:*

- $(1+\varepsilon)$-*Approximate min-sum subset convolution can be solved in time $\widetilde{O}(2^{cn}/\mathsf{poly}(\varepsilon))$,*
- *Exact min-max subset convolution can be solved in time $\widetilde{O}(2^{cn})$.*

4.3 Improved Strongly-Polynomial Approximation Algorithm

We used the $\widetilde{O}(2^{\frac{3n}{2}}/\varepsilon)$-time algorithm for the approximate min-sum subset convolution (Lemma 2) as a basis for the equivalence between *approximate* min-sum subset convolution and *exact* min-max subset convolution. In the following, we design an algorithm with an improved running time of $\widetilde{O}(2^{\frac{3n}{2}}/\sqrt{\varepsilon})$. Our key insight is that the new algorithm can be analyzed using a toolbox similar to that of Bringmann et al. [7]. This will be the running time we use for the applications in Sect. 5 and in the appendix of the full paper.

We split the main result in the corresponding subcases, namely (i) *distant* summands and (ii) *close* summands. The main theorem, Thorem 2, will follow from these.

Lemma 3. *Given set functions f, g on the subset lattice of order n and a parameter $\varepsilon > 0$, let $h = f \star g$ be their min-sum subset convolution. We can compute in time $O(2^{\frac{3n}{2}}\mathsf{polylog}(\frac{2^n}{\varepsilon}))$ a set function $\widetilde{h}$ such that for any $S \subseteq [n]$ we have*

(i) $h(S) \leq \widetilde{h}(S)$, and
(ii) if there is $T \subseteq S$ with $h(S) = f(T) + g(S \setminus T)$ and $\frac{f(T)}{g(S\setminus T)} \notin [\frac{\varepsilon}{4}, \frac{4}{\varepsilon}]$, then $\widetilde{h}(S) \leq (1+\varepsilon)h(S)$.

Briefly, we view the set functions f, g as vectors in $\mathbb{R}_+^{2^n}$. Hence, we can use Distant Covering from the framework [7, Cor. 5.10] on f, g with $\varepsilon' := \frac{\varepsilon}{4}$ and obtain $f^{(1)}, \ldots, f^{(s)}, g^{(1)}, \ldots, g^{(s)}$ with $s = O\left(\mathsf{polylog}(\frac{2^n}{\varepsilon})\right)$.[11] For each $l \in [s]$, we compute $h^{(l)} = f^{(l)} \oslash g^{(l)}$ and scale the entry-wise minimum by $\frac{1}{1-2\varepsilon'}$.

The case of *close* summands is covered by the following result:

Lemma 4. *Given set functions f, g on the subset lattice of order n and a parameter $\varepsilon > 0$, let $h = f \star g$ be their min-sum subset convolution. We can compute in time $\widetilde{O}(2^{\frac{3n}{2}}/\sqrt{\varepsilon})$ a set function $\widetilde{h}$ such that for any $S \subseteq [n]$ we have*

(i) $h(S) \leq \widetilde{h}(S)$, and

---

[11] For completeness, we provide [7, Cor. 5.10] in the appendix of the full paper.

(ii) if $T \subseteq S$ with $h(S) = f(T) + g(S \setminus T)$ and $\frac{f(T)}{g(S \setminus T)} \in [\frac{\varepsilon}{4}, \frac{4}{\varepsilon}]$, then $\widetilde{h}(S) \leq (1+\varepsilon)h(S)$.

We point out one aspect that a careful reader will notice, namely that the for-loop of the algorithm does indeed use the largest value M.[12] At first glance, this seems to contradict our assumption that we are trying to avoid running times that include M. The key insight to understanding this, used by Bringmann et al. [7] in the sequence setting, is the fact that q grows geometrically, and thus the number of entries of f_q and g_q that are not set to ∞ by SCALE is bounded by $O(2^n \log \frac{1}{\varepsilon})$, as we analyze in the proof of Lemma 4. Finally, we obtain:

Theorem 2. $(1+\varepsilon)$-*Approximate min-sum subset convolution can be solved in* $\widetilde{O}(2^{\frac{3n}{2}}/\sqrt{\varepsilon})$-*time.*

Proof. We run both algorithms for the distant and close summands, respectively, and take the entry-wise minimum. Correctness and running time follow from those of Lemma 3 and Lemma 4.

5 Applications

Armed with the novel strongly-polynomial algorithm for $(1+\varepsilon)$-approximate min-sum subset convolution running in time $\widetilde{O}(2^{\frac{3n}{2}}/\sqrt{\varepsilon})$, we can now develop $(1+\varepsilon)$-approximation schemes[13] for a plethora of convolution-like NP-hard optimization problems. In particular, we target all problems that are "scapegoats" of the approach employing min-sum subset convolution on bounded input. In other words, these are all exact algorithms with a dependence on M in their running time. Our goal is to enable out-of-the-box $(1+\varepsilon)$-approximation schemes in time $\widetilde{O}(2^{\frac{3p}{2}}/\sqrt{\varepsilon})$, where p is problem-specific (as a rule of thumb, it is always the exponent in the running time $O(3^p)$ of the exact evaluation). The following problem is intended to demonstrate this technique.

5.1 Minimum-Cost k-Coloring

In their book on parameterized algorithms, Cygan et al. [10] propose a variant of k-coloring, which they entitle minimum-cost k-coloring, and devise an $\widetilde{O}(2^n M)$-time algorithm for it [11, Thorem 10.18]. We are given an undirected graph G, an integer k, and a cost function $c : V(G) \times [k] \to \{-M, \ldots, M\}$. The cost of a coloring $\chi : V(G) \to [k]$ is defined as $\sum_{v \in V(G)} c(v, \chi(v))$, i.e., coloring vertex v with color i incurs cost $c(v, i)$. The task is to determine the minimum cost of a

---

[12] We kindly refer the reader to the appendix of the full paper.
[13] Or $(1-\varepsilon)$-approximation schemes where the *max-sum* subset convolution is applicable.

k-coloring of G (if such a coloring exists). To this end, it is handy to introduce a function $s_i : 2^{V(G)} \to \mathbb{Z} \cup \{+\infty\}$ such that for every $X \subseteq V(G)$ we have

$$s_i(X) = \begin{cases} \sum_{x \in X} c(x,i), & \text{if } X \text{ is an independent set,} \\ +\infty, & \text{otherwise.} \end{cases}$$

Then, one can compute the minimum cost of a k-coloring of $G[X]$ as $(s_1 \star \ldots \star s_k)(X)$. This reduces to simply performing $k-1$ min-sum subset convolutions. To this end, their proposed algorithm runs in $\widetilde{O}(2^n M)$-time, by applying the standard min-sum subset convolution for bounded input. We show how to obtain an $(1+\varepsilon)$-approximation in $\widetilde{O}(2^{\frac{3n}{2}}/\sqrt{\varepsilon})$-time; the running time is independent of M:

Theorem 7. *If $(1+\varepsilon)$-Approximate min-sum subset convolution runs in $T(n, \varepsilon)$-time, then an $(1+\varepsilon)$-approximate minimum-cost k-coloring can be found in time $O(T(n, \frac{\varepsilon}{k-1}))$.*

Proof. Consider the evaluation of the min-sum subset convolution between two set functions f and g at each step. Setting a relative error $\delta > 0$ for each convolution call, we obtain a cumulative relative error bounded by $(1+\delta)^{k-1}$. By setting $\delta = \Theta(\frac{\varepsilon}{k-1})$, we obtain a relative error of at most ε.

As a corollary, Thorem 7 implies Thorem 5. Referring to Ref. [20,26], Fomin et al. [21] point out that if certain reasonable complexity conjectures hold, then k-coloring itself is hard to approximate within $n^{1-\epsilon}$, for any $\epsilon > 0$. It is interesting to ask whether our time bound is optimal for the $(1+\varepsilon)$-approximation. We leave this as future work.

Other Applications. We provide many other applications, such as the prize-collecting Steiner tree and two other applications in computational biology, in the appendix of the full paper. Note that in all applications, we can always replace the strongly-polynomial approximation algorithm (Theorem 2) for the min-sum subset convolution with the weakly-polynomial one (Thorem 1) and alternatively obtain $\widetilde{O}(2^p \log M/\varepsilon)$-time approximation schemes, where p is the problem-specific parameter, e.g., p is equal to n in the minimum-cost k-coloring problem.

6 Discussion

There seemed to be an unyielding "isthmus" between the results on sequence convolutions and subset convolutions on semi-rings. In the following, we outline several future work directions, inspired by research on the sequence setting.

Polynomial Speedups. Currently, the fastest known algorithm for min-plus sequence convolution runs in time $n^2/2^{\Omega(\sqrt{\log n})}$, by combining the reduction to Min-Plus Matrix Product by Bremner et al. [6] and an algorithm for the latter due to Williams [41], subsequently derandomized by Chan and Williams [9]. This has been the culmination of a long line of research starting with the $O(n^2/\log n)$-time algorithm due to Bremner et al. [6]. This leads us to ask whether such speedups can be generalized to the subset context as well:

Are there polynomial-factor speedups for the min-sum subset convolution?

In particular, we are not aware of any $O(3^n/n)$-time algorithm.

Min-Sum Subset Convolution Conjecture. The lack of faster algorithms for this problem leads us to conjecture that a similar scenario as in the case of the min-plus sequence convolution [12,28] is also present in the subset setting:

Conjecture 1. There is no $O((3-\delta)^n \mathsf{polylog}(M))$-time exact algorithm for min-sum subset convolution, with $\delta > 0$.

Indeed, an interesting future work is to find out whether both conjectures are equivalent.

Exploiting Kernelization. Certain NP-hard problems accept polynomial size kernels via the well-known Frank-Tardos' framework [22]. We discuss it in the context of our work and leave as future work the possibility of using lossy kernels [29] for the applications we treated in this paper.

Acknowledgements. The author thanks Petru Pascu, Radu Vintan, and Altan Birler for their helpful feedback and the anonymous reviewers whose detailed comments improved the overall presentation.

References

1. Backurs, A., Indyk, P., Schmidt, L.: Better approximations for tree sparsity in nearly-linear time. In: Proceedings of the Twenty-Eighth Annual ACM-SIAM Symposium on Discrete Algorithms, pp. 2215–2229. SIAM (2017)
2. Bansal, N., Chalermsook, P., Laekhanukit, B., Nanongkai, D., Nederlof, J.: New tools and connections for exponential-time approximation. Algorithmica **81**, 3993–4009 (2019)
3. Björklund, A., Husfeldt, T., Kaski, P., Koivisto, M.: Fourier meets möbius: fast subset convolution. In: Proceedings of the Thirty-Ninth Annual ACM Symposium on Theory of Computing, pp. 67–74 (2007)
4. Björklund, A., Husfeldt, T., Koivisto, M.: Set partitioning via inclusion-exclusion. SIAM J. Comput. **39**(2), 546–563 (2009). https://doi.org/10.1137/070683933
5. Bourgeois, N., Escoffier, B., Paschos, V.T.: Approximation of max independent set, min vertex cover and related problems by moderately exponential algorithms. Discret. Appl. Math. **159**(17), 1954–1970 (2011)
6. Bremner, D., et al.: Necklaces, convolutions, and x+y. In: Algorithms–ESA 2006: 2006 Proceedings of the 14th Annual European Symposium, Zurich, Switzerland, 11–13 September, pp. 160–171. Springer (2006)
7. Bringmann, K., Künnemann, M., Węgrzycki, K.: Approximating APSP without scaling: equivalence of approximate min-plus and exact min-max. In: Proceedings of the 51st Annual ACM SIGACT Symposium on Theory of Computing, pp. 943–954 (2019)

8. Cartis, C., Thompson, A.: An exact tree projection algorithm for wavelets. IEEE Sig. Process. Lett. **20**(11), 1026–1029 (2013). https://doi.org/10.1109/LSP.2013.2278147
9. Chan, T.M., Williams, R.: Deterministic APSP, orthogonal vectors, and more: quickly derandomizing Razborov-Smolensky. In: Proceedings of the Twenty-Seventh Annual ACM-SIAM Symposium on Discrete Algorithms, pp. 1246–1255. SIAM (2016)
10. Cygan, M., et al.: Parameterized Algorithms, 1st edn. Springer, Heidelberg (2015)
11. Cygan, M., et al.: Algebraic techniques: sieves, convolutions, and polynomials. In: Parameterized Algorithms, pp. 321–355 (2015)
12. Cygan, M., Mucha, M., Węgrzycki, K., Włodarczyk, M.: On problems equivalent to (min,+)-convolution. ACM Trans. Algorithms (TALG) **15**(1), 1–25 (2019)
13. Dreyfus, S.E., Wagner, R.A.: The Steiner problem in graphs. Networks **1**(3), 195–207 (1971)
14. Duan, R., Pettie, S.: Fast algorithms for (max, min)-matrix multiplication and bottleneck shortest paths. In: Proceedings of the Twentieth Annual ACM-SIAM Symposium on Discrete Algorithms, pp. 384–391. SIAM (2009)
15. Duan, R., Pettie, S., Su, H.H.: Scaling algorithms for weighted matching in general graphs. ACM Trans. Algorithms **14**(1) (2018). https://doi.org/10.1145/3155301
16. Escoffier, B., Paschos, V.T., Tourniaire, E.: Super-polynomial approximation branching algorithms. RAIRO Oper. Res. **50**(4–5), 979–994 (2016)
17. Esmer, B.C., Kulik, A., Marx, D., Neuen, D., Sharma, R.: Faster exponential-time approximation algorithms using approximate monotone local search. arXiv preprint arXiv:2206.13481 (2022)
18. Esmer, B.C., Kulik, A., Marx, D., Neuen, D., Sharma, R.: Approximate monotone local search for weighted problems. arXiv preprint arXiv:2308.15306 (2023)
19. Esmer, B.C., Kulik, A., Marx, D., Neuen, D., Sharma, R.: Optimally Repurposing Existing Algorithms to Obtain Exponential-Time Approximations, pp. 314–345 (2024)
20. Feige, U., Kilian, J.: Zero knowledge and the chromatic number. J. Comput. Syst. Sci. **57**(2), 187–199 (1998)
21. Fomin, F.V., Kratsch, D.: Exact Exponential Algorithms, 1st edn. Springer, Heidelberg (2010)
22. Frank, A., Tardos, É.: An application of simultaneous diophantine approximation in combinatorial optimization. Combinatorica **7**, 49–65 (1987). https://api.semanticscholar.org/CorpusID:45585308
23. Gabow, H.N., Tarjan, R.E.: Faster scaling algorithms for general graph matching problems. J. ACM (JACM) **38**(4), 815–853 (1991)
24. Garbow, H.N.: Scaling algorithms for network problems. J. Comput. Syst. Sci. **31**(2), 148–168 (1985)
25. Goldberg, A.V.: Scaling algorithms for the shortest paths problem. SIAM J. Comput. **24**(3), 494–504 (1995)
26. Khot, S., Ponnuswami, A.K.: Better inapproximability results for MaxClique, chromatic number and Min-3Lin-deletion. In: Bugliesi, M., Preneel, B., Sassone, V., Wegener, I. (eds.) ICALP 2006. LNCS, vol. 4051, pp. 226–237. Springer, Heidelberg (2006). https://doi.org/10.1007/11786986_21
27. Kosaraju, S.: Efficient tree pattern matching. In: 30th Annual Symposium on Foundations of Computer Science, pp. 178–183 (1989). https://doi.org/10.1109/SFCS.1989.63475
28. Künnemann, M., Paturi, R., Schneider, S.: On the fine-grained complexity of one-dimensional dynamic programming. arXiv preprint arXiv:1703.00941 (2017)

29. Lokshtanov, D., Panolan, F., Ramanujan, M., Saurabh, S.: Lossy kernelization. In: Proceedings of the 49th Annual ACM SIGACT Symposium on Theory of Computing, pp. 224–237 (2017)
30. Manurangsi, P., Trevisan, L.: Mildly exponential time approximation algorithms for vertex cover, uniform sparsest cut and related problems. arXiv preprint arXiv:1807.09898 (2018)
31. Mucha, M., Węgrzycki, K., Włodarczyk, M.: A subquadratic approximation scheme for partition. In: Proceedings of the Thirtieth Annual ACM-SIAM Symposium on Discrete Algorithms, pp. 70–88. SIAM (2019)
32. Orlin, J.B., Ahuja, R.K.: New scaling algorithms for the assignment and minimum mean cycle problems. Math. Program. **54**, 41–56 (1992). https://doi.org/10.1007/BF01586040
33. Polzin, T., Daneshmand, S.V.: On Steiner trees and minimum spanning trees in hypergraphs. Oper. Res. Lett. **31**(1), 12–20 (2003). https://doi.org/10.1016/S0167-6377(02)00185-2
34. Ponta, O., Hüffner, F., Niedermeier, R.: Speeding up dynamic programming for some NP-hard graph recoloring problems. In: Agrawal, M., Du, D., Duan, Z., Li, A. (eds.) TAMC 2008. LNCS, vol. 4978, pp. 490–501. Springer, Heidelberg (2008). https://doi.org/10.1007/978-3-540-79228-4_43
35. Rauf, I., Rasche, F., Nicolas, F., Böcker, S.: Finding maximum colorful subtrees in practice. In: Chor, B. (ed.) RECOMB 2012. LNCS, vol. 7262, pp. 213–223. Springer, Heidelberg (2012). https://doi.org/10.1007/978-3-642-29627-7_22
36. Scott, J., Ideker, T., Karp, R.M., Sharan, R.: Efficient algorithms for detecting signaling pathways in protein interaction networks. J. Comput. Biol. **13**(2), 133–144 (2006)
37. Shapira, A., Yuster, R., Zwick, U.: All-pairs bottleneck paths in vertex weighted graphs. In: SODA, vol. 7, pp. 978–985 (2007)
38. Vassilevska, V., Williams, R.R., Yuster, R.: All pairs bottleneck paths and max-min matrix products in truly subcubic time. Theor. Comput. **5**(1), 173–189 (2009)
39. Warme, D.M.: Spanning Trees in Hypergraphs with Applications to Steiner Trees. Ph.D. thesis, USA (1998). aAI9840474
40. Węgrzycki, K.: Provably Optimal Dynamic Programming (2019). https://books.google.de/books?id=UmYfzwEACAAJ
41. Williams, R.: Faster all-pairs shortest paths via circuit complexity. In: Proceedings of the Forty-Sixth Annual ACM Symposium on Theory of Computing, pp. 664–673 (2014)
42. Zwick, U.: All pairs shortest paths using bridging sets and rectangular matrix multiplication. J. ACM **49**(3), 289–317 (2002). https://doi.org/10.1145/567112.567114

Online String Attractors
And Their Relation to the Lempel-Ziv Factorization

Philip Whittington[✉][iD]

ETH Zurich, Zurich, Switzerland
`philip.whittington@inf.ethz.ch`

Abstract. In today's data-centric world, fast and effective compression of data is paramount. To measure success towards the second goal, Kempa and Prezza [STOC2018] introduce the string attractor, a combinatorial object unifying dictionary-based compression. Given a string $T \in \Sigma^n$, a string attractor (k-attractor) is a set of positions $\Gamma \subseteq [1, n]$, such that every distinct substring (of length at most k) has at least one occurrence that contains one of the selected positions. String attractors are shown to be approximated by and thus measure the quality of many important dictionary compression algorithms such as Lempel-Ziv 77, the Burrows-Wheeler transform, straight line programs, and macro schemes.

In order to handle massive amounts of data, compression often has to be achieved in a streaming fashion. Thus, practically applied compression algorithms, such as Lempel-Ziv 77, have been extensively studied in an online setting. To the best of our knowledge, there has been no such work, and therefore are no theoretical underpinnings, for the string attractor problem. We introduce a natural online variant of both the k-attractor and the string attractor problem.

First, we show that the Lempel-Ziv factorization corresponds to the best online algorithm for this problem, resulting in an upper bound of $\mathcal{O}(\log(n))$ on the competitive ratio. On the other hand, we consider prefixes of the Fibonacci word, and show that any online algorithm has a cost growing with the length of the prefix, for a matching lower bound of $\Omega(\log(n))$. For the online k-attractor problem, we show tight (strict) k-competitiveness.

Keywords: String attractors · Dictionary compression · Online algorithms

1 Introduction

Data is the key element of current technical advances. Large amounts of data are being collected and analysed in every context of our lives. In order to handle these, informatiion needs to be stored in a compressed form. Kempa and

---

A full version of this paper is available at https://arxiv.org/abs/2407.15599.

© The Author(s), under exclusive license to Springer Nature Switzerland AG 2025
M. Bieńkowski and M. Englert (Eds.): WAOA 2024, LNCS 15269, pp. 213–227, 2025.
https://doi.org/10.1007/978-3-031-81396-2_15

Prezza [10] introduced string attractors as a tool to better understand the compression quality of compression methods. They show that any dictionary compression algorithms, such as the Lempel-Ziv factorization, the Burrows-Wheeler transform, and straight-line programs, can be understood as approximation algorithms for the string attractor problem. Kempa and Saha [11] add the LZ-End compression algorithm proposed by Kreft and Navarro [14] to this list. String attractors also support more involved string operations and structures, such as universal data structures [10], fast indexed queries [18], and locating and counting indices [2], which underlines their position as a versatile and important tool to describe the theoretical backbone of dictionary compression.

For a string of length n, an attractor is a set of positions $\Gamma \subseteq [1,n]$ covering all distinct substrings, that is, every distinct substring has an occurrence crossing at least one of the selected positions.

Definition 1 (k-attractor [10]). *A set $\Gamma \subseteq [1,n]$ is a k-attractor of a string $T \in \Sigma^n$ if every substring $T[i \ldots j]$ with $i \leq j < i+k$ has an occurrence $T[i' \ldots j']$ with $j'' \in [i' \ldots j']$ for some $j'' \in \Gamma$.*

A solution Γ is called a *string attractor* or simply *attractor* if $k = n$. The corresponding optimization and decision problems are the minimum-k-attractor problem and the k-attractor problem, respectively. The optimal attractor is denoted by Γ^* and has size γ^*.

In their seminal paper, Kempa and Prezza [10] also show that computing the smallest k-attractor is NP-complete for any $k \geq 3$, by giving a reduction from k-set cover, and extend this proof to non-constant k, especially for $k = n$. On the other hand, the problem is trivially solvable in polynomial time for $k = 1$ by a greedy algorithm. The gap for $k = 2$ was closed by Fuchs and Whittington [5], who show that the 2-attractor problem is also NP-complete.

Variants of the problem have been introduced, such as the sharp k-attractor problem [9], which only considers distinct substrings of length exactly k, and the circular attractor problem [17] defined on the substrings of circular words.

Mantaci et al. [17] study the behavior of attractors under combinatorial operations and obtain results for commonly studied families of words, such as Sturmian words, Thue-Morse words, and de Bruijn words. The results on Thue-Morse words build on the work of Kutsukake et al. [15]. Schaeffer and Shallit [22] investigate string attractors for automatic sequences and compute attractors for finite prefixes of infinite words, including Fibonacci and Thue-Morse sequences. Further work on attractors of such classes has been done by Restivo, Romana, and Sciortino [21], Gheeraert, Romana, and Stipulanti [6], and Dvořáková [3].

Building on string attractors, Kociumaka, Navarro and Prezza [12] introduce the *relative substring complexity* measure δ, which counts the number of different substrings of length l and scales it by l. It is smaller than γ^* by at most a logarithmic factor and efficiently computable.

1.1 Our Contributions

We consider the string attractor and k-attractor problems in an online setting. The online string attractor problem is closely tied to the Lempel-Ziv factorization, as it turns out that Lempel-Ziv is the optimal online algorithm for this problem. This observation, combined with the results of Kociumaka, Navarro and Prezza [12], yields an upper bound of $\mathcal{O}(\log(n))$. A matching lower bound of $\log(n)$ for this problem is achieved by studying Fibonacci and Thue-Morse words for which we show that any online algorithm induces a cost that is at least logarithmic in the input length, whereas it is known that there exist constant-size attractors. For the k-attractor problem, we show k to be both an upper and lower bound on the competitive ratio in Sect. 5. The first result is achieved by an induction that bounds the online performance to the relative substring complexity, which is itself a bound to the optimal attractor. For the second result we introduce a technique for an adversary to construct hard strings for the given online algorithm. This is done by singling out a small fraction of all substrings of length k and using them as delimiters, such that the online algorithm needs to put a marking for each of the remaining substrings. We then show that an offline algorithm has a much better performance on de Bruijn sequences, and combine these two results for the desired competitive ratio.

2 Lempel-Ziv Compression

Lempel-Ziv is a seminal algorithm for the field of dictionary compression, named an IEEE Milestone in 2004. It is of high theoretical importance, but also practically applied. Although it was developed in 1977, it is still part of the foundation for commonly used compression data types such as ZIP, PNG, and GIF.

The Lempel-Ziv factorization subdivides an input string into *phrases* that correspond to substrings seen at earlier positions, which are called *sources*. The string can then be compressed by having phrases link back to their earlier sources using their starting position (or distance) and length, which for large phrases reduces to logarithmic size.

Definition 2 (Lempel Ziv Factorization [15,16]). *For any string T, the Lempel-Ziv factorization of T is the sequence of phrases $f_1, \ldots, f_z$ of non-empty strings such that $T = f_1 \ldots f_z$, and for any $1 \leq i \leq z$, f_i is the longest prefix of $f_i \ldots f_z$ which has at least two occurrences in $f_1 \ldots f_i$, or $|f_i| = 1$ otherwise. The earlier occurrence constitutes the source of f_i. We call z the size of the Lempel-Ziv factorization.*

We discuss two common variants of the Lempel-Ziv factorization. First, it is common to forbid overlap between phrases and sources [10]. In the above definition, the phrase f_i then needs to have at least one occurrence in $f_1 \ldots f_{i-1}$, that is, the source must be fully contained in the already factorized prefix. The variant in our initial definition is called *self-referencing*.

Second, a phrase can be defined to be one position longer and include the first new position, except for the case $|f_i| = 1$ in which a new element of the alphabet is introduced. Then, f_i is the shortest prefix of $f_i \ldots f_z$ which has only one occurrence in $f_1 \ldots f_i$, or none in $f_1 \ldots f_{i-1}$ if it also not self-referencing. Note that in this case, the last phrase f_z might not fit this definition. That way, each phrase (except possibly the last one) is a novel substring, that is, a first occurrence in T. We are not aware of a common name to distinguish this variant, so we call the augmented version *novel*. We show the differences between the variants on the initial example used by Lempel and Ziv in Table 1.

Note that the factorization as originally defined by Lempel and Ziv is both self-referencing and novel. The other variants shown can only increase the size of the factorization. Kempa and Prezza [10] show, for any string T, the relations $\gamma^* \leq z \leq \mathcal{O}(\gamma^* \log^2(n/\gamma^*))$ between the size z of the Lempel-Ziv factorization of T and the size γ^* of the optimal string attractor of T, which was later improved by Kociumaka, Navarro and Prezza [12] to $z \leq \mathcal{O}(\gamma^* \log(n/\delta))$.

Table 1. Behavior of Lempel-Ziv Variants.

self-ref.	novel	$T = aaabbabaabaaabab$	z
✓	✓	a|aab|ba|baa|baaa|bab	5
✓	✗	a|aa|b|b|ab|aab|aaab|ab	7
✗	✓	a|aa|b|ba|baa|baaa|bab	6
✗	✗	a|a|a|b|b|ab|aab|aaab|ab	8

3 Online Attractors

In real-world settings, problems often need to be solved as soon as they occur, and partial solutions need to be committed before the entire scope of the problem is known. This is especially true in the field of data compression where data may be too large to completely load into a computer's cache to compress it, or the data might need to be compressed as soon as it is created in a streaming fashion. The theoretical framework to describe these settings is the area of *online algorithms*.

An instance I of an *online problem* is a finite sequence of *requests* $x_1, \ldots, x_n$ and the corresponding solution S is a finite sequence of *answers* $y_1, \ldots, y_n$. The online algorithm outputs the answers, where y_i may only depend on $x_1, \ldots, x_i$ and $y_1, \ldots, y_{i-1}$ [13].

Often, offline algorithms outperform online algorithms in the quality of the computed solution. The relation between the costs of their results is called the *competitive ratio*. For any online minimization problem, let **I** be the set of all instances of that problem and **S** the set of all solutions. Further, let $f: \mathbf{I} \to \mathbf{S}$ be the function describing the output of a fixed online algorithm and $f^*: \mathbf{I} \to \mathbf{S}$ the function describing the output of an optimal offline algorithm. The online algorithm is then *c-competitive* if there is a non-negative constant α such that

$$|f(I)| \leq c|f^*(I)| + \alpha$$

for all instances $I \in \mathbf{I}$ [13, Definition 1.6]. If $\alpha = 0$, we speak of the *strict competitive ratio*. This is commonly modeled by the instances being created by an *adversary* that knows the online algorithm and accordingly chooses the hardest instance.

Definition 3 (The Online k-Attractor Problem). *The input is a word $T = T[1]T[2]\ldots T[n] \in \Sigma^*$, where both n and $\sigma = |\Sigma|$ are not known to the online algorithm. The online algorithm receives T position by position, that is, in step i, $T[i]$ is revealed. It then needs to decide whether to mark this position as part of the k-attractor. After each step j, the current set of markings $\Gamma(j)$ needs to be a valid k-attractor for $T[1,j]$.*

This is the natural online extension, as it guarantees that the output is a valid k-attractor at all times. As before, $k = n$ defines the online string attractor problem.

The straightforward online algorithm for this problem is the greedy algorithm that scans the input string T from left to right and inserts a marking whenever it encounters a new substring of length at most k. This algorithm has been briefly considered for infinite words by Schaeffer and Shallit [22, Section 9]. Let j be the last marked position, or 0 otherwise. At time $i > j$, $T[i]$ is revealed to the online algorithm which has to decide whether to mark this position. If the string $T[j+1,i]$ did not appear before, we call it *novel* and mark position i. Note that although this approach implements a straightforward greedy strategy, any deterministic online algorithm deviating from this strategy needs to place its markings earlier than greedy, that is, greedy chooses the last possible time to put a marking. Thus, a more appropriate name for this algorithm is LAZY.

Theorem 1. *Any deterministic algorithm for the online k-attractor problem cannot improve upon the competitive ratio of the LAZY algorithm. Further, for each deterministic algorithm A that is not LAZY (or equivalent to it), there are instances on which A has higher costs than LAZY.*

Proof. Given any input string T and any online algorithm A, consider the first marked position where the LAZY deviates from the online algorithm. LAZY marks a position x whereas A marked some $y < x$. The position of A has to appear earlier, otherwise the new substring causing LAZY to cover x is not covered by A at some point. LAZY has marked all substrings in $T[1,x]$, so we change the marking y to x in A's output while still covering at least as many substrings. This is done for all differently marked positions of A, so it produces a k-attractor of a size that is at least equal to that computed by LAZY. If we instead introduce a new symbol right after position y, both A and LAZY will include position $y+1$, thus LAZY is better on this instance.

This statement also holds for the online string attractor problem, and both sharp variants. Note that while LAZY is the natural greedy algorithm for the online variant, there are more sophisticated greedy algorithms in the offline case. For example, if the entire word is known, we can first mark positions that cover a symbol uniquely appearing at that position, or choose the next position to mark according to a weight function on its substrings. The resulting attractors of such algorithms are always minimal under removal and therefore k-approximations [9]. Minimality does not hold for LAZY, however we show that it still achieves k-competitiveness.

The initial purpose of string attractors is their ability to serve as lower bounds for famous dictionary compression algorithms such as Lempel-Ziv, as those approximate string attractors. We observe that Lempel-Ziv actually computes an online string attractor, thus any hardness results for online attractors translate to Lempel-Ziv.

Theorem 2. *The greedy algorithm LAZY for the online string attractor problem exactly computes the self-referencing, novel Lempel-Ziv factorization.*

Proof. On a string $T \in \Sigma^n$, LAZY marks a position i when the substring seen since the last marking j is novel, that is, $T[j+1,i]$ does not appear a second time in $T[1,i]$. It is then the shortest such prefix of $T[j+1,n]$ and thus exactly a Lempel-Ziv phrase.

This equivalence immediately yields interesting results for both the online string attractor problem and the Lempel-Ziv compression. First, the efficient online computation of the Lempel-Ziv compression is well studied, see Policriti and Prezza [19,20] who compute Lempel-Ziv in time $\mathcal{O}(n \log(R))$ and space $R \log(n))$ where R corresponds to the number of runs in the reversed Burrows-Wheeler transform of the input, i.e., in linear time and logarithmic space for highly repetitive inputs. Thus, the online string attractor can also be computed efficiently. On the other hand, we can now use the toolbox of competitive online algorithms to better understand the Lempel-Ziv factorization.

In the following, we limit the size σ of the alphabet or the scope k of the attractor. If we were to limit both at the same time, only a finite number of substrings could exist, which implies that the maximum costs are bounded.

4 Limiting the Alphabet

We first study hard instances for the online string attractor problem. More precisely, we want to consider families of words that have a constant-size (offline) attractor, which can only be achieved if the alphabet is of constant size. We especially turn our attention to infinite strings, and families of words that are prefixes of these strings. Lemma 4 tells us that a family of words only has optimal attractors of constant size if its complexity function, that is, the number of factors of length l, grows at most linearly with l.

However, if this number grows less than linearly, we obtain periodic instances that, after a finite prefix, repeat the same string according to the Morse-Hedlund Theorem [1]. Then, after that string is repeated twice, the online attractor also does not incur any further cost. LAZY's cost is then also bounded by a constant and we only obtain a constant lower bound on the competitive ratio.

In the following, we consider infinite strings that are generated by an iterative application of a morphism and are thereby equipped with a linear complexity function. As described in the preliminaries, such morphisms give rise to an infinite family of prefixes of the infinite string. There are already rich results on offline attractors of such families, e.g., Sturmian words [3,6,17] and Thue-Morse words [15,17]. We study the behavior of the LAZY algorithm on these families and compare the results.

Perhaps the most famous sequence of numbers is the *Fibonacci sequence*, where each value after the two initial ones is defined as the sum of its two predecessors. We define the Fibonacci sequence (of numbers) by

$$f_{-2} = 0, f_{-1} = 1, f_m = f_{m-1} + f_{m-2}.$$

Fibonacci numbers are closely connected to the *golden ratio* $\phi = (1+\sqrt{5})/2$, as they can be computed by $f_m = \lfloor \phi^{m-2}/\sqrt{5} \rceil$, which shows their exponential growth.

Fibonacci words follow the same idea, as each Fibonacci word is the concatenation of its two predecessors, to ultimately form the infinite word also called the Fibonacci sequence. Fibonacci words are recursively defined on a binary alphabet $\{a, b\}$, with

$$F_{-2} = \varepsilon, F_{-1} = b, F_0 = a, F_m = F_{m-1}F_{m-2}.$$

They are also the result of iteratively applying the morphism $\varphi(a) = ab, \varphi(b) = a$. Starting at $F_0 = a$, we get $F_m = \varphi^m(a)$, and it holds that F_m is a prefix of F_{m+1} for all $m \geq 0$. The infinite *Fibonacci sequence* F_∞ is defined as the fixed point of this morphism. Due to the analogous definition and the choice of indices, the length of F_m is f_m.

Dvořáková [3] shows that each factor w of an episturmian sequence has an attractor of size σ' where σ' is the number of distinct letters in w. Sturmian sequences correspond to aperiodic binary episturmian sequences, and the Fibonacci sequence is the standard Sturmian sequence with directive sequence $q_i = 1$, for all $i \in \mathbb{N}$. Thus, each Fibonacci word has an attractor of size two.

Table 2. Fibonacci and Kernel words [8].

m	-2	-1	0	1	2	3	4	5
F_m	ε	b	a	ab	aba	abaab	abaababa	abaababaabaab
K_m	ε	a	b	aa	bab	aabaa	babaabab	aabaababaabaa
f_m	0	1	1	2	3	5	8	13

Definition 4 (Kernel word). Let $\delta_m = a$ if m is even and $\delta_m = b$ if m is odd, then δ_m denotes the last letter of F_m. For every Fibonacci word F_m we call $K_m = \delta_{m+1} F_m \delta_m^{-1}$ the m-th kernel word or singular word. The concatenation of δ_m^{-1} denotes the removal of δ_m at the end of F_m.

Table 2, originally from Huang and Wen [8], shows the first Fibonacci and kernel words. Concatenating all kernel words also yields the Fibonacci sequence [23, Theorem 1]. Our main result on Fibonacci words is that kernel words correspond exactly to the novel substrings that incur costs for LAZY.

Lemma 1. *We note the following properties of kernel words:*

1. *Every kernel word $K_m = \delta_{m+1} F_m \delta_m^{-1}$ is a palindrome [23, Property 2.9].*
2. *K_{m-2} is the largest kernel word contained in F_m [23, Property 2.5].*
3. *The largest kernel word in F_m appears only once [7, Proposition 1.8].*

Lemma 2. *For each Fibonacci word F_m, its prefix $F_m \delta_m^{-1} \delta_{m+1}^{-1}$ consisting of all but its last two characters is a palindrome.*

Proof. By the first property of Lemma 1, $\delta_{m+1} F_m \delta_m^{-1}$ is a palindrome, thus removing δ_{m+1} on both ends also yields a palindrome.

In a note from 2015, Fici [4] states a connection between singular words and the Lempel-Ziv factorization of the Fibonacci word that turns out to be equivalent to the next lemma, as we show in Theorem 2. While the necessary preliminaries are stated, no proof is given.

Lemma 3. *On the m-th Fibonacci word $F_m, m \geq 3$, LAZY puts a marking on the second to last position, that is, right after the largest palindromic prefix of F_m.*

Proof. We prove the statement by induction. $F_3 = ab\underline{aa}b$ exhibits the desired behavior. We assume that F_{m-1} is already marked as described and $m \geq 4$. Then, $F_m[f_{m-1}, f_m - 1] = K_{m-2}$ is the substring that has appeared since the last marking. It is the largest kernel word contained in F_m and thus appears only once by properties 2 and 3 of Lemma 1. Therefore, LAZY needs to put a marking within that substring.

On the other hand, the substring $w = F_m[f_{m-1} - 1, f_m - 2] = K_{m-2}\delta_{m-1}^{-1}$ already appears before in F_{m-1}. Thus, LAZY does not put an earlier marking. To see that, we apply Lemma 2 which tells us that $F_m[1, f_m - 2]$ is a palindrome ending in w, thus starting in its reversal $\overleftarrow{w}$. $F_m[1, f_{m-1}-2]$ has length $f_{m-1}-2 = f_{m-2}+f_{m-3}-2$ whereas $w = F_m[f_{m-1}, f_m-2]$ is shorter than $f_m-2-f_{m-1}+1 = f_{m-2} - 1$ because $m \geq 4$ implies $f_{m-3} \geq 2$. Therefore, $F_m[1, f_{m-1} - 2]$ has that same prefix $\overleftarrow{w}$ and is again a palindrome due to Lemma 2. This implies that the suffix of $F_m[1, f_{m-1} - 2]$ is equal to w. Consider Fig. 1 for a visualization of this proof. $F_m[1, f_{m-1} - 2]$ is a prefix of F_{m-1}, thus w and all its substrings are already covered according to our induction hypothesis.

Theorem 3. *For $m \geq 3$, LAZY has a cost of m on F_m. The marked positions are $\{1, 2, 4, 7, 12, \ldots\} = \{1, 2\} \cup \{f_m - 1 \mid m \geq 3\}$. As the optimal solution has size 2 [3, 17], the resulting competitive ratio is $m/2 = f^{-1}(n)/2 \in \mathcal{O}(\log(n))$.*

Proof. $F_3 = ab\underline{aa}b$ has a cost of 3. By Lemma 3, each additional increment of the Fibonacci morphism adds a cost of 1.

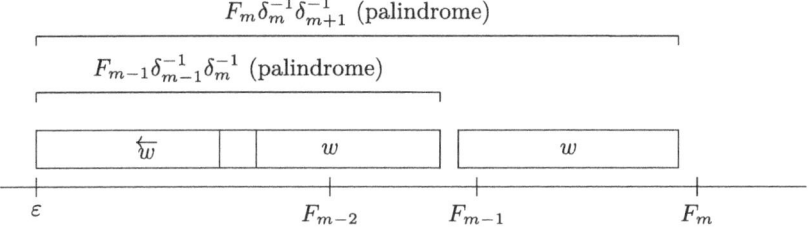

Fig. 1. $w = F_m[f_{m-1}, f_m - 2]$ appears in F_{m-1}.

We now turn towards another family of words that are aperiodic, yet have a constant-size optimal attractor. *Thue-Morse words* are recursively defined on a binary alphabet $\{a, b\}$, with $G_0 = a, G_m = G_{m-1}\overline{G_{m-1}}$ where $\overline{T}$ denotes the negation of the string T. They are also the result of iteratively applying the morphism $\psi(a) = ab, \psi(b) = ba$. Starting at $G_0 = a$, we get $G_m = \psi^m(a)$, and it holds that G_m is a prefix of G_{m+1} for all $m \geq 0$. The infinite *Thue-Morse sequence* G_∞ is defined as the fixed point of this morphism. The length of G_m is 2^m.

The complexity function of the Thue-Morse sequence is slightly larger than that of Sturmian words, but still linear. Kutsukake et al. [15] show that every Thue-Morse word G_m has a string attractor of size 4, and this bound is optimal. A subsection on Thue-Morse words containing the proof of the next theorem can be found in the appendix.

Theorem 4. LAZY *has a cost of* $2m - 2$ *on* G_m *for* $m \geq 3$.

On the Fibonacci words, we obtain a competitive ratio of $\log_\phi(n)/2$. The competitive ratio of LAZY on Thue-Morse words is $(2m-2)/4 = (\log_2(n) - 1)/2$. As $\log_2(n)$ is smaller than $\log_\phi(n)$ by a factor $\log_2(\phi) \approx 0.694$, the Fibonacci sequence yields a stronger bound.

The factors of general Sturmian and episturmian words are not as well understood as for the specific Fibonacci word. We conjecture that for every episturmian sequence, LAZY puts a marking right after each palindromic prefix, after some initial offset to cover the elements of Σ. Directive sequences can be used to show that the family of Fibonacci words grows slower than all other Sturmian words, that is, the Fibonacci sequence has a higher density of palindromic prefixes and would thus yield the best competitive ratio if the conjecture holds.

4.1 Conclusions for the Unrestricted Case

We combine Theorems 2 and 3 and Corollary 3.15 from Kempa and Prezza [10] for a tight bound on the competitive ratio.

Theorem 5. *The greedy algorithm* LAZY *for the online string attractor problem is* $\Theta(\log(n))$-*competitive*.

We also obtain a lower bound on the performance of Lempel-Ziv compression as an approximation to the string attractor problem. Note that this result is based on the Fibonacci word with an optimal attractor of constant size, so a bound taking γ^* into account as in Theorem 5 might be more accurate. In fact, it is not possible for the lower bound to hold for large optimal attractors as this would exceed the upper bound.

Corollary 1. *The approximation guarantee of the Lempel-Ziv compression algorithm for the string attractor problem is not better than $\Omega(\log(n))$.*

We have seen evidence indicating that the performance of LAZY/Lempel-Ziv depends on the size of the optimal attractor, and that the Fibonacci word is a particularly hard instance with a small attractor. This leads to the conjecture that the Fibonacci word is actually the overall worst-case instance, and we get better competitive ratios for larger attractors, such as $\mathcal{O}(\log(n/\gamma^*))$.

5 Limiting the Scope

In this section, we study the online k-attractor problem and the online sharp k-attractor problem for constant k. We first consider how these problems differ from each other in their definition. In a set cover interpretation, any marking in an optimal solution for k-attractor covers at most $k(k+1)/2$ elements, or k elements for sharp k-attractor. As each marking done by LAZY covers at least one new element, this immediately yields these values as upper bounds for the respective competitive ratios.

In the sharp setting, the adversary has fewer substrings to force costs for the online algorithm. The optimal solution is of course also subject to fewer conditions, but that does not help much as an efficient k-attractor is close to a sharp k-attractor anyway. Our first result is to show that the competitive ratio of k-attractor is also bounded by k, thus the adversary has no significant gain from the additional, shorter substrings.

Next, we construct instances for online k-attractor on which LAZY has a competitive ratio converging to k, showing that this bound is tight. Furthermore, this can also be achieved using only substrings of length k, thus yielding the same result for the online sharp k-attractor problem. Thus, while the problems are different in their definition, their online versions behave similarly and have the same competitive ratio.

The main result of this section is given by the following theorem, and the remainder of the subsection is devoted to proving it.

Theorem 6. *The greedy algorithm LAZY for the online k-attractor problem for constant k has a strict competitive ratio of k, and is not $k - \varepsilon$-competitive for any constant $\varepsilon > 0$.*

5.1 Upper Bound

Kempa et al. [9] show that minimal k-attractors are at most a factor k away from the size of an optimal solution. While the online k-attractor computed by LAZY is easily seen to be not necessarily minimal, e.g., on the string $ab\underline{aa}$, we still show the same performance guarantee.

We introduce two helpful observations before proving the upper bound. The following Lemma 4 has already appeared in different contexts [2,12] to show the relation between a string's attractor size and complexity function. We rephrase it for our purposes.

Lemma 4. *For any length l and string T, if T contains x many different substrings of length l, any sharp l-attractor and thus any l'-attractor for T with $l' \geq l$ has size at least $\lceil x/l \rceil$.*

Proof. Any single attractor position i can cover at most l different substrings of length l, namely the substrings $T[i-l+1,i], T[i-l+2,i+1], \ldots, T[i,i+l-1]$. Thus, any attractor that asks to cover x different substrings of length l needs at least $\lceil x/l \rceil$ many attractor positions.

Lemma 5. *Let T be the input to the online k-attractor problem, Γ_k the output by LAZY for online k-attractor and Γ_{k-1} the output by LAZY for online $(k-1)$-attractor. If Γ_k has no occurrence of at least $k-1$ unmarked positions between two markings or the last marking and the end of string, $\Gamma_k = \Gamma_{k-1}$.*

Proof. Each time LAZY puts a marking to produce Γ_k, it is due to a novel substring which is as long as the distance to the last marking. As there are at most $k-2$ unmarked positions between any two markings, every novel substring has length at most $k-1$ and therefore also induces a marking when producing Γ_{k-1}. Clearly, Γ_{k-1} will also not have an earlier marking than Γ_k.

Lemma 6. *The LAZY algorithm is strictly k-competitive for the k-attractor problem.*

Proof. We show the statement by induction on k. The case $k=1$ corresponds to a trivial coverage of symbols from Σ, thus LAZY is optimal. Assume $(k-1)$-attractor is $(k-1)$-competitive. Let T be the input to the online k-attractor problem, Γ_k the output by LAZY and Γ_k^* an optimal solution. We first assert that LAZY produces an output Γ_k where at some point there are at least $k-1$ unmarked positions either between two markings or the last marking and the end of string.

If this does not hold, we apply Lemma 5 and obtain $|\Gamma_k| = |\Gamma_{k-1}|$. It always holds that $|\Gamma_{k-1}^*| \leq |\Gamma_k^*|$. Thus, we compute the competitive ratio on T by

$$\frac{|\Gamma_k|}{|\Gamma_k^*|} \leq \frac{|\Gamma_{k-1}|}{|\Gamma_{k-1}^*|} \leq k-1$$

where the second inequality holds due to our induction hypothesis.

With this assumption, we are now ready to count k-substrings and uniquely assign them to markings. Every marking chosen by LAZY is due to a novel substring $w_i = T[x, x + k_i - 1]$ of length $k_i \leq k$ appearing for the first time. Note that multiple novel substrings can appear at once, so we fix w_i to be the longest of those. This substring can always be extended either to the right or to the left to length k. We present a technique of extending them such that each marking is uniquely assigned a k-substring appearing in T. The number of different k-substrings in T is then at least $|\Gamma|$, and an optimal offline k-attractor has cost at least $|\Gamma|/k$, according to Lemma 4.

We define $LE(w_i, k) = LE(T[x, x+k_i-1], k) = T[x-(k-k_i), x+k_i-1]$ as the leftward extension and $RE(w_i, k) = RE(T[x, x+k_i-1], k) = T[x, x+k-1]$ as the rightward extension of $w_i = T[x, x+k_i-1]$. By our assumption, there exists at least one set of $k-1$ consecutive unmarked positions. We construct a set S of substrings of length k. For all w_i to the left of the first such set, add the rightward extension to S, and for all w_i to the right of the first such set, add the leftward extension to S. We claim that for every w_i, a unique substring of length k is added, which implies $|S| = |\Gamma|$. First, observe that all these extensions actually exist, that is, they do not cross the start or end of the string. Further, each extended string is different from all other extended strings. Assume the extensions of two novel substrings w_i, w_j with $i < j$ are the same. If the extensions are both rightward, w_j is a prefix of $RE(w_i, k)$, thus not a novel substring at its position and already covered, contradicting its definition. The mirrored argument applies if both extensions are leftward. If w_i is extended rightwards and w_j is extended leftwards, we make use of the fact that we switch from rightward to leftward extensions at a gap of at least $k-1$ unmarked positions. That way, $RE(w_i, k)$ starts at a position x that is strictly smaller than the position y at which $LE(w_j, k)$ starts, thus also w_j appears in $RE(w_i, k)$ and is not novel.

5.2 Lower Bound

This subsection is devoted to creating strings that induce a high cost for the online greedy algorithm LAZY. The key idea is to single out a small set of substrings, such that a vastly higher number of other substrings can be separated using the small set as delimiters. On the other hand, we analyse how to make use of the fact that attractors are not monotone to create a much better solution for the offline algorithm, that is, what strings allow the most efficient coverage of a set of substrings?

For space reasons, we only give the main result and an overview of the proof ideas. The lemmata in this subsection and their proofs mostly consist of bookkeeping, calculations, and observations. A full version of this subsection and an excursion to de Bruijn sequences can be found in the full version of the paper.

The online algorithm is presented two strings T_1, T_2 successively as input T. Both strings include the same set of substrings of length up to k, and their overlap, that is, the end of T_1 combined with the start of T_2, does not generate any additional substrings. For the examples presented here, this always means all possible substrings of length up to k, but this is not necessary, otherwise the proof just needs to be executed more carefully. As both include the same set of substrings, the online algorithm induces a cost only on T_1. Conversely, there is an optimal solution (or a bound on it) that only uses T_2. We can then compute a lower bound on the online algorithm LAZY's competitive ratio c by

$$c = \frac{\text{cost}(\text{LAZY}(T))}{\text{cost}(\text{OPT}(T))} \geq \frac{\text{cost}(\text{LAZY}(T_1))}{\text{cost}(\text{OPT}(T_2))}$$

Definition 5 (Spoon-feeding Substrings). *For $l \geq 3$ and a fixed alphabet Σ, we define the set of all spoon-feeding substrings $\mathcal{W}(l)$ to contain exactly all strings $w \in \Sigma^l$ with $w[1] \neq w[2]$ and $w[l-1] \neq w[l]$. Thus,*

$$\mathcal{W}(l) = \Sigma^l - \{x \in \Sigma^l \mid x = \alpha^2 x' \text{ or } x = x' \alpha^2 \text{ for some } \alpha \in \Sigma, x' \in \Sigma^{l-2}\}$$

The core idea is that this set of spoon-feeding substrings can be written, using repeated symbols as delimiters, in a way such that the first occurrence of each such substring is at least k positions apart from every other first occurrence. This will incur costs for the LAZY algorithm equal to the size of the set, which we show to be asymptotically equal to the size of the set of all substrings. On the other hand, we show that de Bruijn sequences can be used to list all substrings of a given length in a very dense fashion, saving a factor of k (asymptotically) for the optimal solution. This enables us to prove Lemma 7, which, combined with Lemma 6, yields Theorem 6.

Lemma 7. *The competitive ratio of LAZY for the online k-attractor problem for constant k on the string*

$$T_1 T_2 = \text{SF}(k) \prod_{i=1}^{k} \text{dB}(i, \Sigma)$$

converges to k.

6 Conclusion

In this paper, we produced the first explicit results on the respective online versions of the string attractor and k-attractor problems.

The online string attractor problem is closely related to the Lempel-Ziv factorization as we showed that this algorithm actually describes the optimal online algorithm. We thus obtained a bound on the performance of Lempel-Ziv on the families of Fibonacci and Thue-Morse words. The competitive ratio of the LAZY algorithm for the string attractor problem is in $\Theta(\log(n))$, although the lower bound is only proven when the optimal attractor has constant size. For the limited scope, we show that online k-attractor is k-competitive and that this bound is tight.

Going further into this topic, we ask how the competitive ratio depends on the size of the optimal solution, and what worst-case instances for each optimal cost look like. We conjecture that the competitive ratio decreases when the size of an optimal string attractor increases, that is, the cost of LAZY scales sublinearly with the size of the optimal solution. In future work, this approach could be used to obtain better approximation guarantees for string attractors, as one either obtains a 'bad' approximation of a small attractor, which is still small relative to the length of the input, or a better approximation of a large attractor.

Acknowledgements. I want to thank my colleagues and supervisors Janosch Fuchs, Dennis Komm, Moritz Stocker and Walter Unger for their helpful comments and suggestions that have improved this paper.

Disclosure of Interests. The authors have no competing interests to declare that are relevant to the content of this article.

References

1. Cassaigne, J., Nicolas, F.: Factor complexity. In: Combinatorics, automata, and number theory, pp. 163–217. Cambridge: Cambridge University Press (2010). https://doi.org/10.1017/CBO9780511777653.005
2. Christiansen, A.R., Ettienne, M.B., Kociumaka, T., Navarro, G., Prezza, N.: Optimal-time dictionary-compressed indexes. ACM Trans. Algorithms **17**(1) (dec 2021). https://doi.org/10.1145/3426473
3. Dvořáková, L.: String attractors of episturmian sequences. arXiv **abs/2211.01660v2** (2022). https://doi.org/10.48550/arXiv.2211.01660
4. Fici, G.: Factorizations of the fibonacci infinite word. arXiv **abs/1508.06754** (2015). https://doi.org/10.48550/arXiv.1508.06754
5. Fuchs, J., Whittington, P.: The 2-Attractor Problem Is NP-Complete. In: Beyersdorff, O., Kanté, M.M., Kupferman, O., Lokshtanov, D. (eds.) 41st International Symposium on Theoretical Aspects of Computer Science (STACS 2024), vol. 289, pp. 35:1–35:13. Schloss Dagstuhl – Leibniz-Zentrum für Informatik (2024). https://doi.org/10.4230/LIPIcs.STACS.2024.35
6. Gheeraert, F., Romana, G., Stipulanti, M.: String attractors of fixed points of k-bonacci-like morphisms. arXiv **abs/2302.13647** (2023). https://doi.org/10.48550/arXiv.2302.13647
7. Huang, Y., Wen, Z.: The sequence of return words of the fibonacci sequence. Theoret. Comput. Sci. **593**, 106–116 (2015). https://doi.org/10.1016/j.tcs.2015.05.048
8. Huang, Y., Wen, Z.: The numbers of repeated palindromes in the fibonacci and tribonacci words. Discret. Appl. Math. **230**, 78–90 (2017). https://doi.org/10.1016/j.dam.2017.06.012
9. Kempa, D., Policriti, A., Prezza, N., Rotenberg, E.: String attractors: Verification and optimization. In: Azar, Y., Bast, H., Herman, G. (eds.) 26th Annual European Symposium on Algorithms, ESA 2018, August 20-22, 2018, Helsinki, Finland. LIPIcs, vol. 112, pp. 52:1–52:13. Schloss Dagstuhl - Leibniz-Zentrum für Informatik (2018). https://doi.org/10.4230/LIPIcs.ESA.2018.52
10. Kempa, D., Prezza, N.: At the roots of dictionary compression: string attractors. In: Diakonikolas, I., Kempe, D., Henzinger, M. (eds.) Proceedings of the 50th Annual ACM SIGACT Symposium on Theory of Computing, STOC 2018, Los Angeles, CA, USA, June 25-29, 2018. pp. 827–840. ACM (2018). https://doi.org/10.1145/3188745.3188814

11. Kempa, D., Saha, B.: An upper bound and linear-space queries on the lz-end parsing. In: Naor, J.S., Buchbinder, N. (eds.) Proceedings of the 2022 ACM-SIAM Symposium on Discrete Algorithms, SODA 2022, Virtual Conference / Alexandria, VA, USA, January 9 - 12, 2022. pp. 2847–2866. SIAM (2022). https://doi.org/10.1137/1.9781611977073.111
12. Kociumaka, T., Navarro, G., Prezza, N.: Towards a definitive measure of repetitiveness. In: Kohayakawa, Y., Miyazawa, F.K. (eds.) LATIN 2020: Theoretical Informatics - 14th Latin American Symposium, São Paulo, Brazil, January 5-8, 2021, Proceedings. Lecture Notes in Computer Science, vol. 12118, pp. 207–219. Springer (2020). https://doi.org/10.1007/978-3-030-61792-9_17
13. Komm, D.: An Introduction to Online Computation - Determinism, Randomization, Advice. Springer, Heidelberg (2016). https://doi.org/10.1007/978-3-319-42749-2
14. Kreft, S., Navarro, G.: Lz77-like compression with fast random access. In: 2010 Data Compression Conference, pp. 239–248 (2010). https://doi.org/10.1109/DCC.2010.29
15. Kutsukake, K., Matsumoto, T., Nakashima, Y., Inenaga, S., Bannai, H., Takeda, M.: On repetitiveness measures of thue-morse words. In: Boucher, C., Thankachan, S.V. (eds.) String Processing and Information Retrieval. Springer International Publishing, Cham (2020). https://doi.org/10.1007/978-3-030-59212-7_15
16. Lempel, A., Ziv, J.: On the complexity of finite sequences. IEEE Trans. Inf. Theory **22**(1), 75–81 (1976). https://doi.org/10.1109/TIT.1976.1055501
17. Mantaci, S., Restivo, A., Romana, G., Rosone, G., Sciortino, M.: A combinatorial view on string attractors. Theor. Comput. Sci. **850**, 236–248 (2021). https://doi.org/10.1016/j.tcs.2020.11.006
18. Navarro, G., Prezza, N.: Universal compressed text indexing. Theoret. Comput. Sci. **762**, 41–50 (2019). https://doi.org/10.1016/j.tcs.2018.09.007
19. Policriti, A., Prezza, N.: Fast online lempel-ziv factorization in compressed space. In: Iliopoulos, C., Puglisi, S., Yilmaz, E. (eds.) String Processing and Information Retrieval. pp. 13–20. Springer International Publishing, Cham (2015). https://doi.org/10.1007/978-3-319-23826-5_2
20. Policriti, A., Prezza, N.: Computing lz77 in run-compressed space. In: 2016 Data Compression Conference (DCC), pp. 23–32 (2016). https://doi.org/10.1109/DCC.2016.30
21. Restivo, A., Romana, G., Sciortino, M.: String attractors and infinite words. CoRR **abs/2206.00376** (2022). https://doi.org/10.48550/arXiv.2206.00376
22. Shallit, J.O., Schaeffer, L.: String attractors for automatic sequences. CoRR **abs/2012.06840** (2020). https://doi.org/10.48550/arXiv.2012.06840
23. Zhi-Xiong, W., Zhi-Ying, W.: Some properties of the singular words of the fibonacci word. Eur. J. Comb. **15**(6), 587–598 (1994). https://doi.org/10.1006/eujc.1994.1060

Author Index

A
Aarts, Sander 1
Amelia, Afrouz Jabal 89
Amouzandeh, Aflatoun 16

B
Blom, Danny 89

C
Cabello, Sergio 31
Chitnis, Rajesh 46

D
Dentes, Jacob 1

G
Giannopoulos, Panos 31

H
Hartmann, Tim A. 61
Høgemo, Svein 104
Hougardy, Stefan 76
Hyatt-Denesik, Dylan 89

J
Janßen, Tom 61

K
Kuo, Tung-Wei 119

L
Lu, Kefu 135

M
Marchetti, Mason 135
Miltenburg, Steven 183

N
Nutov, Zeev 151

O
Oosterwijk, Tim 183

S
Shapley, Richard 167
Shmoys, David B. 1, 167
Sitters, René 183
Smeulders, Bart 89
Stoian, Mihail 198

T
Tammemaa, Karolina 76
Thomas, Samuel 46

V
van Stee, Rob 16

W
Whittington, Philip 213
Wirth, Anthony 46
Wu, Manxi 1

The manufacturer's authorised representative in the EU is Springer Nature Customer Service Centre GmbH, Europaplatz 3, 69115 Heidelberg, Germany. If you have any concerns regarding our products, please contact ProductSafety@springernature.com

Printed and bound by CPI Group (UK) Ltd, Croydon, CR0 4YY

26/03/2026

02078935-0006

The manufacturer's authorised representative in the EU is Springer Nature Customer Service Centre GmbH, Europaplatz 3, 69115 Heidelberg, Germany. If you have any concerns regarding our products, please contact ProductSafety@springernature.com

Printed and bound by CPI Group (UK) Ltd, Croydon, CR0 4YY

26/03/2026

02078935-0006